Orchard Management under Climate Change

Orchard Management under Climate Change

Guest Editors

Sarita Leonel
Sergio Ruffo Roberto
Simone Rodrigues da Silva

MDPI

Basel • Beijing • Wuhan • Barcelona • Belgrade • Novi Sad • Cluj • Manchester

Guest Editors

Sarita Leonel
School of Agriculture
Department of Crop Science
São Paulo State University
(UNESP)
Botucatu
Brazil

Sergio Ruffo Roberto
Department of Agronomy
Agricultural Research Center
Londrina State University
(UEL)
Londrina
Brazil

Simone Rodrigues da Silva
Luiz de Queiroz College of
Agriculture
University of São Paulo (USP)
Piracicaba
Brazil

Editorial Office
MDPI AG
Grosspeteranlage 5
4052 Basel, Switzerland

This is a reprint of the Special Issue, published open access by the journal *Horticulturae* (ISSN 2311-7524), freely accessible at: https://www.mdpi.com/journal/horticulturae/special_issues/F82593HVXW.

For citation purposes, cite each article independently as indicated on the article page online and as indicated below:

Lastname, A.A.; Lastname, B.B. Article Title. *Journal Name* **Year**, *Volume Number*, Page Range.

ISBN 978-3-7258-3337-5 (Hbk)
ISBN 978-3-7258-3338-2 (PDF)
https://doi.org/10.3390/books978-3-7258-3338-2

Cover image courtesy of Marco Antonio Tecchio

Contents

About the Editors . . . . **vii**

Sarita Leonel, Sergio Ruffo Roberto and Simone Rodrigues da Silva
Orchard Management Under Climate Change
Reprinted from: *Horticulturae* **2025**, *11*, 98, https://doi.org/10.3390/horticulturae11010098 . . . **1**

Ricardo Bordignon Medina, Yane Caroline dos Anjos Bezerra, Ellen Rayssa Oliveira, Ricardo Alfredo Kluge and Marcel Bellato Spósito
Cropping and Pruning Systems of Primocane Raspberries in the Subtropical Climate
Reprinted from: *Horticulturae* **2024**, *10*, 1197, https://doi.org/10.3390/horticulturae10111197 . . **6**

Francisco José Domingues Neto, Marco Antonio Tecchio, Cristine Vanz Borges, João Domingos Rodrigues, Elizabeth Orika Ono, Giuseppina Pace Pereira Lima, et al.
Yield Performance and Quality Assessment of Brazilian Hybrid Grapes Influenced by Rootstocks and Training Systems
Reprinted from: *Horticulturae* **2024**, *10*, 909, https://doi.org/10.3390/horticulturae10090909 . . . **17**

Stanko Vršič, Borut Pulko and Andrej Perko
Structure and Trends in Climate Parameters of Wine-Growing Regions in Slovenia
Reprinted from: *Horticulturae* **2024**, *10*, 854, https://doi.org/10.3390/horticulturae10080854 . . . **37**

Roxana Mihaela Filimon, Claudiu Ioan Bunea, Răzvan Vasile Filimon, Florin Dumitru Bora and Doina Damian
Long-Term Evolution of the Climatic Factors and Its Influence o n G rape Q uality in Northeastern Romania
Reprinted from: *Horticulturae* **2024**, *10*, 705, https://doi.org/10.3390/horticulturae10070705 . . . **54**

Stanko Vršič, Borut Pulko, Tadeja Vodovnik-Plevnik and Andrej Perko
The Impact of Climatic Warming on Earlier Wine-Grape Ripening in Northeastern Slovenia
Reprinted from: *Horticulturae* **2024**, *10*, 611, https://doi.org/10.3390/horticulturae10060611 . . . **76**

Gabriel Cássia Fortuna, Caio Scardini Neves, Olivia Pak Campos, Jordany Aparecida Oliveira Gomes, Júlio César Rodrigues Lopes Silva, Amauri Alves Souza, et al.
Hop Tropicalization: Chemical Compositions of Varieties Grown under Organic and Conventional Systems in Subtropical Conditions
Reprinted from: *Horticulturae* **2023**, *9*, 855, https://doi.org/10.3390/horticulturae9080855 . . . . **93**

Francisco José Domingues Neto, Adilson Pimentel Junior, Lenon Romano Modesto, Mara Fernandes Moura, Fernando Ferrari Putti, Carmen Silvia Fernandes Boaro, et al.
Photosynthesis, Biochemical and Yield Performance of Grapevine Hybrids in Two Rootstock and Trellis Height
Reprinted from: *Horticulturae* **2023**, *9*, 596, https://doi.org/10.3390/horticulturae9050596 . . . . **105**

Rafaelly Calsavara Martins, Sarita Leonel, Jackson Mirellys Azevedo Souza, Giuseppina Pace Pereira Lima, Magali Leonel, Fernando Ferrari Putti, et al.
Profile of Bioactive Compounds in Orange Juice Related tothe Combination of Different Scion/Rootstocks, Packagingand Storage
Reprinted from: *Horticulturae* **2023**, *9*, 347, https://doi.org/10.3390/horticulturae9030347 . . . . **118**

Sarita Leonel, Magali Leonel, Paulo Ricardo Rodrigues de Jesus, Marco Antonio Tecchio, Marcelo de Souza Silva, Hebert Teixeira Cândido, et al.
Achievements of Banana (*Musa* sp.)-Based Intercropping Systems in Improving Crop Sustainability
Reprinted from: *Horticulturae* **2024**, *10*, 956, https://doi.org/10.3390/horticulturae10090956 . . . **136**

Saleem Jaffar, Syed Arif Hussain Rizvi and Yongyue Lu
Understanding the Invasion, Ecological Adaptations, and Management Strategies of *Bactrocera dorsalis* in China: A Review
Reprinted from: *Horticulturae* **2023**, *9*, 1004, https://doi.org/10.3390/horticulturae9091004 . . . **156**

About the Editors

Sarita Leonel

Sarita Leonel, graduated in the field of Agronomy, is a Full Professor in the Department of Crop Sciences at the College of Agricultural Sciences at São Paulo State University (UNESP), focused on the topic of fruit crops. Her research projects are centered on the physiology, production and post-harvest handling of fruit.

Sergio Ruffo Roberto

Sergio Ruffo Roberto, graduated in Agronomy, is an Associate Professor in the Department of Agronomy, Agricultural Research Center, Londrina State University (UEL). He holds the title of Professor of Fruit Crops and Fruit Production Technology. His research projects are focused on the physiology, production and post-harvest handling of fruit and hops.

Simone Rodrigues da Silva

Simone Rodrigues da Silva, graduated in Agronomy, is an Associate Professor in the Department of Crop Sciences, Luiz de Queiroz College of Agriculture, University of São Paulo (USP). As Professor of Fruit Crops and Plant Propagation, his research projects focus on the management and evaluation of cultivars, and on fruit propagation.

horticulturae

MDPI

Editorial

Orchard Management Under Climate Change

Sarita Leonel [1,*], Sergio Ruffo Roberto [2] and Simone Rodrigues da Silva [3]

1 Department of Horticulture, School of Agriculture, São Paulo State University (UNESP), Avenida Universitária, 3780, Botucatu 18610-307, SP, Brazil
2 Department of Agronomy, Agricultural Research Center, Londrina State University (UEL), Celso Garcia Cid Road, P.O. Box 10.011, Londrina 86057-970, PR, Brazil; sroberto@uel.br
3 Luiz de Queiroz College of Agriculture, University of São Paulo (USP), Avenida Pádua Dias, n° 11, P.O. Box 9, Piracicaba 13418-900, SP, Brazil; srsilva@usp.br
* Correspondence: sarita.leonel@unesp.br

Received: 14 January 2025
Accepted: 15 January 2025
Published: 16 January 2025

Citation: Leonel, S.; Roberto, S.R.; da Silva, S.R. Orchard Management Under Climate Change. *Horticulturae* **2025**, *11*, 98. https://doi.org/10.3390/horticulturae11010098

Health-conscious consumers are looking for healthier foods that promote health and well-being. Fruit stands out and is recognized for its nutritional value, with considerable levels of fibre, minerals, phenolic content, other antioxidant substances, vitamins and other nutrients [1]. Expanding fruit crops can help increase food production and consequently reduce food insecurity and hunger around the world [2], as well as generating employment and income along the production supply chain.

Hundreds of fruit species are spread across continents all over the world. Dozens of the best fruits, produced in regions with specific climates and soils and distinct vocations, supply the global market and find their place in competitive international trade [3]. Almost every type of fruit on the planet is harvested in the tropics and temperate zones, which explains how much of the land of micro, small, medium and large commercial farms is identified as orchards [4]. Like agribusiness as a whole, this sector is cutting costs wherever possible and planning for extreme efficiency in operations, but harvests are only viable if plantations are well looked after. Meanwhile, export sectors benefit from the exchange rate, turning crises in some sectors into opportunities for growth in others [4].

Climate is one of the most important limiting factors for fruit production and climate change is currently the biggest threat to the environment [5,6]. A shortage or excess of rainfall, extreme temperatures, hail and frost are affecting fruit crops all over the world. In this context, it is necessary to continually assess the long-term evolution of climatic factors and their influence on the production and quality of fruit worldwide [7].

Climate change is a very topical and relevant issue for orchard management and has been addressed in different ways in this Special Issue. A series of articles are presented on various topics of cultural practices in orchards under different climatic conditions and with different fruit trees and the impact of climate warming on vineyard crops. Eight research articles and two reviews were briefly described in this Editorial. The main aim was to motivate readers to further explore the articles.

Raspberries (*Rubus idaeus* L.) are originally grown in temperate climate regions. Increased market demand for fresh raspberries, however, has led to the development of new strategies to supply the market with this fruit all year round, with production outside the main harvest season and the expansion of growing areas to warm climate regions [8]. Medina et al. (contribution 1) evaluated the cultivation and pruning systems for primocane raspberries in a subtropical climate in the state of São Paulo, Brazil, where raspberry production is limited to high latitude cool temperate areas due to the need for low temperatures for flowering and fruiting of most cultivars. Nevertheless, primocane cultivars, which are less demanding in terms of cold conditions, represent a possible alternative that adapts to subtropical climate regions. The results show that the Heritage raspberry cv. Primocana can be grown in the subtropical Cwa climate with sequential pruning, allowing the production of commercial fruit with harvests distributed over the year, without reducing the post-harvest quality of the fruit produced.

Ongoing studies are needed to ensure that wine production in Brazil is optimized. The relationship between the variety, rootstock and trellis height is important for vine management, especially for the production of new varieties of grapes for juice and wine in new wine-growing regions with high production potential. The choice of training system depends on the architecture and physiological characteristics of the vines, affecting the production and quality of the grapes [9]. Domingues Neto et al. (contribution 2) studied the production and quality of Brazilian hybrid grapes according to the rootstock and training system in the state of São Paulo, Brazil. The high trellis system provided the overall most effective results, increasing the photosynthetic rate, improving water use efficiency, increasing vine production and improving fruit quality. The authors recommended its use for training vines. With regard to rootstocks, the best compatibility between scion and rootstock was found between hybrid vines and 'IAC 766 Campinas'.

Following the same research, the effects on photosynthesis, biochemistry and production performance of grapevine hybrids on two rootstocks and trellis heights were evaluated in the Cfa climate of São Paulo state, Brazil, by Domingues Neto et al. (contribution 3). The rootstock and trellis height combination had a positive effect on the variables evaluated. In summary, under subtropical conditions, better photosynthetic, biochemical and productive performance was observed when the cultivars IAC 138-22 Maximo and BRS Violeta were grafted onto the rootstock 'IAC 766'. 'IAC 138-22 Maximo' was trained at 2.0 m and grafted onto 'IAC 766' rootstock, increasing grape production and photosynthesis efficiency. In addition to this, this variety was more productive than 'BRS Violeta'.

The sustainability of citrus crops is a global matter of interest. Brazil is among the countries that are vulnerable to cultivating a reduced number of citrus genotypes, which leads to greater susceptibility to pests and diseases, as well as lower economic competitiveness [10]. New combinations of scion/scion cultivars are a constant need for citrus growers and are also aimed at satisfying the preferences of consumers, who are increasingly demanding in terms of the quality attributes of the fruit and the orange juice consumed. There is currently a growing demand for nutritious food and many attempts have been made to maximize nutrient retention during storage and during processing. Martins et al. (contribution 4) explored the profile of bioactive compounds in orange juice as a result of using different rootstocks, packaging and storage. The research insights can contribute to the diversification of scion/rootstock cultivars in hopes of increasing orchard variety by choosing the best combinations for pasteurized orange juice with the highest nutritional value.

An important topic presented studies on the tropicalization of hops in subtropical conditions (contribution 5). The interest in hop production in Brazil, motivated by the third position in the world ranking of beer producers and the growth of the craft brewery business, justifies the intensification of studies on its adaptation to local cultivation conditions. Considering the high internal demand and expansion of the national beer market with interest in hops with peculiar phytochemical profiles, studies that promote the expansion of new cultivation zones and that guarantee quality are very desirable. Fortuna et al. (contribution 5) investigated the chemical compositions of varieties grown in organic and conventional systems in the state of São Paulo in Brazil. This study contributes the first report of the chemical profiles of hops grown in subtropical conditions.

Climate change is threatening wine production everywhere, especially in regions with hot, dry climates. The risks of frost and drought during the growing season are common problems in viticulture. Therefore, maintaining viticulture also requires adapting to climate change, and the evaluation of adaptation strategies needs to be more precise and multidisciplinary and adapted to specific local conditions [4].

Vrši˘c et al. (contribution 6) examined the structure and trends of climatic parameters important for grape production over seventy years in Slovenian wine regions. Mean and extreme temperature and precipitation data from six meteorological stations in three wine regions were divided into annual and growing seasons. The trends show a decrease in total annual precipitation.

Vrši˘c et al. (contribution 7) also reported on the impact of climate warming on the early ripening of wine grapes in north-eastern Slovenia. In this study, the development trends of bioclimatic parameters recorded over seventy years and the dynamics of grape ripening in early, medium and late ripening grape varieties were investigated. Based on the data on soluble solid content, total acidity and the recommended harvest date per year, the trends in the reduction in the growing period of the vines were calculated. Temperature changes were more pronounced and the number of so-called hot days (with a maximum T > 30 °C) increased the most, which has the greatest impact on other bioclimatic parameters, for example, average temperature and growing degree days. Total annual rainfall and rainfall in the growing season show downward trends.

Filimon et al. (contribution 8) highlighted the changes in the main climatic elements during the last five decades (1971–2020) in north-eastern Romania and their impact on grape quality, as part of precision viticulture strategies and efficient management of vine orchards. The main outcomes indicated a significant increase in the values of the bioclimate indicators, requiring the reclassification of the viticulture area into higher classes of favorability, increasing the opportunity to grow cultivars more suited to warmer climates, ensuring the efficiency of the vineyard and meeting the current consumer demands.

In this Special Issue, selected topics on the perspectives, challenges and sustainability of orchard management under climate change are covered. Sustainable agricultural practices need to be continually sought so that a greater number of producers can adopt them, taking into account, above all, the food security scenario, land-use efficiency and climate change.

Leonel et al. (contribution 9) reviewed recent findings on banana-based intercropping systems. The authors provided an overview of studies on intercropping banana plantations, focusing on the contextualization of land use, monoculture and intercropping, and evaluating intercropping indicators, as well as the benefits, risks and disadvantages discussed in the literature and the main results of banana-based intercropping systems. The main conclusions relate to the use of combined crops with aromatic species and preliminary reports on the contributions of intercropping to the suppression of Fusarium wilt disease.

Jaffar et al. (contribution 10) presented an important review article that provides an overview of the invasion history of B. dorsalis in China, its ecological and physiological mechanisms that facilitate invasion, and the progress made in understanding its main biological characteristics. Bactrocera dorsalis (Hendel, 1912) (Diptera: Tephritidae), commonly known as the oriental fruit fly, is a highly destructive pest that infests fruit and vegetables worldwide, resulting in economic losses every year. The main B. dorsalis management approaches that have been or are likely to be implemented in China were presented, including quarantine measures, monitoring procedures, physical controls, biological controls, the sterile insect technique, RNA interference and CRISPR-Cas-9.

Given the great importance and relevance of the theme of this Special Issue, ongoing studies should be encouraged and rapidly publicized, allowing readers to find research in global and national contexts that provides a more complete overview of the field of research into climate change affecting the growth, yield and qualitative performance of fruit orchards.

Author Contributions: S.L., S.R.R. and S.R.d.S. wrote the Editorial. All authors have read and agreed to the published version of the manuscript.

Acknowledgments: As guest editors of the Special Issue "Orchard Management Under Climate Change", we would like to extend our sincere thanks to all the authors who contributed their valuable articles to the success of this Special Issue.

Conflicts of Interest: The authors declare no conflicts of interest.

List of Contributions:

1. Medina, R.B.; Bezerra, Y.C.S.; Oliveira, E.R.; Kluge, R.A.; Spósito, M.B. Cropping and Pruning Systems of Primocane Raspberries in the Subtropical Climate. *Horticulturae* **2024**, *10*, 1197. https://doi.org/10.3390/horticulturae10111197.
2. Domingues Neto, F.J.; Tecchio, M.A.; Borges, C.V.; Rodrigues, J.D.; Ono, E.O.; Lima, G.P.P.; Moura, M.F.; Hernandes, J.L.; Silva, M.d.S.; Leonel, M. Yield Performance and Quality Assessment of Brazilian Hybrid Grapes Influenced by Rootstocks and Training Systems. *Horticulturae* **2024**, *10*, 909. https://doi.org/10.3390/horticulturae10090909.
3. Domingues Neto, F.J.; Pimentel Junior, A.; Modesto, L.R.; Moura, M.F.; Putti, F.F.; Boaro, C.S.F.; Ono, E.O.; Rodrigues, J.D.; Tecchio, M.A. Photosynthesis, Biochemical and Yield Performance of Grapevine Hybrids in Two Rootstock and Trellis Height. *Horticulturae* **2023**, *9*, 596. https://doi.org/10.3390/horticulturae9050596.
4. Martins, R.C.; Leonel, S.; Souza, J.M.A.; Lima, G.P.P.; Leonel, M.; Putti, F.F.; Monteiro, G.C.; Züge, P.G.Ü.; Napoleão, G.M.; Figueira, R.; et al. Profile of Bioactive Compounds in Orange Juice Related to the Combination of Different Scion/Rootstocks, Packaging and Storage. *Horticulturae* **2023**, *9*, 347. https://doi.org/10.3390/horticulturae9030347.
5. Fortuna, G.C.; Neves, C.S.; Campos, O.P.; Gomes, J.A.O.; Silva, J.C.R.L.; Souza, A.A.; Funari, C.S.d.; Marques, M.O.M.; Bonfim, F.P.G. Hop Tropicalization: Chemical Compositions of Varieties Grown under Organic and Conventional Systems in Subtropical Conditions. *Horticulturae* **2023**, *9*, 855. https://doi.org/10.3390/horticulturae9080855.
6. Vršič, S.; Pulko, B.; Perko, A. Structure and Trends in Climate Parameters of Wine-Growing Regions in Slovenia. *Horticulturae* **2024**, *10*, 854. https://doi.org/10.3390/horticulturae10080854.
7. Vršič, S.; Pulko, B.; Vodovnik-Plevnik, T.; Perko, A. The Impact of Climatic Warming on Earlier Wine-Grape Ripening in Northeastern Slovenia. *Horticulturae* **2024**, *10*, 611. https://doi.org/10.3390/horticulturae10060611.
8. Filimon, R.M.; Bunea, C.I.; Filimon, R.V.; Bora, F.D.; Damian, D. Long-Term Evolution of the Climatic Factors and Its Influence on Grape Quality in Northeastern Romania. *Horticulturae* **2024**, *10*, 705. https://doi.org/10.3390/horticulturae10070705.
9. Leonel, S.; Leonel, M.; Jesus, P.R.R.d.; Tecchio, M.A.; Silva, M.d.S.; Cândido, H.T.; Molha, N.Z.; Ouros, L.F.d. Achievements of Banana (*Musa* sp.)-Based Intercropping Systems in Improving Crop Sustainability. *Horticulturae* **2024**, *10*, 956. https://doi.org/10.3390/horticulturae10090956.
10. Jaffar, S.; Rizvi, S.A.H.; Lu, Y. Understanding the Invasion, Ecological Adaptations, and Management Strategies of Bactrocera dorsalis in China: A Review. *Horticulturae* **2023**, *9*, 1004. https://doi.org/10.3390/horticulturae9091004.

References

1. Petsakos, A.; Orager, S.; Gonzalez, C.E.; Gama, A.C.; Sulser, T.B.; Gbegbelegbe, S.; Kikulwe, E.M.; Hareau, G. Understanding the Consequences of Changes in the Production Frontiers for Roots, Tubers, and Bananas. *Glob. Food Secur.* **2019**, *20*, 180–188. [CrossRef]
2. Cândido, H.T.; Marzullo, Y.O.T.; Leonel, M. Green Banana Flour Technology: From Raw Material to Sensory Acceptance of Products Made with Green Banana Flour in the Brazilian Scenario. *Braz. Arch. Biol. Technol.* **2023**, *66*, e23210543. [CrossRef]
3. Boteon, M. *HortiFruit Brasil Yearbook*; Editora CEPEA/USP: Piracicaba, Brazil, 2024; n. 240; 42p, ISSN 1981-1837. Available online: https://www.hfbrasil.org.br/en/magazine/access/completo/hf-brasil-yearbook-2023-retrospective-2024-outlook.aspx (accessed on 6 January 2025).
4. Kist, B.B. *Brazilian Horti & Fruit Yearbook*; Editora Gazeta Santa Cruz: Santa Cruz do Sul, Brazil, 2022; 88p, Available online: https://editoragazeta.com.br/produto/anuario-brasileiro-de-horti-fruti-2022/ (accessed on 7 January 2025).

5. Naulleau, A.; Gary, C.; Prévot, L.; Hossard, H. Strategies for adaptation to climate change in grapevine production—A systematic review. *Front. Plant Sci.* **2021**, *11*, 607859. [CrossRef] [PubMed]
6. Fraga, H. Climate change: A new challenge for the winemaking sector. *Agronomy* **2020**, *10*, 1465. [CrossRef]
7. Venios, X.; Korkas, E.; Nisiotou, A.; Banilas, G. Grapevine Responses to Heat Stress and Global Warming. *Plants* **2020**, *9*, 1754. [CrossRef] [PubMed]
8. Heide, O.M.; SØNsteby, A. Physiology of flowering and dormancy regulation in annual- and biennial-fruiting red raspberry (*Rubus idaeus* L.)—A review. *J. Hortic. Sci. Biotechnol.* **2011**, *86*, 433–442. [CrossRef]
9. Simonetti, L.M.; Sousa, M.C.; Moura, M.F.; Nunes, J.G.d.S.; Diamante, M.S.; da Silva, M.B.; da Silva, M.J.R.; Callili, D.; Lima, G.P.P.; Tecchio, M.A. The influence of different training systems and rootstocks on Sauvignon Blanc grapes. *Bragantia* **2021**, *80*, e2021. [CrossRef]
10. Carvalho, L.M.; Carvalho, H.W.L.; Barros, I.; Martins, C.R.; Soares Filho, W.S.; Girard, E.A.; Passos, O.S. New scion-rootstock combinations for diversification of sweet orange orchards in tropical hardsetting soils. *Sci. Hortic.* **2019**, *243*, 169–176. [CrossRef]

horticulturae

Article

Cropping and Pruning Systems of Primocane Raspberries in the Subtropical Climate

Ricardo Bordignon Medina [1], Yane Caroline dos Anjos Bezerra [2], Ellen Rayssa Oliveira [1,*], Ricardo Alfredo Kluge [1] and Marcel Bellato Spósito [2,*]

1 Department of Biological Sciences, Luiz de Queiroz College of Agriculture, University of São Paulo, Piracicaba 13418-900, Brazil; ricardobmedina@gmail.com (R.B.M.); rakluge@usp.br (R.A.K.)
2 Department of Plant Production, Luiz de Queiroz College of Agriculture, University of São Paulo, Piracicaba 13418-900, Brazil; yanecaroline.bezerra@gmail.com
* Correspondence: ellen.rayssa@usp.br (E.R.O.); mbsposito@usp.br (M.B.S.); Tel.: +55-19-983716439 (M.B.S.)

Abstract: Raspberry production is limited to cold temperate areas of high latitude due to the requirement of low temperatures for flowering and fruiting from most cultivars. However, primocane cultivars, as they are less demanding in cold conditions, represent a possible alternative that suits regions with a subtropical climate. The cultivar Heritage primocane raspberry was investigated in the Cwa climate, in three production systems (PS), during two crop cycles. In PS1, canes were hard pruned at ground level after primocane fruiting. In PS2, canes were tipped to promote subapical bud break for a second harvest. In PS3, canes were tipped again after the second harvest to induce a third harvest. PS1 had the lowest yield, however, after two cycles; in plants of this system it was observed the highest root weight, and starch content. Raspberries subjected to subapical pruning show lower carbohydrate storage in the root system. The production systems had little influence on fruit qualities, in both cycles. The cultivation of cv. Heritage raspberry primocane, in the subtropical Cwa climate can be carried out with sequential pruning, allowing for the production of commercial fruits with harvests distributed over the months, without any reduction in the postharvest quality of the fruits produced.

Keywords: *Rubus idaeus*; cultivation; warm regions; harvest; fruit quality

Citation: Medina, R.B.; Bezerra, Y.C.d.A.; Oliveira, E.R.; Kluge, R.A.; Spósito, M.B. Cropping and Pruning Systems of Primocane Raspberries in the Subtropical Climate. *Horticulturae* **2024**, *10*, 1197. https://doi.org/10.3390/horticulturae10111197

Academic Editor: Cristina Moniz Oliveira

Received: 27 September 2024
Revised: 4 November 2024
Accepted: 9 November 2024
Published: 14 November 2024

1. Introduction

Raspberry (*Rubus idaeus* L.) is traditionally cultivated in regions of temperate climates. Raspberries are cultivated worldwide in an area of 116,393 hectares and yield a production of 947,852 tons [1]. The majority of global production takes place in the Northern Hemisphere, led by Russia, Mexico, Serbia, Poland, and the United States of America, collectively contributing to 72.12% of the total production [1]. In South America, Chile stands out as the largest producer with a production of 11, 775 tons. In Brazil, raspberries are cultivated on 40 hectares, producing 240 tons annually, representing just 0.025% of global production [2]. Nevertheless, the increase in the market demand for fresh raspberries has driven the development of new strategies to supply the market for this fruit during the whole year, with productions outside the traditional harvest season and the expansion of cultivation areas for regions of hot climate [3–5].

Raspberry cultivars are classified into primocanes and floricanes based on their fruiting habits, with primocanes producing fruit from the apical nodes on current-year canes. These cultivars do not need low temperatures nor to undergo dormancy period for the induction of flower buds to occur [6]. Floricane cultivars present a biannual cycle. The canes develop vegetatively during spring and summer, and they need to undergo periods of low temperatures to enter dormancy during winter, bloom, and bear fruit in the spring and summer of the second year [3].

The use of primocane raspberry cultivars represents an alternative that is economically feasible for production in regions without chilling weather since the plants have the potential to produce fruit during the whole year under conditions of protected cultivation [5]. Raspberry cultivars, such as Heritage, Autumn Bliss, Autumn Britten, Caroline, Himbo Top, Polka, and Sugana, are some of the primocane cultivars planted in several regions of the world [7].

One of the primary objectives of fruiting pruning in raspberry cultivation is to remove the harvested inflorescences and optimize cane density for the upcoming production cycle. However, pruning strategies vary according to the cultivar group. In primocane cultivars, fruiting first occurs at the apex of newly developed canes. These canes are pruned following the initial harvest, as they retain the potential for additional fruiting. After their second production phase, the canes desiccate and must be replaced by new canes, which will support the subsequent harvest cycle [2,8].

Previous studies indicate that primocane raspberry pruning impacts root carbohydrate reserves, yield, and the duration of the production cycle [9–11]. Therefore, it is essential to conduct investigations that deepen the understanding of the influence of pruning on the physiological aspects of the raspberry plant, aiming to optimize fruit production and quality. In this context, studies on the pruning systems and the resulting performance of primocane raspberries are necessary, aiming at the expansion of the cultivation to areas of warm climate.

Thus, this study aims to assess production systems with various pruning types and their effects on the development, production, and fruit quality of cv. Heritage primocane raspberries cultivated in a subtropical climate (Cwa).

2. Materials and Methods

2.1. Plant Material and Description of the Study Area

Primocane raspberry (*Rubus idaeus* L.) rooted cuttings, cv. Heritage, were grown in 30 L pots filled with a substrate composed of coconut fiber (Golden Mix Misto 98, Amafibra, São Paulo, Brazil) and *Sphagnum* peat (Jiffy TPS, Jiffy Group, Zwijndrecht, The Netherlands) at a 2:1 ratio. Weekly nutrient applications were administered through fertigation (N: Ammonium Sulfate 1000 mg L^{-1}; Calcium Nitrate 500 mg L^{-1}; P: Mono Ammonium Phosphate 150 mg L^{-1}; K: Potassium Sulfate 500 mg L^{-1}). The study was conducted in Piracicaba, Brazil (22°42′27.7″ S 47°37′47.3″ W, altitude of 554 m), within a Cwa climate according to the Köppen and Geiger classification [12]. Raspberry pots were placed in a greenhouse, externally covered with a 150 µm low-density polyethylene (LDPE) diffuser film, and internally equipped with a gray heat-reflective screen (Freshnet®—providing 65% nominal shading). One week post-transplantation, two canes per pot were selected, and any additional emerging buds were removed weekly to control cane growth and evaluate production.

2.2. Treatments and Experimental Design

The treatments under examination comprised distinct production systems (PS) for each cane: PS1 involved harvesting at the apical nodes, followed by hard pruning, after which another cane would be used in the next production cycle (Figure 1A); PS2 entailed harvesting at the apical nodes, followed by tipping to induce subapical bud break, allowing a subsequent harvest on the same cane, and concluded with hard pruning (Figure 1B); PS3 encompassed harvesting at the apical nodes, followed by tipping to promote subapical bud break, a subsequent harvest, further tipping to encourage new sprouting, and a third harvest on the same cane, ultimately concluding with hard pruning (Figure 1C).

The plants underwent evaluation for two complete production cycles, with hard pruning of the canes performed at the end of each cycle. The duration from cane emission to the last harvest of each treatment defined a complete production cycle. Following the first harvest in the apical nodes of the canes during the first cycle, two new canes per plot were selected, initiating the second production cycle (Figure 1A–C).

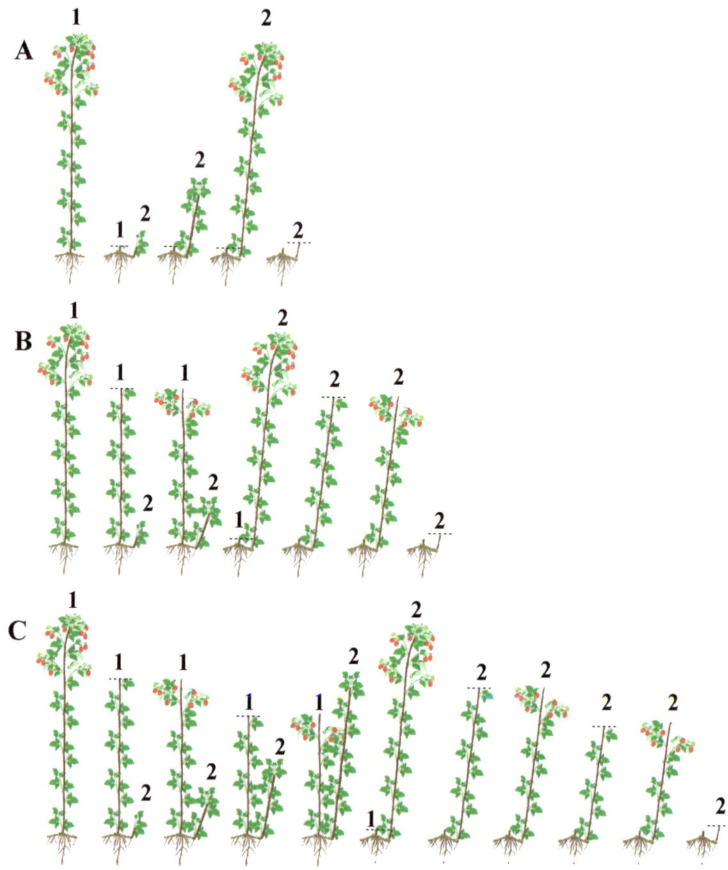

Figure 1. Production systems of cv. Heritage raspberries primocane cultivated in Piracicaba, Brazil, for two cycles (1 and 2). (**A**): Production system PS1, single harvest in the apical nodes of the canes, followed by hard pruning; (**B**): Production system PS2, harvest in the apical nodes of the canes, followed by tipping to induce the second harvest, and subsequent hard pruning; (**C**): Production system PS3, harvest in the apical nodes of the canes, followed by tipping to induce the second harvest, second tipping to induce the third harvest, and finally, hard pruning.

The experimental design employed a randomized block approach in a double factorial scheme (3 treatments × 2 production cycles) with four blocks. Each block included three pots from each treatment, with two canes in each pot, resulting in a total of 12 pots per treatment and 72 canes evaluated per cycle.

2.3. *Vegetative Development, Aspects of Production, and Postharvest Quality*

The vegetative development was evaluated by the growth of the canes in each treatment. Over two vegetative cycles, the length of two canes per pot was measured weekly, considering the distance from the base to the apical meristem of each cane. This evaluation continued until the onset of the reproductive period, marked by flowering, at which point the canes ceased their vegetative development.

To determine production per cane, harvests were conducted three times a week, assessing both the number and weight of harvested fruit per cane. The fruits were harvested based on the developmental stages of raspberry fruit, using the color scale for cv. Heritage, specifically at the pink (P) stage, when the drupelets detach easily from the receptacle [13].

Data were recorded in grams per cane per week. Fresh fruit mass was measured using an analytical scale, model AG 200 (Gehaka, São Paulo, Brazil), immediately after each harvest. Soluble solids content was determined using a digital refractometer, model Palette 101 (Atago, Tokyo, Japan), and expressed in °Brix. Titratable acidity (TA) and pH were measured using an automatic titrator (Model 848 Titrino Plus, Metrohm, Herisau, Switzerland) with four replicates of ten fruit for each treatment, and results were expressed as a percentage of citric acid. For anthocyanin content, extracts were obtained from 20 mg of freeze-dried raspberries and 10 mL of extraction solution (85% ethanol P.A. and 15% 1.5 N HCl), following the spectrophotometric method [14]. Absorbance readings were taken with a spectrophotometer (Model Libra S22, Biochrom, Cambridge, UK) at 535 nm. The analyses were performed in periods in which there was fruit in all treatments, encompassing the period from Aug to Dec in the first cycle and from Mar to June in the second cycle, in the Southern Hemisphere.

2.4. Root Biomass and Starch Accumulation

The dry and fresh root masses, along with the starch content, were determined after the completion of the second production cycle for each system used. Root collection took place upon the conclusion of each system's production. Both dry and fresh root masses were measured using an analytical scale (Model AG 200, Gehaka, São Paulo, Brazil). Fresh mass data were collected after the conclusion of the second production cycle, coinciding with the hard pruning of the plants. Subsequently, roots were thoroughly washed in running water until complete substrate removal. After measuring the fresh mass, the roots were placed in identified paper bags and dried in an oven at 55 °C until weight stabilization.

The determination of starch in root samples was conducted following a previously established protocol, with modifications [15]. Soluble sugars were extracted using 200 mg of root samples subjected to three consecutive extractions in 70% ethanol at 60 °C and two extractions with 37% perchloric acid. The extract comprised the recovery of supernatants post-centrifugation at 500 rpm. Glucose measurement employed the phenol sulfuric acid method with a reaction mixture of 50 μL of extract, 450 μL of H_2O, and 500 μL of phenol reagent (5% in water). After vortexing, 2 mL of concentrated sulfuric acid were added, followed by further agitation. Readings were taken with a spectrophotometer at 490 nm. A glucose standard curve (2 to 80 μg) facilitated the calculation of the glucose amount released from perchloric acid digestion. The data were expressed in mg of starch per g dry mass of root.

2.5. Statistical Analysis

The obtained data were subjected to Analysis of Variance (ANOVA) using R Studio software (R Core Team, 2018—Version 1.2.5033), and mean comparisons were performed with the Scott-Knott test ($p \leq 0.05$).

3. Results

3.1. Vegetative Development, Aspects of Production, and Postharvest Quality

The final cane growth was not influenced by the production system. The mean cane length at the end of the first cycle was 130.9 ± 3.4 cm and 153.5 ± 5.0 cm in the second cycle. During the first cycle, the vegetative period of the plants extended for 104 days. In contrast, the second cycle experienced a reduced vegetative development period of 90 days, from the beginning of cane development until the first harvest.

Harvest timing was similar for treatment PS1, as well as for PS2 and PS3, initiating in early Aug and extending until early Dec (Figure 2). In PS1, the first harvest was followed by hard pruning. Subsequent harvests started only from Mar of the subsequent year, extending until early June in the new canes that developed from Dec to Apr (Figure 2A). The first production cycle for PS2 occurred from Aug to Feb, when the plants ceased production in the subapical buds, and a hard pruning of the canes was conducted. The second production cycle of PS2 began in late Mar with canes that developed from Dec to Apr, producing fruit in the apical nodes until June, which was followed by cane tipping that promoted subapical bud break and resulted in a second harvest from June to Sept (Figure 2B). In PS3, a third harvest was subsequently obtained from mid-Feb to late Mar (Figure 2C).

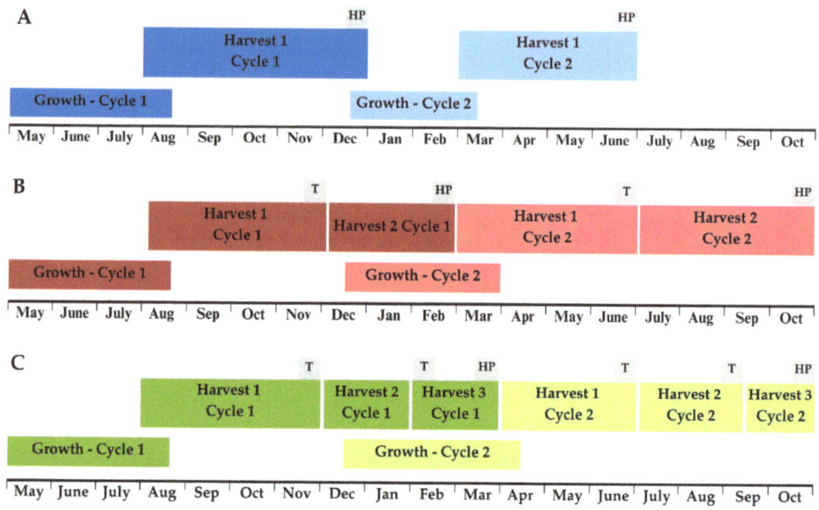

Figure 2. Three production systems in two production cycles for a single cane, showing different harvest strategies. Each system includes two cycles, with additional harvests in systems PS2 and PS3. (**A**): Production system PS1, with a single harvest at the apical nodes of the canes, followed by hard pruning (HP); (**B**): Production system PS2, with an initial harvest at the apical nodes, followed by tipping (T) to induce a second harvest, and subsequent hard pruning; (**C**): Production system PS3, featuring an initial harvest at the apical nodes, followed by tipping to induce a second harvest, a second tipping to promote a third harvest, and ending with hard pruning.

During the first cycle, the total production per cane in PS1 was 168 g, 252 g in PS2, and 313 g in PS3 (Table 1). Differences were also observed in the second cycle, where PS1 continued to produce less per cane (164 g), PS3 more (204 g) and PS2 was an intermediary between them (189 g). In the second production cycle in all treatments, the production was lower (Table 1). There were declines in production between the first and second cycles for all three treatments, with the smallest decrease of 2.7% for PS1, followed by 24.8% for PS2 and 34.6% for PS3 (Table 1). According to the results for the area under the progress curve, there was no interaction between the factors' production system and cultivation cycles (Table 1). Production systems PS2 and PS3 were equivalent to each other, and superior to PS1. The distribution of yields from Mar to Oct in the second cycle exhibited a comparable pattern between PS2 and PS3, but significantly differed from PS1. PS1 concluded its entire harvest distribution by July, while PS2 and PS3 extended their harvest until Oct in the second cycle.

The physicochemical parameters of the raspberries were minimally affected by the variations in production systems. The soluble solids contents, titrable acidity, and anthocyanins varied throughout the months of harvest but did not show significant variations as a function of the treatments (Tables 2–4). For the soluble solids contents, in the first cycle, a clear trend for higher values occurring in Aug and Sept is observed, with a noticeable decrease from Oct in the first cycle to Mar in the second cycle (Table 2). In the second cycle, the differences were less pronounced from Mar to Aug, with a reduction observed in Sept and Oct of that year (Table 2).

Table 1. Production per cane of cv. Heritage primocane raspberry, in three production systems (PS1, PS2, and PS3) across two production cycles, in Piracicaba, Brazil. The first cycle began in August and ended in March, while the second cycle started in March and concluded in October. PS1: single harvest at apical nodes followed by hard pruning; PS2: initial harvest at apical nodes, followed by tipping to induce a second harvest and subsequent hard pruning; PS3: initial harvest at apical nodes, followed by tipping to induce a second harvest, a second tipping to promote a third harvest, and final hard pruning.

Systems	Production per Cane (g)	
	1st Cycle	2nd Cycle
PS 1	168.5 bA	163.9 bB
PS 2	252.3 aA	189.7 aB
PS 3	313.2 aA	204.7 aB

Different lowercase letters in the columns and capital letters in the lines indicate significant differences ($p \leq 0.05$) according to the Scott-Knott test.

Table 2. Soluble solids content (°Brix) in fruits of cv. Heritage primocane raspberries from monthly harvests over two production cycles across three production systems (PS1, PS2, and PS3), in Piracicaba, Brazil. PS1: single harvest at apical nodes followed by hard pruning; PS2: initial harvest at apical nodes, followed by tipping to induce a second harvest and subsequent hard pruning; PS3: initial harvest at apical nodes, followed by tipping to induce a second harvest, a second tipping to promote a third harvest, and final hard pruning.

Soluble Solids Content (°Brix)								
				1st Cycle				
System	Aug	Sept	Oct	Nov	Dec	Jan	Feb	Mar
PS 1	10.6 aA	10.2 aA	8.4 aB	8.6 aB	8.1 bB	-	-	-
PS 2	10.4 aA	10.3 aA	8.6 aB	8.7 aB	9.9 aA	9.1 aB	7.8 aC	7.6 bC
PS 3	10.8 aA	10.1 aB	8.2 aE	8.5 aD	9.5 aC	8.9 aD	7.8 aE	8.4 aD
				2nd Cycle				
System	Mar	Apr	May	June	July	Aug	Sept	Oct
PS 1	9.1 aA	9.7 aA	9.3 aA	8.6 bB	-	-	-	-
PS 2	9.0 aC	9.8 aB	9.1 aC	9.4 aC	10.5 aA	8.7 bC	7.6 bD	7.7 aD
PS 3	8.0 bC	9.7 aA	9.0 aB	8.8 bB	9.4 bA	9.5 aA	8.7 aB	7.9 aC

Different lowercase letters in the columns and capital letters in the lines indicate significant differences ($p \leq 0.05$) according to the Scott-Knott test.

Table 3. Titrable acidity in fruits of cv. Heritage primocane raspberries from monthly harvests over two production cycles across three production systems (PS1, PS2, and PS3), in Piracicaba, Brazil. A: Production system PS1: single harvest in the apical nodes of the canes, followed by hard pruning; B: Production system PS2, harvest in the apical nodes of the canes, followed by tipping to induce the second harvest, and subsequent hard pruning; C: Production system PS3, harvest in the apical nodes of the canes, followed by tipping to induce the second harvest, second tipping to induce the third harvest, and finally, hard pruning.

Titrable Acidity (% of Citric Acid)								
				1st Cycle				
System	Aug	Sept	Oct	Nov	Dec	Jan	Feb	Mar
PS 1	1.87 aA	1.79 aA	1.47 aC	1.59 aB	1.81 aA	-	-	-
PS 2	1.78 aB	1.63 bC	1.49 aC	1.56 aC	1.69 bB	1.54 aC	1.77 aB	1.92 bA
PS 3	1.85 aB	1.71 aC	1.47 aD	1.53 aD	1.81 aB	1.55 aD	1.86 aB	2.18 aA
				2nd Cycle				
System	Mar	Apr	May	June	July	Aug	Sept	Oct
PS 1	1.76 aB	1.84 aB	1.95 aA	2.02 aA	-	-	-	-
PS 2	1.59 bC	1.80 aB	1.88 aA	1.88 bA	1.79 bB	1.94 aA	1.62 aC	1.41 aD
PS 3	1.60 bC	1.79 aB	1.92 aA	1.91 bA	1.95 aA	1.88 aA	1.65 aC	1.26 bD

Different lowercase letters in the columns and capital letters in the lines indicate significant differences ($p \leq 0.05$) according to the Scott-Knott test.

Table 4. Anthocyanin content in fruits of cv. Heritage primocane raspberries from monthly harvests across two production cycles (cycle 1: Aug to Dec; cycle 2: Mar to June) in three production systems (PS1, PS2, and PS3) in Piracicaba, Brazil. PS1: single harvest at the apical nodes of the canes, followed by hard pruning; PS2: harvest at the apical nodes, followed by tipping to stimulate a second harvest, and then hard pruning; PS3: harvest at the apical nodes, followed by a first tipping to induce a second harvest, a second tipping for a third harvest, and finally, hard pruning.

Anthocyanins (mg g^{-1} of Freeze-Dried Fruit)					
	1st Cycle				
System	Aug	Sept	Oct	Nov	Dec
PS 1	28.9 aB	22.1 aB	39.3 aA	35.8 aA	38.0 aA
PS 2	28.4 aB	21.2 aB	36.1 aA	33.9 aA	36.6 aA
PS 3	24.2 aB	26.4 aB	36.4 aA	33.5 aA	39.7 aA
	2nd Cycle				
System	Mar	Apr	May	June	
PS 1	37.0 aA	33.5 aB	33.9 aB	32.6 aB	
PS 2	39.8 aA	34.6 aB	34.3 aB	35.6 aB	
PS 3	39.5 aA	34.3 aB	33.2 aB	34.7 aB	

Different lowercase letters in the columns and capital letters in the lines indicate significant differences ($p \leq 0.05$) according to the Scott-Knott test.

The titrable acidity of the cv. Heritage raspberries, in both the first and second cycles, ranged from 1.47% to 2.18% of citric acid in the first cycle and 1.26% to 2.02% in the second cycle (Table 3). During the first cycle, there was a tendency for the highest values to occur in Aug and Sept, and in the second cycle, in Feb and Mar, with the lowest values recorded from Oct in the first cycle to Jan in the second cycle. In the second cycle, the highest values were observed between Apr and Aug, while the lowest values occurred in Mar, Sept, and Oct of the same year (Table 3).

In both cycles, seasonal differences in anthocyanin concentrations were observed. During the first cycle, the highest anthocyanin levels were recorded in Oct, Nov, and Dec, while the lowest levels occurred in Aug and Sept. In the second cycle, the peak was in Mar, whereas the lowest contents were noted in Apr, May, and June (Table 4).

3.2. Root Biomass and Starch Accumulation

After the termination of the harvests of the second cycle, the fresh and dry masses were evaluated, as well as the starch content in the roots of the plants. The values for PS2 and PS3 are very close, not differing from each other for these three parameters (Table 5). Nevertheless, they were around 30% inferior to PS1 (Table 5).

Table 5. Fresh mass, dry mass, and starch content in the roots of cv. Heritage raspberries subjected to different production systems for two complete cycles, in Piracicaba, Brazil. Values represent the average of all harvests throughout each cycle. A: Production system PS1: single harvest in the apical nodes of the canes, followed by hard pruning; B: Production system PS2, harvest in the apical nodes of the canes, followed by tipping to induce the second harvest, and subsequent hard pruning; C: Production system PS3, harvest in the apical nodes of the canes, followed by tipping to induce the second harvest, second tipping to induce the third harvest, and finally, hard pruning.

Systems	**Fresh Mass (g)**	**Dry Mass (g)**	**Starch (mg g^{-1})**
PS 1	117.8 a	28.9 a	4.65 a
PS 2	81.7 b	19.0 b	3.14 b
PS 3	82.7 b	19.5 b	3.23 b

Different lowercase letters in the columns indicate significant differences ($p \leq 0.05$) according to the Scott-Knott test.

4. Discussion

The raspberry plants were subjected to different pruning systems over two production cycles. Due to the different periods of the year in which these cycles started, differences were observed in both the time required to start production and the size of the canes. In the first cycle, the reproductive period began, on average, 104 days after transplanting, in canes measuring 130 cm. In the second cycle, the start of the reproductive period was shorter, averaging 90 days, with taller canes reaching 153 cm. This difference likely occurred because, in the first cycle, vegetative development took place from May to September, with milder temperatures and lower global solar radiation in the Southern Hemisphere. In the second cycle, the canes developed from December to April, experiencing higher temperatures and greater radiation, resulting in canes that grew, on average, taller than those of the previous cycle. Similar seasonal characteristics influencing cane growth have been observed in the primocane raspberry cv. Autumn Bliss, where higher cultivation temperatures resulted in an increased growth rate of the canes [16].

The pruning systems used influenced fruit production per cane (g), regardless of the production cycle. The systems involving one subapical pruning (PS2) or two subapical prunings (PS3) yielded similar production levels within each cycle, both of which were higher compared to the hard pruning without subapical prunings (PS1). However, systems with subapical prunings on the same cane may present additional challenges for the labor responsible for this cultural practice in raspberry cultivation. These findings are consistent with a study in which the primocane raspberry production over three consecutive years was higher when subjected to a double-cropping system that included harvesting from the apical nodes of the canes, followed by a second harvest from the subapical buds, as compared to production only from the apical nodes of the canes [17]. It should be noted that the present experiment compared production across two complete cycles of different durations. The tippings and subsequent harvests in PS2 and PS3 increased fruit production but prolonged the production cycle, whereas PS1 completed both production cycles three and a half months before PS2 and five months before PS3. During this period, it would have been possible to initiate a third cycle and the beginning of a new harvest, based on the time elapsed until a new harvest began for PS1 in the first and second cycles.

High levels of soluble solids and acidity are important parameters in fruit farming, often associated with a better flavor profile and organoleptic quality of the fruit [18], while polyphenols and anthocyanins are more closely related to health benefits [19]. The various production systems had minimal impact on fluctuations in the soluble solids content of the raspberries. This parameter, along with the fruit's titratable acidity, is significantly influenced by factors such as the cultivation region, specific cultivar, and production period [20,21]. The reduction in soluble solids content throughout the production cycle may be associated with climatic factors, as this decrease is more evident in months typically characterized by high temperatures. Previous studies suggest that elevated temperatures impact soluble solids content in raspberry fruits, likely due to an increase in plant respiration rate, which heightens carbohydrate consumption and modifies the source–sink relationship, ultimately resulting in lower soluble solids content levels [22]. Despite this reduction during the production cycle, fruit quality for commercialization is not adversely affected, as the minimum values observed in the present study remain acceptable and comparable to those reported for the cv. Heritage in Brazil, which show approximately 6.0 °Brix in fruits of this cultivar [22,23].

Notably, the fruit exhibited higher soluble solids content at the onset of the cycles, followed by a reduction in sugar accumulation during the harvest period. The production systems also did not significantly affect the titratable acidity of the fruit. The highest acidity levels occurred in the colder months of the cycles, while in the second production cycle, the lowest acidity levels were observed in the warmer months. Lower acidity might be related to metabolic changes and the consumption of organic acids during warmer periods [24].

The anthocyanin content of the raspberries was not affected by the different production systems. However, environmental factors are key determinants of antioxidant compound accumulation in raspberries [19,25,26]. In this study, the time of year (months) during which production occurred significantly influenced the anthocyanin content. Higher anthocyanin levels were recorded in October, November, and December in the first cycle, and in March in the second cycle, coinciding with the months of higher global solar radiation during each evaluated period. Anthocyanins play an important role in protecting plants from the damage of high solar radiation and are more abundant under high light intensity [27].

The pruning systems with subapical prunings (PS2 and PS3), which included sequential harvests that promoted higher production per cane, led to lower root system masses and reduced carbohydrate reserves in the roots compared to the hard pruning system (PS1). System PS1, which had the lowest mean fruit production per cane at the end of the two cycles, showed the highest root mass and greatest starch content, indicating greater reserve accumulation and lower depletion in plants subjected to this system. Sequential prunings (subapical prunings) may result in greater reserve use, consequently leading to greater depletion of the plants for future production [28]. In this experiment, the severe pruning system (PS1) experienced a minimal decline in production of approximately 2.73% from the first cycle to the second cycle, likely due to the carbohydrate reserves available to the plant. In treatments with two harvests (PS2) and three harvests (PS3), the production reduction from one harvest to the next was 24.8% and 34.6%, respectively, likely due to the depletion of reserves from the sequential prunings.

Starch is the most abundant storage carbohydrate in woody tissues, accumulating in plants during periods of high photosynthetic activity and depleting when carbohydrate usage exceeds production [29,30]. Studies on the raspberry cv. Titan have shown that raspberry roots serve as major carbon sinks, utilized during fruit maturation when carbohydrate synthesis sources are limited [31]. In the present study, during and after apical production, the canes were undergoing lignification, and leaf senescence occurred. In systems PS2 and PS3, where lignified canes were retained for subsequent harvests, the carbohydrate source for fruit maturation may have come from root system reserves, potentially explaining the lower starch content and root mass in these systems compared to PS1.

5. Conclusions

This study provides evidence that under subtropical climate conditions (Cwa), the cv. Heritage primocane raspberry, cultivated with subapical prunings and subjected to two or three harvests on the same cane, exhibits higher production over the corresponding period than the system with only one harvest. This approach prolongs the uninterrupted harvest season for two complete production cycles, spanning 14 months. Furthermore, it does not adversely affect the postharvest quality of the fruit, as observed in anthocyanin content, soluble solids, and titratable acidity. However, raspberries subjected to subapical pruning show lower carbohydrate storage in the root system compared to hard pruning, which may impact the longevity of these plants. For future studies, it is recommended to conduct a socioeconomic analysis of labor costs across different production systems to assess the economic feasibility of management practices in primocane raspberry cultivation under a subtropical climate.

Author Contributions: Conceptualization: R.B.M. and M.B.S. Data Curation: R.B.M. and Y.C.d.A.B. Formal analysis: R.B.M., Y.C.d.A.B., R.A.K. and M.B.S. Methodology: R.B.M., R.A.K. and M.B.S. Writing—review and editing: R.B.M., Y.C.d.A.B., E.R.O., R.A.K. and M.B.S. All authors have read and agreed to the published version of the manuscript.

Funding: This work was financially supported by Coordination for the Improvement of Higher Education Personnel (CAPES, Brazil, Finance Code 001) and the National Council for Scientific and Technological Development (CNPq, Brazil).

Data Availability Statement: The data presented in this study are available on request from the corresponding author.

Acknowledgments: The authors express gratitude for the technical assistance of Rodrigo Veraldi Ismael in plant cultivation.

Conflicts of Interest: The authors declare no conflict of interest.

References

1. FAOSTAT. *Crops and Livestock Products*; Food and Agriculture Organization of the United Nations: Rome, Italy, 2023. Available online: https://www.fao.org/faostat/ (accessed on 2 March 2024).
2. Tezotto-Uliana, J.V.; Kluge, R.A. *Framboesa: Cultura Alternativa para Pequenas Propriedades Rurais em Regiões Subtropicais*; ESALQ—Divisão de Biblioteca: Piracicaba, Brazil, 2013; Série Produtor Rural, No. 55; p. 33.
3. Gambardella, M.; Contreras, E.; Alcalde, J.; Neri, D. Phenotyping primocane fruiting trait in raspberry (*Rubus idaeus*). *Acta Hortic.* **2016**, *1133*, 67–74. [CrossRef]
4. Heide, O.M.; SØNsteby, A. Physiology of flowering and dormancy regulation in annual- and biennial-fruiting red raspberry (*Rubus idaeus* L.)—A review. *J. Hortic. Sci. Biotechnol.* **2011**, *86*, 433–442. [CrossRef]
5. Dale, A.; Pirgozliev, S.; King, E.M.; Sample, A. Scheduling primocane-fruiting raspberries (*Rubus idaeus* L.) for year-round production in greenhouses by chilling and Summer-pruning of canes. *J. Hortic. Sci. Biotechnol.* **2005**, *80*, 346–350. [CrossRef]
6. Vasilakakis, M.D.; McCown, B.H.; Dana, M.N. Low temperature and flowering of primocane-fruiting red Raspberries. *HortScience* **1980**, *15*, 750–751. [CrossRef]
7. Weber, C. Cultivar development and selection. In *Raspberries*; CABI: Wallingford, CT, USA, 2013; pp. 55–72. [CrossRef]
8. Bortolini, A.J. Avaliação de Diferentes Sistemas de Condução para a Cultura da Framboeseira no Planalto Sul Catarinense. Master's Dissertation, Mestrado em Produção Vegetal, Centro de Ciências Agroveterinárias, Universidade do Estado de Santa Catarina (UDESC), Lages, Brazil, 2024.
9. Parra-Quezada, R.Á.; Guerrero-Prieto, V.M.; Arreola-Avila, J.G. Efecto de Fecha y Tipo de Poda en Frambuesa Roja 'Malling Autumn Bliss'. *Rev. Chapingo Ser. Hortic.* **2007**, *13*, 201–206. [CrossRef]
10. Zorenc, Z.; Veberic, R.; Koron, D.; Mikulic-Petkovsek, M. Impact of Raspberry (*Rubus idaeus* L.) Primocane Tipping on Fruit Yield and Quality. *Not. Bot. Horti Agrobot. Cluj-Napoca* **2017**, *45*, 543–550. [CrossRef]
11. Koron, D.; Weber, N. Delay of Raspberry Harvesting Season with Successive Primocane Pruning. *Acta Hortic.* **2019**, *1265*, 223–228. [CrossRef]
12. Beck, H.E.; Zimmermann, N.E.; McVicar, T.R.; Vergopolan, N.; Berg, A.; Wood, E.F. Present and future Köppen-Geiger climate classification maps at 1-km resolution. *Sci. Data* **2018**, *5*, 180214. [CrossRef]
13. Álvarez, F.; Moya, M.; Rivera-Mora, C.; Zúñiga, P.E.; Jara-Cornejo, K.; Muñoz, P.; Ayala-Raso, A.; Munné-Bosch, S.; Figueroa, C.R.; Figueroa, N.E.; et al. Abscisic Acid Synthesis and Signaling during the Ripening of Raspberry (*Rubus idaeus* 'Heritage') Fruit. *Plants* **2023**, *12*, 1882. [CrossRef]
14. Francis, F.J. Anthocyanins of 'Durham' and 'Heritage' Raspberry Fruits. *HortScience* **1972**, *7*, 398. [CrossRef]
15. Dubois, M.; Gilles, K.A.; Hamilton, J.K.; Rebers, P.A.; Smith, F. Colorimetric method for determination of sugars and related substances. *Anal. Chem.* **1956**, *28*, 350–356. [CrossRef]
16. Carew, J.G.; Mahmood, K.; Darby, J.; Hadley, P.; Battey, N.H. The effect of temperature, photosynthetic photon flux density, and photoperiod on the vegetative growth and flowering of 'Autumn Bliss' raspberry. *J. Am. Soc. Hortic. Sci.* **2003**, *128*, 291–296. [CrossRef]
17. Hanson, E.; Crain, B.; Moses, J. Cropping on primocanes and floricanes increases yields of organic raspberries in high tunnels. *HortScience* **2019**, *54*, 459–462. [CrossRef]
18. Wozniak, W.; Radajewska, B.; Reszelska-Sieciechowicz, A.; Dejwor, I. Sugars and acid content influence organoleptic evaluation of fruits of six strawberry cultivars from controlled cultivation. *Acta Hortic.* **1997**, *439*, 333–336. [CrossRef]
19. Rao, A.V.; Snyder, D.M. Raspberries and Human Health: A Review. *J. Agric. Food Chem.* **2010**, *58*, 3871–3883. [CrossRef]
20. Maro, L.A.C.; Pio, R.; Guedes, M.N.S.; de Abreu, C.M.P.; Moura, P.H.A. Environmental and genetic variation in the post-harvest quality of raspberries in subtropical areas in Brazil. *Acta Sci. Agron.* **2014**, *36*, 323. [CrossRef]
21. Mazur, S.P.; Nes, A.; Wold, A.-B.; Remberg, S.F.; Aaby, K. Quality and chemical composition of ten red raspberry (*Rubus idaeus* L.) genotypes during three harvest seasons. *Food Chem.* **2014**, *160*, 233–240. [CrossRef]
22. Fagundes, C.d.M.; Biasi, L.A.; Tofanelli, M.B.D.; Rodrigues, A.P.S.; Franciscon, V.G.; Siqueira, A.d.F.C. Production, quality, bioactive compounds, and phenology of raspberry cultivars under an organic cropping system in a subtropical region of Brazil. *Sci. Agric.* **2023**, *81*, e20230009. [CrossRef]
23. Moura, P.H.A.; Campagnolo, M.A.; Pio, R.; Curi, P.N.; Assis, C.N.; Silva, T.C. Fenologia e produção de cultivares de framboeseiras em regiões subtropicais no Brasil. *Pesq. Agropecu. Bras.* **2012**, *47*, 1714–1721. [CrossRef]
24. Conradie, W.J.; Carey, V.A.; Bonnardot, V.; Saayman, D.; Van Schoor, L.H. Effect of different environmental factors on the performance of Sauvignon Blanc grapevines in the Stellenbosch/Durbanville districts of South Africa. I. Geology, soil, climate, phenology and grape composition. *S. Afr. J. Enol. Vitic.* **2002**, *23*, 78–91.
25. Frías-Moreno, M.N.; Parra-Quezada, R.Á.; Ruíz-Carrizales, J.; González-Aguilar, G.A.; Sepulveda, D.; Molina-Corral, F.J.; Jacobo-Cuellar, J.L.; Olivas, G.I. Quality, bioactive compounds and antioxidant capacity of raspberries cultivated in northern Mexico. *Int. J. Food Prop.* **2021**, *24*, 603–614. [CrossRef]
26. Moore, P.P.; Perkins-Veazie, P.; Weber, C.A.; Howard, L. Environmental effect on antioxidant content of ten raspberry cultivars. *Acta Hortic.* **2008**, *777*, 499–504. [CrossRef]

27. Jaakola, L.; Määttä-Riihinen, K.; Kärenlampi, S.; Hohtola, A. Activation of flavonoid biosynthesis by solar radiation in bilberry (*Vaccinium myrtillus* L.) leaves. *Planta* **2004**, *218*, 721–728. [CrossRef] [PubMed]
28. Sartori, I.A.; Koller, O.C.; Theisen, S.; de Souza, P.V.D.; Bender, R.J.; Marodin, G.A.B. Efeito da poda, raleio de frutos e uso de fitorreguladores na produção de tangerinas (*Citrus deliciosa* Tenore) cv. montenegrina. *Rev. Bras. Frutic.* **2007**, *29*, 5–10. [CrossRef]
29. Kozlowski, T.T. Carbohydrate sources and sinks in woody plants. *Bot. Rev.* **1992**, *58*, 107–222. [CrossRef]
30. Keller, J.D.; Loescher, W.H. Nonstructural carbohydrate partitioning in perennial parts of sweet cherry. *J. Am. Soc. Hortic. Sci.* **1989**, *114*, 969–975. [CrossRef]
31. Fernandez, G.E.; Pritts, M.P. Carbon supply reduction has a minimal influence on current year's red raspberry (*Rubus idaeus* L.) fruit production. *J. Am. Soc. Hortic. Sci.* **1996**, *121*, 473–477. [CrossRef]

horticulturae

MDPI

Article

Yield Performance and Quality Assessment of Brazilian Hybrid Grapes Influenced by Rootstocks and Training Systems

Francisco José Domingues Neto [1,*], Marco Antonio Tecchio [1], Cristine Vanz Borges [2], João Domingos Rodrigues [3], Elizabeth Orika Ono [3], Giuseppina Pace Pereira Lima [3], Mara Fernandes Moura [4], José Luiz Hernandes [4], Marcelo de Souza Silva [1] and Magali Leonel [1]

[1] School of Agriculture Sciences, Sao Paulo State University (UNESP), Botucatu 18610-034, Sao Paulo, Brazil; marcelo.s.silva@unesp.br (M.d.S.S.); magali.leonel@unesp.br (M.L.)
[2] School of Agriculture Sciences, Alto Vale do Rio do Peixe University (UNIARP), Caçador 89500-199, Santa Catarina, Brazil
[3] Institute of Biosciences, Sao Paulo State University (UNESP), Botucatu 18610-034, Sao Paulo, Brazil; joao.domingos@unesp.br (J.D.R.)
[4] Advanced Fruit Research Center, Agronomic Institute (IAC), Jundiaí 13214-820, Sao Paulo, Brazil
* Correspondence: fjdominguesneto@hotmail.com

Abstract: The choice of training system impacts the architecture and physiological characteristics of grapevines, affecting grape production and quality. Continuous studies are necessary to optimize viticulture production in Brazil. This study aimed to evaluate the effects of rootstocks and different training systems on the production and quality of 'IAC 138-22 Máximo' and 'BRS Violeta' grapevines for juice and wine. The experiment was conducted over two productive cycles (2019/2020 and 2020/2021) in an experimental vineyard at the Advanced Center for Fruit Research as part of the Agronomic Institute (IAC), in Jundiaí, São Paulo, Brazil (23°06′ S, 46°55′ W, and 745 m altitude). For each cultivar, a randomized block design in a 2 × 2 factorial scheme was used, with two rootstocks ('IAC 766 Campinas' and 106-8 'Mgt') and two training systems (low and high trellises), with five blocks of three plants per experimental plot. In both cycles, the gas exchange and grapevine production, the chemical characteristics of the grape juice (must), and the chemical compounds in the berry skins were evaluated. The rootstocks and training systems influenced the variables evaluated in both cultivars, with the high trellis system providing the best results. This approach increased the photosynthetic rate, improved water-use efficiency, elevated grapevine production, and enhanced fruit quality. Therefore, its use is recommended for training grapevines. Regarding rootstocks, the best scion–rootstock affinity was found between hybrid grapevines and 'IAC 766 Campinas'.

Keywords: juice and wine grapes; trellis height; yield and quality; phenolic compounds; anthocyanins; gas exchange; yield; grape juice; 'IAC 138-22 Máximo'; 'BRS Violeta'; 'IAC 766 Campinas'; 106-8 'Mgt'

Citation: Domingues Neto, F.J.; Tecchio, M.A.; Borges, C.V.; Rodrigues, J.D.; Ono, E.O.; Lima, G.P.P.; Moura, M.F.; Hernandes, J.L.; Silva, M.d.S.; Leonel, M. Yield Performance and Quality Assessment of Brazilian Hybrid Grapes Influenced by Rootstocks and Training Systems. *Horticulturae* **2024**, *10*, 909. https://doi.org/10.3390/horticulturae10090909

Academic Editor: Ute Albrecht

Received: 5 August 2024
Revised: 14 August 2024
Accepted: 23 August 2024
Published: 27 August 2024

1. Introduction

The Brazilian viticulture landscape is diverse, driven by the genetic breeding of grapevines for greater productivity and adaptation to the tropical climate [1]. In 2023, Brazil produced 1,719,630 tons of grapes over 76,747 hectares [2]. The regional diversity in national viticulture includes different production cycles, harvest times, cultivars, and management practices, requiring studies on specific cultivation techniques. The Brazilian grape market is divided between fresh consumption (table grapes) and processing (grape juice and wines), with a predominance of non-vinifera grapes (*Vitis labrusca* and hybrids). The diversity of cultivars results in grapes with distinct characteristics, influenced by the polyphenolic content and the search for high-quality antioxidant products.

Genetic breeding has enabled the development of hybrid cultivars, such as 'IAC 138-22 Máximo' and 'BRS Violeta' [3,4], both red and cultivated for processing juice and wines.

These cultivars are aimed at meeting the grape juice and derivatives market, characterized by high productivity and lower sensitivity to major fungal diseases compared to *Vitis vinifera*. However, studies are needed to understand the performance of these cultivars grafted on different rootstocks and grown using different training systems, as well as to evaluate their productive characteristics and fruit quality.

Given climate change, the selection of rootstocks and scion varieties becomes crucial for viticulture, as it allows better adaptation of the grapevines to adverse conditions and increases their resistance to abiotic stresses [5]. This strategic choice optimizes the use of water and nutrients, maintaining grape productivity and quality, which is essential for the sustainability of the viticulture sector [5].

The use of rootstocks in viticulture not only protects against damage caused by Phylloxera (*Daktulosphaira vitifoliae*), but also enables cultivation in situations involving different abiotic factors and adverse soil conditions (the presence of pests and high salinity), which directly affect vegetative development, the duration of the phenological phases, and production quality [6].

The interaction between scion varieties and rootstocks is widely studied under different edaphoclimatic conditions, aiming to optimize vegetative growth, productivity, and grape quality [7–11].

With the expansion of viticulture areas, the choice of the vineyard training system directly influences the architecture and physiological characteristics of grapevines, as well as the quantitative and qualitative aspects of the fruits [12,13]. Although the three-wire trellis system is the most commonly used, new variations have emerged to increase the leaf area, favoring physiological characteristics that increase production and improve fruit quality, especially in terms of phenolic compounds and sugars, which are fundamental to the quality of juices and wines [14].

Among the training system alternatives, there is the high trellis system, with four wires, providing a greater height for the vegetative canopy. The interaction of solar radiation with the grapevine canopy structure directly influences its productivity, affecting the light distribution, carbon assimilation, and water deficit [15]. Adequate foliage is essential for photosynthesis, without creating excessive shading that could hinder grape development and ripening [16,17]. During maturation, proper light exposure is essential for the synthesis and accumulation of anthocyanins and flavanols, which are positively related to solar radiation exposure [18,19].

Studies on training systems used in grape processing have shown significant variations in grape and wine or juice production and quality [12,14,20,21]. However, for the cultivars 'IAC 138-22 Máximo' and 'BRS Violeta', in subtropical conditions, there are no reports in the literature. Grapes are recognized for their antioxidant properties due to phenolic compounds, such as flavonoids and anthocyanins, which play important roles in the human diet and in the prevention of cardiovascular, cancerous, and neurological diseases [22–24]. In addition to improving nutritional quality, these compounds enhance the commercial quality of red grapes, being fundamental for their color and external appearance [25].

In this context, the present work aimed to evaluate the effects of interactions between rootstocks ('IAC 766 Campinas' and 106-8 'Mgt') and training systems (low and high trellises), focusing on improving the productivity and quality of 'IAC 138-22 Máximo' and 'BRS Violeta' grapevine fruits.

2. Materials and Methods

2.1. Installation and Vineyard Management, and Location of the Experimental Area

The experiment was conducted over two productive cycles (2019/2020 and 2020/2021) in a vineyard in its 9th and 10th year of production, located at the Advanced Center for Fruit Research as part of the Agronomic Institute (IAC), Jundiaí, São Paulo, Brazil (23°06′ S, 46°55′ W, with an average altitude of 745 m). According to the Köppen classification, the region's climate is classified as Cfb, with an average annual rainfall of 1400 mm, an average annual temperature of 19.5 °C, and a relative humidity of 70.6%. The soil in the experimental area is classified as Dystrophic Red Cambisol [26].

In both experimental years, short pruning was performed, with one bud, followed by the application of 5% hydrogen cyanamide (Dormex®, BASF, Ludwigshafen am Rhein, Germany). After budburst, only one productive shoot per spur was maintained, with disbudding and tying of the shoots to wires, thinning, and defoliation. Additionally, cultural practices, such as weeding, and applications of herbicides, fungicides, and insecticides, were carried out.

2.2. Experimental Design and Treatments

The cultivars 'IAC 138-22 Máximo' ('Seibel 11342' × 'Syrah') and 'BRS Violeta' ('BRS Rúbea' × 'IAC 1398-21') were evaluated over two productive cycles in an experimental vineyard located at the Advanced Center for Fruit Research as part of the Agronomic Institute (IAC), Jundiaí, São Paulo, Brazil. For each cultivar, a randomized block design was used in a 2 × 2 factorial arrangement, with two rootstocks ('IAC 766 Campinas' and 106-8 'Mgt') and two training systems: low trellis with three wires situated at 1, 1.3, and 1.6 m from the ground, and high trellis with four wires situated at 1, 1.3, 1.6, and 2 m from the ground, with five blocks and three plants per experimental plot.

2.3. Harvest and Evaluations

Harvests were performed according to the physiological maturation of each cultivar, occurring at 140 days for 'BRS Violeta' and 146 days for 'IAC 138-22 Máximo' in the first cycle, and at 153 days for both cultivars in the second cycle (Figure 1).

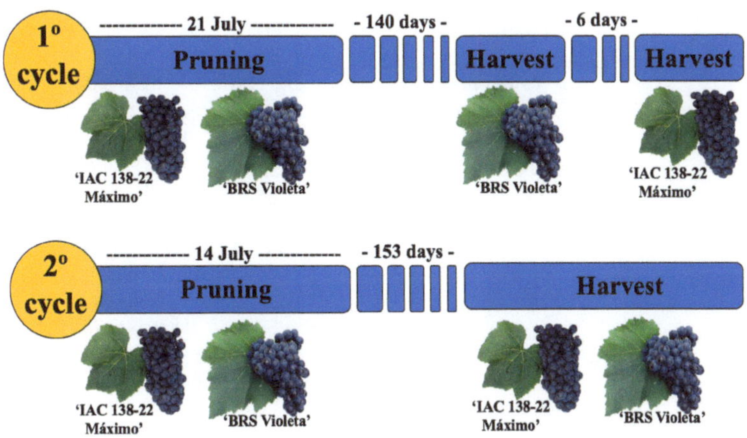

Figure 1. Pruning and harvest times for Brazilian hybrid grapes in two productive cycles, in Jundiaí, São Paulo, Brazil.

2.4. Gas Exchange, Water-Use Efficiency, SPAD Index, and Total Chlorophyll

The gas exchange was evaluated at the full flowering stage of the vines, using an open photosynthesis system, with an infrared CO_2 and water vapor analyzer (Infrared Gas Analyzer—IRGA, model LI-6400, Li-Cor). During the evaluations, the ambient CO_2 concentration was used as a reference, along with data collection on the temperature and relative humidity. The CO_2 assimilation rate (A, µmol CO_2 m^{-2} s^{-1}), transpiration rate (E, mmol H_2O m^{-2} s^{-1}), internal CO_2 concentration (Ci), and stomatal conductance (gs) were calculated according to the method proposed by [27].

The water-use efficiency (WUE, µmol CO_2 (mmol $H_2O)^{-1}$) was determined by the ratio of CO_2 assimilation to the transpiration rate. The electron transport rate (ETR) was calculated following the method by [28]. The SPAD index was evaluated on four leaves per plant, with readings taken at three points per leaf using a SPAD meter (Model 502, Konica Minolta, Tokyo, Japan). The total chlorophyll concentration was determined using 100 g of fresh leaf material [29].

2.5. Number of Bunches per Plant, Productivity, and Physicochemical Composition of Grape Must

After harvesting, the grape bunches were counted and weighed to determine the number of bunches per plant, the fresh mass of the bunches, and the yield per plant. Subsequently, based on the number and fresh mass of the bunches and the cultivation stand, the productivity (t ha^{-1}) was estimated.

For the physicochemical analyses of grape must, 250 berries per experimental plot (n = 250) were used. The must was obtained by pressing the berries, and the soluble solids (SS) content was evaluated by direct refractometry, using an Atago® digital refractometer (Schmidt Haensch, Berlin, Germany) with automatic temperature compensation, with results expressed in °Brix. The pH was measured by a direct reading of the must using a Micronal B-274 pH meter (Mettler Toledo, Barueri, Brazil). The titratable acidity (TA) was determined by titration, using 5 g of must, diluted in 100 mL of distilled water, titrating with a standardized 0.1 N sodium hydroxide solution, using phenolphthalein as an indicator, until the color change endpoint, with results expressed as a percentage of tartaric acid. The reducing sugars were determined by the colorimetric method of Somogyi-Nelson, based on a standard glucose curve and readings taken at 510 nm in a BEL Photonics® SP 2000 UV/vis spectrophotometer [30], with the results expressed as a percentage of reducing sugars per mL of must.

2.6. Biochemical Composition of Grape Skins

For the analyses of the total phenolic compounds [31] and total monomeric anthocyanins [32], as well as the antioxidant activity via DPPH (2,2-diphenyl-1-picrylhydrazyl) [33], modified by [34] and FRAP [35], 200 berries per experimental plot (n = 1000 berries per treatment) were collected. The berries were halved, the seeds were removed, and the skins and pulp were separated and frozen in liquid nitrogen. Approximately 100 g of frozen fresh skin was manually pulverized in a porcelain mortar with a pestle and stored at −80 °C until the analyses. All the analyses were performed in triplicate.

2.7. Statistical Analysis

The means obtained from the two production cycles for each cultivar were subjected to analysis of variance (ANOVA) to determine the effects of the rootstocks and training systems and their interactions. Subsequently, the means were compared using the Tukey test ($p < 0.05$), utilizing the Sisvar 6.0 software [36]. Additionally, to characterize the interaction between the rootstocks and the training systems and to assess the correlation of the variables, principal component analysis (PCA) was performed using Statistical Analysis Software 4.0 (SAS).

3. Results

3.1. Impact of Rootstocks and Training Systems on 'IAC 138-22 Maximo' Grape Variety: Physiological, Biochemical Parameters and Yield

Significant interactions were observed between rootstocks and training systems concerning the electron transport rate (ETR), stomatal conductance (gs), transpiration rate (E), water-use efficiency (WUE), assimilation rate (A), and internal carbon concentration (Ci) in the 'IAC 138-22 Máximo' grape variety (Table 1).

A higher ETR was achieved with the combination of the rootstock 106-8 'Mgt' and the high espalier, as well as with the rootstock 'IAC 766 Campinas' and the low espalier. This suggests that these specific combinations are more efficient in capturing and utilizing light energy for electron transport, leading to increased production of ATP and NADPH, which are essential for the synthesis of photoassimilates.

Table 1. Electron transport rate (ETR), stomatal conductance (gs), transpiration rate (E), water-use efficiency (WUE), CO_2 assimilation rate (A), and internal carbon concentration (Ci) of the 'IAC 138-22 Máximo' vine on different rootstocks and training systems.

Variables	Cycle	Trellis System	Rootstocks		CV (%)
			'IAC 766 Campinas'	106-8 'Mgt'	
ETR (µmol electrons $m^{-2}s^{-1}$)	I	Low	130.34 ± 3.0	119.90 ± 3.4	3.71
		High	119.49 ± 1.9	149.39 ± 4.3	
	II	Low	166.82 ± 4.9	147.82 ± 3.4	
		High	126.57 ± 4.4	166.91 ± 4.3	
	Aver.	Low	148.58 ± 7.07 aA	133.86 ± 5.51 bB	
		High	123.03 ± 3.16 bB	158.15 ± 9.93 aA	
gs (mol $m^{-2}s^{-1}$)	I	Low	0.18 ± 0.01	0.20 ± 0.01	4.12
		High	0.12 ± 0.01	0.17 ± 0.01	
	II	Low	0.28 ± 0.01	0.30 ± 0.01	
		High	0.22 ± 0.01	0.27 ± 0.01	
	Aver.	Low	0.23 ± 0.01 aB	0.25 ± 0.01 aA	
		High	0.17 ± 0.001 bB	0.22 ± 0.01 bA	
E (mmol water vapor $m^{-2}s^{-1}$)	I	Low	4.09 ± 0.3	4.69 ± 0.4	4.26
		High	3.19 ± 0.9	3.94 ± 0.3	
	II	Low	10.03 ± 0.4	10.49 ± 0.2	
		High	10.45 ± 0.3	8.99 ± 0.3	
	Aver.	Low	7.06 ± 0.42 aB	7.59 ± 0.30 aA	
		High	6.82 ± 0.24 aA	6.47 ± 0.17 bB	
WUE (µmol CO_2 (mmol $H_2O)^{-1}$)	I	Low	3.90 ± 0.3	3.80 ± 0.3	4.09
		High	3.80 ± 0.2	4.19 ± 0.4	
	II	Low	5.16 ± 0.1	5.04 ± 0.3	
		High	5.32 ± 0.3	7.13 ± 0.4	
	Aver.	Low	4.53 ± 0.25 aA	4.42 ± 0.11 bA	
		High	4.56 ± 0.18 aB	5.66 ± 0.15aA	
A (µmol CO_2 $m^{-2}s^{-1}$)	I	Low	39.49 ± 3.4	42.89 ± 1.8	5.59
		High	34.47 ± 3.0	40.10 ± 1.1	
	II	Low	29.01 ± 3.4	34.75 ± 1.8	
		High	24.55 ± 3.4	32.24 ± 1.9	
	Aver.	Low	34.75 ± 1.38 aA	38.82 ± 2.48 aA	
		High	29.51 ± 1.51 bB	36.17 ± 1.25 aA	
Ci (µmolCO_2 mol^{-1} air)	I	Low	189.90 ± 4.9	139.39 ± 3.4	3.72
		High	180.03 ± 4.4	100.12 ± 3.9	
	II	Low	169.96 ± 3.9	157.95 ± 4.0	
		High	170.59 ± 4.0	106.30 ± 3.9	
	Aver.	Low	179.93 ± 3.74 aA	148.67 ± 5.70 aB	
		High	175.31 ± 5.76 aA	103.21 ± 6.11 bB	

Mean followed by the same lowercase letter in the column and an uppercase letter in the row do not differ from each other, according to Tukey's test ($p < 0.05$).

Stomatal conductance (gs) is another vital parameter, referring to the rate of CO_2 influx and water vapor efflux through the stomata. An increase in gs, as observed in vines grafted onto the rootstock 106-8 'Mgt' with the low espalier training system, allows for greater CO_2 intake, thereby increasing the assimilation rate (A). It was also noted that the transpiration rate (E), which measures water loss through the stomata, increased as well, potentially aiding in maintaining leaf temperature and ensuring continuous water and nutrient flow.

The increased gs and E in vines grafted onto the rootstock 106-8 'Mgt' may have contributed to the lower water-use efficiency (WUE) in this combination of rootstock and training system, indicating that, despite the increased photosynthesis, water was not used as efficiently, which may result in lower biomass accumulation.

In both training systems, vines grafted onto the rootstock 'IAC 766 Campinas' exhibited a higher internal carbon concentration (Ci), suggesting that despite the high stomatal conductance (gs), CO_2 fixation may not have been as efficient due to lower Rubisco activity and reduced water-use efficiency (WUE). Another hypothesis is that mesophyll cells consume CO_2 during photosynthetic assimilation, consequently resulting in lower CO_2 concentration in the intercellular airspace compared to the ambient air outside the leaf. The internal carbon concentration (Ci) reflects the concentration of CO_2 within the leaves and indicates the balance between CO_2 fixation and its entry through the stomata.

Thus, it was observed that the correct choice of rootstock and training system can optimize photosynthesis and water use in vines, resulting in more efficient and productive plants. These interactions directly influence physiological parameters, such as the ETR, gs, E, WUE, A, and Ci, and consequently, the agronomic performance of the 'IAC 138-22 Máximo' vines.

No interaction was observed between the rootstocks and training systems for the SPAD index and total chlorophyll content in the 'IAC 138-22 Máximo' vine, with only rootstocks showing a significant isolated effect. Higher SPAD indices and chlorophyll contents were obtained with the use of the rootstock 106-8 'Mgt' (Table 2). Vines with higher SPAD indices exhibit greater green coloration intensity, as a result of higher concentrations of photosynthetic pigments, particularly chlorophylls. This aspect is particularly important for grapes intended for processing, where a high SPAD index and chlorophyll content enhance light utilization during photosynthesis, increasing carbohydrate accumulation, which will be converted into sugars in the berries and energy for the next production cycle.

Table 2. SPAD index and total chlorophyll of the 'IAC 138-22 Máximo' vine on different rootstocks and training systems.

Variables	Cycle	Trellis System		Rootstocks		CV (%)
		Low	High	'IAC 766 Campinas'	106-8 'Mgt'	
SPAD index	I	30.03 ± 1.49	31.04 ± 2.04	30.09 ± 3.49	34.10 ± 3.10	4.83
	II	36.91 ± 1.92	34.12 ± 2.03	31.85 ± 3.39	36.06 ± 3.12	
	Aver.	33.47 ± 2.79 a	32.58 ± 2.38 a	30.97 ± 1.63 b	35.08 ± 1.27 a	
Total chlorophyll (mg 100 g^{-1} of leaves)	I	44.03 ± 4.49	42.89 ± 3.90	40.12 ± 4.10	49.90 ± 3.48	11.94
	II	51.15 ± 3.90	44.47 ± 3.89	41.90 ± 3.90	50.62 ± 3.92	
	Aver.	47.59 ± 8.10 a	43.68 ± 5.61 a	41.01 ± 5.12 b	50.26 ± 5.54 a	

Mean followed by the same lowercase letter within the same factor do not differ from each other, according to Tukey's test ($p < 0.05$).

Associating these results with photosynthetic parameters, it can be inferred that vines with a higher chlorophyll content and SPAD index, as observed with the rootstock 106-8 'Mgt', exhibit greater photosynthetic efficiency. This is reflected in a higher ETR and CO_2 assimilation rate (A), promoting better photosynthetic performance.

No significant interaction was observed between rootstocks and training systems for the productive and physicochemical variables of the 'IAC 138-22 Máximo' grape must. Therefore, the isolated effect of the variables was assessed. Significant differences were found for the production, yield, cluster fresh weight, soluble solids, and pH, among the training systems. Except for pH, all the variables showed higher results when the 'IAC 138-22 Máximo' vine was trained with a high espalier (Table 3).

The yield of the vines trained with a high espalier was 38.52% higher than those trained with a low espalier, demonstrating greater compatibility of the cultivar with this training system, possibly due to its high vigor. The 'IAC 138-22 Máximo' vine requires a system that allows for better distribution of the branches and the vegetative canopy, thus providing more efficient leaf distribution and greater solar radiation capture. This benefits the photosynthetic process, increasing photoassimilate production and, consequently, the yield. Additionally, this training system resulted in heavier clusters, which directly contributes to the observed higher yield.

Table 3. Number and fresh weight of clusters, yield, productivity, soluble solids, pH, titratable acidity (TA), SS/TA ratio, and reducing sugars of the 'IAC 138-22 Máximo' vine on different rootstocks and training systems.

Variable	Cycle	Trellis System		Rootstocks		CV (%)
		Low	High	'IAC 766 Campinas'	106-8 'Mgt'	
Number of clusters per plant	I	22.33 ± 3.4	23.00 ± 3.4	26.80 ± 3.1	28.17 ± 3.1	26.04
	II	21.31 ± 2.3	23.34 ± 3.2	16.88 ± 3.4	18.13 ± 3.4	
	Aver.	21.82 ± 7.18 a	23.17 ± 8.16 a	21.84 ± 6.64 a	23.15 ± 8.60 a	
Fresh cluster weight (g)	I	176.82 ± 30.4	266.58 ± 31.3	175.12 ± 34.4	246.45 ± 33.4	31.99
	II	218.14 ± 32.3	281.64 ± 30.3	333.56 ± 34.3	188.07 ± 34.9	
	Aver.	197.48 ± 37.72 b	274.11 ± 23.69 a	254.34 ± 27.96 a	217.26 ± 32.91 a	
Yield (kg per plant)	I	3.65 ± 1.2	5.40 ± 1.9	6.34 ± 1.3	4.58 ± 1.3	28.02
	II	4.67 ± 1.4	6.12 ± 1.9	3.80 ± 1.2	5.12 ± 1.2	
	Aver.	4.16 ± 1.14 b	5.76 ± 1.47 a	5.07 ± 1.57 a	4.85 ± 1.52 a	
Productivity (t/ha)	I	14.59 ± 2.3	23.74 ± 1.9	14.61 ± 1.8	13.65 ± 1.3	28.02
	II	18.69 ± 2.4	22.36 ± 1.8	25.95 ± 1.9	25.19 ± 1.9	
	Aver.	16.64 ± 4.54 b	23.05 ± 5.89 a	20.28 ± 6.28 a	19.42 ± 6.10 a	
Soluble solids (°Brix)	I	14.22 ± 1.2	15.03 ± 1.9	15.00 ± 1.8	14.39 ± 1.0	9.18
	II	14.16 ± 1.3	15.47 ± 1.4	15.26 ± 1.2	14.23 ± 1.2	
	Aver.	14.19 ± 1.36 b	15.25 ± 1.55 a	15.13 ± 1.63 a	14.31 ± 1.36 a	
pH	I	3.34 ± 0.08	3.30 ± 0.09	3.25 ± 0.03	3.32 ± 0.04	2.04
	II	3.44 ± 0.09	3.32 ± 0.02	3.45 ± 0.04	3.40 ± 0.03	
	Aver.	3.39 ± 0.09 a	3.31 ± 0.08 b	3.35 ± 0.08 a	3.36 ± 0.11 a	
TA (% tartaric acid)	I	1.02 ± 0.1	1.01 ± 0.1	1.02 ± 0.4	1.08 ± 0.6	13.76
	II	0.94 ± 0.2	0.87 ± 0.3	0.78 ± 0.4	0.96 ± 0.9	
	Aver.	0.98 ± 0.16 a	0.94 ± 0.16 a	0.90 ± 0.18 b	1.02 ± 0.11 a	
SS/TA	I	13.94 ± 1.9	14.88 ± 1.3	14.70 ± 1.1	13.32 ± 1.3	20.07
	II	15.88 ± 1.8	18.68 ± 1.3	20.34 ± 1.2	15.02 ± 1.9	
	Aver.	14.91 ± 3.15 a	16.78 ± 4.31 a	17.52 ± 4.40 a	14.17 ± 2.25 b	
Reducing sugars (%)	I	9.32 ± 1.3	9.44 ± 1.4	9.34 ± 1.3	9.40 ± 1.3	10.76
	II	11.80 ± 1.4	12.52 ± 1.9	11.82 ± 1.3	12.52 ± 1.4	
	Aver.	10.56 ± 1.89 a	10.98 ± 1.70 a	10.58 ± 1.53 a	10.96 ± 2.04 a	

Mean followed by the same lowercase letter within the same factor do not differ from each other, according to Tukey's test ($p < 0.05$).

The high espalier system promotes an increase in the soluble solids in the must of the 'IAC 138-22 Máximo' grape (15.25 °Brix). Although the training system did not influence the titratable acidity and the SS/TA ratio, the low espalier system produced fruit with an acidity of 0.98% tartaric acid and an SS/TA ratio of 14.91, values below the requirements in Brazilian legislation, which stipulates a maximum acidity of 0.9% tartaric acid and a minimum SS/TA ratio of 15 for grapes intended for processing. This result was also observed when the vines were grafted onto the rootstock 106-8 'Mgt', which had a titratable acidity of 1.02% tartaric acid and an SS/TA ratio of 14.17.

When a grape cultivar exhibits high acidity, low soluble solids content, or any other undesirable chemical characteristic, it becomes crucial to diversify cultivars for blending purposes. This aims to balance the potential limitations of each cultivar, thus meeting the required standards for beverage production and enhancing the quality of juices and wines.

The high espalier system, by providing better canopy distribution, supports a higher ETR and assimilation rate, optimizing light utilization and increasing ATP and NADPH production. This not only improves photosynthesis, but also enhances the efficiency of photoassimilate production, resulting in a higher soluble solids content and better grape must quality. Therefore, choosing an appropriate training system, such as the high espalier, in combination with efficient rootstocks, is crucial for maximizing the productivity and quality of the 'IAC 138-22 Máximo' vine.

A well-distributed canopy allows more leaves to receive direct light, increasing the CO_2 assimilation rate and light-use efficiency. Increased photosynthetic activity promotes a higher concentration of carbohydrates, which are utilized for the growth and development of the clusters, explaining the greater cluster fresh weight and higher soluble solids concentration. Thus, the high espalier offers optimal conditions to maximize the photosynthetic efficiency and productivity of the 'IAC 138-22 Máximo' vine.

The rootstock 'IAC 766 Campinas' provided the 'IAC 138-22 Máximo' grape must with lower titratable acidity, increasing the SS/TA ratio and improving the grape flavor.

Improvements to the photosynthetic efficiency and carbohydrate production of grapevines are essential for fruit growth and development. Consequently, the increased absorption and availability of nutrients, common to vigorous rootstocks, may lead to reduced titratable acidity, balancing the SS/TA ratio and enhancing the flavor and quality of the must.

Moreover, the greater vigor of the 'IAC 766 Campinas' rootstock may be associated with its better adaptation to various environmental and soil conditions, promoting more balanced and healthy vine growth. This results in a more efficient distribution of resources within the plant, contributing to a more uniform development of the aerial parts and, particularly, the clusters, leading to better overall fruit quality.

Significant interactions were observed between rootstocks and training systems for the total phenolic compounds, monomeric anthocyanins, and antioxidant activity (DPPH and FRAP) of the 'IAC 138-22 Máximo' grapevine (Table 4). The combination of the high espalier with the 'IAC 766 Campinas' rootstock resulted in higher levels of monomeric anthocyanins and total phenolic compounds in the grapes, increasing the antioxidant activity as expressed by DPPH and FRAP values (Table 4).

The presence of high levels of phenolic compounds, anthocyanins, and antioxidant activity in the berry skins of 'IAC 138-22 Máximo' grapes, especially when combined with the 'IAC 766 Campinas' rootstock and the high espalier system, offers numerous health benefits. Anthocyanins and flavonoids are positively related to vine exposure to solar radiation, which justifies the high antioxidant activity observed in grapes cultivated with this combination of training system and rootstock.

The high antioxidant activity observed, as measured by the DPPH and FRAP methods, indicates that these grapes have a significant capacity to combat cellular damage caused by reactive oxygen species.

The higher concentration of these compounds in the combination of the high espalier and the 'IAC 766 Campinas' rootstock suggests an increased synthesis of flavonoids.

The combination of the high espalier and the 'IAC 766 Campinas' rootstock proved effective in maximizing the levels of these bioactive compounds in the grape skins. This training system and rootstock favor light exposure and canopy distribution, optimizing photosynthesis [14] and secondary metabolite production.

Consequently, the increase in phenolic compounds and anthocyanins not only improves the functional and sensory quality of the grapes, but also enhances the health benefits, making these grapes highly beneficial for both fresh consumption and for the production of juices and wines. Therefore, the choice of appropriate training systems and rootstocks is crucial for maximizing the beneficial chemical and biochemical attributes of 'IAC 138-22 Máximo' grapes for consumers.

Table 4. Total phenolic compounds, total monomeric anthocyanins, and antioxidant activity (DPPH and FRAP) in the skin of 'IAC 138-22 Máximo' grapes on different rootstocks and training systems.

Variables	Cycle	Trellis System	Rootstocks		CV (%)
			'IAC 766 Campinas'	106-8 'Mgt'	
Total phenolic compounds (mg 100 g^{-1} of berry skins)	I	Low	903.80 ± 1.3	890.14 ± 1.2	1.07
		High	989.19 ± 1.9	919.12 ± 1.4	
	II	Low	957.98 ± 3.4	918.54 ±1.8	
		High	1130.15 ± 3.4	1054.48 ± 1.9	
	Aver.	Low	930.89 ± 3.64 bA	904.34 ± 1.65 bB	
		High	1059.67 ± 18.30 aA	986.80 ± 8.89 aB	
Total monomeric anthocyanins (mg 100 g^{-1} of berry skins)	I	Low	642.39 ± 3.0	400.39 ± 3.4	9.35
		High	649.39 ± 1.9	600.89 ± 3.4	
	II	Low	649.81 ± 1.8	471.17 ± 3.9	
		High	656.95 ± 1.4	610.89 ± 3.4	
	Aver.	Low	646.10 ± 35.12 aA	435.78 ± 23.45 bB	
		High	653.17 ± 38.79 aA	605.89 ± 30.56 aA	
DPPH (Mmol g^{-1} of berry skins)	I	Low	12.19 ± 0.9	17.39 ± 1.8	8.84
		High	19.39 ± 0.8	19.49 ± 1.2	
	II	Low	16.35 ± 0.3	27.49 ± 1.2	
		High	31.77 ± 0.4	29.49 ± 1.9	
	Aver.	Low	14.27 ± 0.69 bB	22.44 ± 2.00 bA	
		High	30.90 ± 0.74 aA	24.49 ± 3.11 aB	
FRAP (Mmol Fe kg^{-1} of berry skins)	I	Low	190.39 ± 3.4	189.90 ± 3.4	3.35
		High	290.44 ± 3.3	290.34 ± 1.2	
	II	Low	371.21 ± 3.9	325.22 ± 3.4	
		High	318.08 ± 4.9	271.70 ± 3.9	
	Aver.	Low	280.80 ± 25.85 bA	257.56 ± 11.89 bB	
		High	304.26 ± 4.26 aA	281.02 ± 27.15 aB	

Mean followed by the same lowercase letter within the column and a capital letter within the row do not differ from each other, according to Tukey's test ($p < 0.05$).

3.2. *Impact of Rootstock and Training Systems on 'BRS Violeta' Grape Variety: Physiological, Biochemical Parameters and Yield*

For the variables, namely the electron transport rate (ETR), stomatal conductance (gs), CO_2 assimilation rate (A), internal carbon concentration (Ci), and SPAD index of the 'BRS Violeta' vine, a significant interaction between the rootstocks and training systems was observed (Table 5). The combination of 'IAC 766 Campinas' with the low espalier resulted in reduced energy dissipation and a higher ETR, indicating a less stressful condition for the scion cultivar and resulting in less wear on the vines. This combination also increased the CO_2 assimilation rate and reduced the internal carbon concentration, reinforcing the benefits of this combination for vine cultivation (Table 5).

The lower energy dissipation observed with the combination of 'IAC 766 Campinas' and a low espalier suggests that this interaction optimizes the capture of light energy, directing it more efficiently towards photosynthetic processes.

Furthermore, the reduction in the internal carbon concentration (Ci) observed with this interaction suggests that the available CO_2 is rapidly fixed during photosynthesis, preventing CO_2 accumulation in leaf cells and potentially reducing oxidative stress [27,28].

For the SPAD index, a significant interaction was observed, with the 'IAC 766 Campinas' rootstock providing higher SPAD values in 'BRS Violeta' vines, regardless of whether it was combined with a low or high espalier. The SPAD index is an indirect indicator of chlorophyll content, with high SPAD values being associated with greater photosynthetic capacity of the plant.

Table 5. Electron transport rate (ETR), stomatal conductance (gs), assimilation rate (A), internal carbon concentration (Ci), and SPAD index of the 'BRS Violeta' vine on different rootstocks and training systems.

Variables	Cycle	Trellis System	Rootstocks		CV (%)
			'IAC 766 Campinas'	106-8 'Mgt'	
ETR (µmol electrons $m^{-2}s^{-1}$)	I	Low	119.39 ± 1.4	100.39 ± 1.4	1.49
		High	112.34 ± 1.2	100.12 ± 1.2	
	II	Low	129.37 ± 1.2	127.89 ± 3.4	
		High	89.06 ± 1.3	100.10 ± 3.4	
	Aver.	Low	124.38 ± 3.50 aA	114.14 ± 5.64 aB	
		High	100.70 ± 5.96 bA	100.11 ± 4.87 bA	
gs (mol $m^{-2}s^{-1}$)	I	Low	0.14 ± 0.01	0.10 ± 0.01	2.24
		High	0.19 ± 0.01	0.12 ± 0.01	
	II	Low	0.42 ± 0.01	0.18 ± 0.01	
		High	0.33 ± 0.01	0.14 ± 0.01	
	Aver.	Low	0.28 ± 0.01 aA	0.14 ± 0.001 aB	
		High	0.26 ± 0.001 bA	0.13 ± 0.001 aB	
A (µmol CO_2 $m^{-2}s^{-1}$)	I	Low	30.30 ± 1.2	30.01 ± 1.4	3.63
		High	34.03 ± 1.4	34.39 ± 1.9	
	II	Low	41.96 ± 1.3	32.79 ± 1.3	
		High	27.97 ± 1.2	25.09 ± 1.2	
	Aver.	Low	36.13 ± 1.27 aA	31.40 ± 1.06 aB	
		High	31.00 ± 0.38 bA	29.74 ± 1.30 aA	
Ci (µmolCO_2 mol^{-1} air)	I	Low	139.44 ± 3.4	130.44 ± 3.4	4.76
		High	148.44 ± 3.9	144.44 ± 3.4	
	II	Low	119.14 ± 3.4	134.42 ± 3.4	
		High	176.52 ± 3.4	158.74 ± 3.4	
	Aver.	Low	129.29 ± 8.21 bA	132.43 ± 3.20 bA	
		High	162.48 ± 3.37 aA	151.59 ± 7.21 aA	
SPAD index	I	Low	17.12 ± 1.2	17.89 ± 0.3	4.90
		High	19.39 ± 1.2	19.90 ± 0.3	
	II	Low	41.10 ± 1.4	25.25 ± 1.2	
		High	41.29 ± 1.4	33.60 ± 1.2	
	Aver.	Low	29.11 ± 2.72 aA	21.57 ± 0.57 bB	
		High	30.34 ± 0.83 aA	26.75 ± 1.35 aB	

Mean followed by the same lowercase letter in the column and a capital letter in the row indicate that they do not differ, according to Tukey's test ($p < 0.05$).

The presence of higher chlorophyll content suggests that the 'IAC 766 Campinas' rootstock is particularly effective in enhancing the photosynthetic efficiency of 'BRS Violeta', providing optimal conditions for the production of ATP and NADPH, necessary for CO_2 assimilation.

The combination of the 'IAC 766 Campinas' rootstock with both training systems (low or high espalier) maximizes the SPAD index, contributing to a higher ETR, gs, and A. The increased ETR facilitates electron transfer between the photosystems, while a higher gs enhances the gas exchange and CO_2 uptake. The combination of these physiological and biochemical characteristics optimizes photosynthesis, promoting more vigorous growth of 'BRS Violeta' vines.

There were no significant interactions between the training systems and rootstocks in terms of the transpiration rate (E), water-use efficiency (WUE), and chlorophyll content; therefore, these variables were analyzed separately (Table 6). For the photosynthetic pigments, no significant differences were observed in any of the evaluated variables.

Table 6. Transpiration rate (E), water-use efficiency (WUE), and total chlorophyll of the 'BRS Violeta' vine on different rootstocks and training systems.

Variables	Cycle	Trellis System		Rootstocks		CV (%)
		Low	High	'IAC 766 Campinas'	106-8 'Mgt'	
E (mmol water vapor $m^{-2}s^{-1}$)	I	6.49 ± 1.3	7.02 ± 0.9	8.71 ± 0.3	6.27 ± 0.9	4.64
	II	9.03 ± 1.9	9.72 ± 1.0	10.79 ± 0.4	6.49 ± 1.0	
	Aver.	7.76 ± 2.00 b	8.37 ± 1.66 a	9.75 ± 0.38 a	6.38 ± 0.56 b	
WUE (μmol CO_2 (mmol $H_2O)^{-1}$)	I	3.39 ± 0.9	2.89 ± 0.8	4.34 ± 1.1	4.39 ± 1.0	5.26
	II	5.45 ± 0.8	4.77 ± 0.9	2.98 ± 1.4	4.81 ± 1.2	
	Aver.	4.42 ± 0.62 a	3.83 ± 0.44 b	3.66 ± 0.29 b	4.60 ± 0.44 a	
Total chlorophyll (mg 100 g^{-1} of leaves)	I	53.39 ± 3.8	69.03 ± 1.9	64.12 ± 1.8	60.03 ± 1.0	9.89
	II	69.61 ± 3.4	63.97 ± 1.8	67.82 ± 1.4	64.03 ± 1.2	
	Aver.	61.50 ± 8.96 a	66.50 ± 4.21 a	65.97 ± 6.70 a	62.03 ± 4.21 a	

Mean followed by the same lowercase letter in the column indicate that they do not differ, according to Tukey's test ($p < 0.05$).

The 'BRS Violeta' hybrid grown with a low espalier has higher water-use efficiency (WUE), particularly when combined with the 106-8 'Mgt' rootstock, which also resulted in a lower transpiration rate (E). Since the WUE indicates the plant's capacity to assimilate a greater amount of carbon dioxide with less water lost through transpiration, a higher water-use efficiency implies a lower E, which can contribute to better carbohydrate synthesis efficiency and reduced vulnerability to water stress.

The increased WUE observed with the low espalier and the 106-8 'Mgt' rootstock suggests an optimization in the plant's water balance, allowing efficient CO_2 assimilation without excessive water loss through transpiration (E). This trait is particularly advantageous in environments with limited water resources, where the ability to maintain efficient photosynthesis and carbohydrate production with reduced water consumption is crucial. Therefore, selecting training systems and rootstocks that maximize the WUE can significantly improve the sustainability and productivity of the 'BRS Violeta' vine under water stress conditions.

No significant interactions were observed between rootstocks and training systems for any of the productive or physicochemical variables of the 'BRS Violeta' grape must. Therefore, the factors were evaluated individually (Table 7). Training systems only influenced the soluble solids (SS) content and titratable acidity (TA) of the must.

The highest soluble solids (SS) content (16.32 °Brix) and the lowest titratable acidity (TA) (0.68% tartaric acid) were obtained from grapes of vines trained on a high espalier. The SS content in grapes is primarily composed of sugars (glucose and fructose) and, along with the TA, are direct indicators of fruit quality, closely linked to photosynthetic metabolism and the plant's resource-use efficiency. The higher SS content achieved with a high espalier suggests better photosynthetic efficiency, as the high espalier configuration allows for improved light distribution over the canopy, enhancing light capture and CO_2 assimilation (A). This efficiency is highlighted by the elevated values of the electron transport rate (ETR), stomatal conductance (gs), and SPAD index, which together contribute to increased sugar production in the fruits (Table 5).

The lower TA observed in grapes from a high espalier can be explained by the increased CO_2 assimilation and greater Rubisco enzyme activity, which channels more carbon into sugar production, reducing the relative concentration of organic acids in the must. This combination is ideal for producing grapes with a more balanced and attractive sensory profile for winemaking.

The 'IAC 766 Campinas' rootstock was noted for promoting higher production and productivity in the 'BRS Violeta' vine. The increase in these variables can be attributed to several physiological advantages provided by this rootstock; hence, it is recognized for its superior vigor, translating into greater water and nutrient absorption capacity from the soil.

This is crucial to meet the high metabolic demands associated with photosynthesis and plant growth.

Table 7. Number and fresh weight of the clusters, yield, productivity, soluble solids, pH, titratable acidity (TA), SS/TA ratio, and reducing sugars of the 'BRS Violeta' vine on the different rootstocks and training systems.

Variables	Cycle	Trellis System		Rootstocks		CV (%)
		Low	High	'IAC 766 Campinas'	106-8 'Mgt'	
Number of clusters per plant	I	8.64 ± 1.2	8.52 ± 1.8	8.91 ± 1.7	7.49 ± 1.9	1.87
	II	7.92 ± 1.9	7.80 ± 1.9	8.87 ± 1.8	7.61 ± 1.2	
	Aver.	8.28 ± 2.49 a	8.16 ± 3.87 a	8.89 ± 3.02 a	7.55 ± 3.34 a	
Fresh cluster weight (g)	I	173.00 ± 24.3	183.00 ± 30.1	190.30 ± 31.3	164.39 ± 19.8	27.97
	II	202.86 ± 19.3	197.5 ± 14.9	213.42 ± 14.8	188.25 ± 12.1	
	Aver.	187.93 ± 40.55 a	190.25 ± 48.05 a	201.86 ± 52.05 a	176.32 ± 30.06 a	
Yield (kg per plant)	I	1.60 ± 0.7	2.02 ± 1.1	2.30 ± 1.3	1.49 ± 0.9	0.36
	II	1.44 ± 0.3	1.08 ± 0.9	1.16 ± 1.0	1.21 ± 1.2	
	Aver.	1.52 ± 0.48 a	1.55 ± 0.81 a	1.73 ± 0.60 a	1.35 ± 0.68 b	
Productivity (t/ha)	I	5.69 ± 1.3	5.91 ± 1.4	6.34 ± 1.9	5.49 ± 1.2	1.47
	II	6.49 ± 1.1	6.55 ± 1.2	7.50 ± 1.8	5.31 ± 1.0	
	Aver.	6.09 ± 1.94 a	6.23 ± 3.25 a	6.92 ± 2.39 a	5.40 ± 2.73 b	
Soluble solids (°Brix)	I	15.61 ± 2.1	16.19 ± 1.9	16.10 ± 1.3	15.80 ± 1.9	0.31
	II	16.37 ± 1.9	16.45 ± 1.8	16.54 ± 1.2	16.18 ± 1.3	
	Aver.	15.99 ± 0.49 b	16.32 ± 0.53 a	16.32 ± 0.51 a	15.99 ± 0.52 b	
pH	I	3.47 ± 0.8	3.45 ± 0.9	3.49 ± 0.1	3.44 ± 0.9	0.07
	II	3.69 ± 0.9	3.65 ± 0.8	3.75 ± 0.1	3.60 ± 0.9	
	Aver.	3.58 ± 0.16 a	3.55 ± 0.12 a	3.62 ± 0.16 a	3.52 ± 0.10 b	
TA (% tartaric acid)	I	0.73 ± 0.3	0.69 ± 0.10	0.71 ± 0.3	0.70 ± 0.4	0.06
	II	0.77 ± 0.2	0.67 ± 0.9	0.75 ± 0.1	0.70 ± 0.3	
	Aver.	0.75 ± 0.15 a	0.68 ± 0.09 b	0.73 ± 0.15 a	0.70 ± 0.11 a	
SS/TA	I	20.19 ± 3.9	21.90 ± 3.8	22.90 ± 2.7	21.89 ± 2.3	1.53
	II	26.37 ± 3.2	27.10 ± 3.1	24.86 ± 1.1	25.91 ± 1.9	
	Aver.	23.28 ± 2.86 a	24.50 ± 3.09 a	23.88 ± 3.46 a	23.90 ± 2.56 a	
Reducing sugars (%)	I	10.39 ± 1.3	11.34 ± 1.4	12.49 ± 1.3	11.89 ± 1.9	0.80
	II	14.89 ± 1.2	15.40 ± 1.9	14.05 ± 1.4	13.59 ± 1.9	
	Aver.	12.64 ± 1.38 a	13.37 ± 1.42 a	13.27 ± 1.73 a	12.74 ± 1.04 a	

Mean followed by the same lowercase letter on the line, within the same factor, do not differ from each other, according to Tukey's test ($p < 0.05$).

Thus, the greater photosynthetic efficiency of vines grafted onto 'IAC 766 Campinas' is evidenced by the higher values of the ETR, gs, and CO_2 assimilation rate (A) (Table 5). These parameters indicate that the vines have a greater capacity to capture and utilize solar light for photosynthesis, resulting in higher production of ATP and NADPH, which are essential for carbohydrate synthesis and increased vine production.

Vines grafted onto this rootstock showed higher SS and pH values, indicating an optimal balance between sugar accumulation and must pH. The superior capacity of 'IAC 766 Campinas' to support cultivation conditions and maximize carbohydrate production makes it a recommended choice for the 'BRS Violeta' vine. The higher CO_2 assimilation rate (A) observed with the use of this rootstock indicates that more carbon is being fixed and converted into sugars and other photoassimilates, which are used for fruit growth and development.

Moreover, the improved regulation of must pH with 'IAC 766 Campinas' indicates a balance between sugar synthesis and organic acid accumulation, resulting in higher quality fruit for wine and juice production. Specifically concerning pH, studies suggest that the effects of rootstocks may be related to their capacity for potassium extraction from the soil.

In both grapes and beverages, such as grape juice and wine, pH measurement is crucial as it is directly related to anthocyanin stability, which affects the color intensity of these beverages. Therefore, the choice of rootstock not only impacts vine productivity and health, but also the final quality of the grape-derived products, highlighting the importance of 'IAC 766 Campinas' in optimizing the production and quality of 'BRS Violeta' grapes.

There were interactions among the factors evaluated for the secondary metabolites in the skin of 'BRS Violeta' grapes. The combination of using the 'IAC 766 Campinas' rootstock with the high espalier resulted in higher levels of total phenolic compounds and monomeric anthocyanins, as well as greater antioxidant activity, using both methods (Table 8).

Table 8. Total phenolic compounds, total monomeric anthocyanins, and antioxidant activity (DPPH and FRAP) in the skin of 'BRS Violeta' grape in different rootstocks and training systems.

Variables	Cycle	Trellis System	Rootstocks		CV (%)
			'IAC 766 Campinas'	106-8 'Mgt'	
Total phenolic compounds (mg 100 g^{-1} of berry skins)	I	Low	1059.67 ± 18.3	1130.15 ± 3.4	1.81
		High	1390.90 ± 12.3	1300.39 ± 3.4	
	II	Low	1977.87 ± 10.1	1030.73 ± 1.9	
		High	1809.78 ± 12.1	1711.55 ± 1.8	
	Aver.	Low	1518.77 ± 23.45 bA	1080.44 ± 8.80 bB	
		High	1600.34 ± 20.77 aA	1505.97 ± 40.27 aB	
Total monomeric anthocyanins (mg 100 g^{-1} of berry skins)	I	Low	989.19 ± 11.2	930.89 ± 12.4	6.88
		High	1089 ± 10.9	1300.39 ± 12.3	
	II	Low	1672.35 ± 10.2	700.45 ± 10.2	
		High	1732.62 ± 10.3	1292.75 ± 10.3	
	Aver.	Low	1330.77 ± 49.62 bA	815.67 ± 13.85 bB	
		High	1410.81 ± 46.89 aA	1296.57 ± 47.90 aB	
DPPH (Mmol g^{-1} of berry skins)	I	Low	600.37 ± 6.4	500.13 ± 10.3	1.36
		High	600.39 ± 9.3	500.69 ± 4.4	
	II	Low	674.43 ± 3.4	527.47 ± 1.9	
		High	679.35 ± 3.4	638.79 ± 1.9	
	Aver.	Low	637.40 ± 6.49 bA	513.80 ± 10.63 bB	
		High	639.87 ± 9.51 aA	569.74 ± 3.32 aB	
FRAP (Mmol Fe kg^{-1} of berry skins)	I	Low	39.89 ± 3.4	30.30 ± 3.0	1.39
		High	40.39 ± 1.9	34.34 ± 3.4	
	II	Low	42.09 ± 1.2	43.78 ± 1.2	
		High	47.29 ± 1.2	45.44 ± 1.2	
	Aver.	Low	40.99 ± 7.26 bA	37.04 ± 5.82 bB	
		High	43.84 ± 3.11 aA	39.89 ± 4.06 aB	

Mean followed by the same lowercase letter in the column and a capital letter in the row do not differ, according to Tukey's test ($p < 0.05$).

The higher concentration of phenolic compounds obtained with the high espalier can be attributed to better sunlight distribution, which promotes increased photosynthetic activity and synthesis of secondary metabolites, such as phenolic compounds.

This same interaction between the rootstock and the training system also resulted in increased anthocyanin content in the grape skins (Table 8). Biochemically, anthocyanins are pigments responsible for grape coloration and play a significant role in antioxidant activity.

Greater light exposure with a high espalier can induce the expression of genes related to anthocyanin biosynthesis, resulting in higher concentrations of these pigments. Enhanced photosynthetic efficiency, evidenced by the high ETR and increased production of ATP and NADPH (Table 5), also contributes to the increased synthesis of anthocyanins.

Anthocyanins, along with other phenolic components present in 'BRS Violeta' grapes, are of extreme importance in the production of red wines and, particularly, juices. Therefore, the correct choice of training system and rootstock, which have greater affinity with the 'BRS Violeta' cultivar, is essential to maximize the accumulation potential of antioxidant compounds in grapes.

Vines trained with a high espalier and grafted onto the 'IAC 766 Campinas' rootstock exhibited higher antioxidant activity (DPPH: 639.87 μmol g^{-1} of grape skin; and FRAP: 43.84 μmol Fe kg^{-1} of grape skin). These results indicate that the increased synthesis of phenolic compounds and anthocyanins as a result of a high espalier not only enhances the nutritional quality of the grapes, but also boosts their antioxidant capacity.

Therefore, the choice of training system and rootstock has a significant impact on the phytochemical composition and antioxidant activity of 'BRS Violeta' grapes. The high espalier, combined with the 'IAC 766 Campinas' rootstock, proved to be the most effective combination for maximizing the concentration of phenolic compounds, anthocyanins, and antioxidant activity. This combination not only improves the nutritional and sensory quality of the grapes, but also enhances the photosynthetic efficiency and resistance to environmental stresses, promoting more sustainable and high-quality production.

3.3. *Principal Component Analysis (PCA)*

Principal Component Analysis (PCA) of the 'IAC 138-22 Máximo' vine reveals that the first two principal components (PCA 1 and PCA 2, Figure 2A) account for 85.03% of the total variance in the data. It is observed that parameters related to photosynthetic efficiency, such as the SPAD index, internal carbon concentration (Ci), assimilation rate (A), and electron transport rate (ETR), are positively correlated and associated with the 'IAC 766 Campinas' rootstock and the high espalier. This indicates that this combination promotes better conditions for photosynthesis, resulting in higher production of photoassimilates and higher levels of phenolic compounds and anthocyanins, which are crucial for the nutritional and sensory quality of the grapes and wines.

Parameters, such as the pH, yield, productivity, and soluble solids, are also positively correlated with a high espalier, highlighting the importance of this training system in maximizing the quality of grape must. The 106-8 'Mgt' rootstock, associated with the low espalier, showed lower efficiency in regard to all these parameters, suggesting reduced physiological and biochemical performance of the 'IAC 138-22 Máximo' vines under these conditions.

In the PCA of the 'BRS Violeta' vine (Figure 2B), the first two principal components explain 91.62% of the total variance in the data. Similar to 'IAC 138-22 Máximo', the photosynthetic efficiency parameters (SPAD, Ci, A, ETR) are strongly associated with the 'IAC 766 Campinas' rootstock combined with the high espalier. This rootstock demonstrated notable superiority in maximizing the photosynthetic efficiency and, consequently, the production of phenolic compounds, anthocyanins, and antioxidant activity.

Additionally, must quality parameters, such as the soluble solids (SS) content, yield, and titratable acidity (TA) are positively correlated with the high espalier and the 'IAC 766 Campinas' rootstock. This indicates that this combination not only improves the photosynthetic efficiency, but also optimizes the physical and chemical qualities of the grapes. In contrast, the low espalier and the 106-8 'Mgt' rootstock are associated with lower levels of phenolic compounds, anthocyanins, and antioxidant activity (DPPH and FRAP), reflecting the reduced physiological and biochemical performance of the grapes from this combination.

Figure 3 summarizes how different combinations of rootstocks and training systems affect the production and quality of the grapes from the 'IAC 138-22 Máximo' and 'BRS Violeta' cultivars. The use of the 'IAC 766 Campinas' rootstock and the high espalier resulted in increased productivity and improved physiological, physicochemical, and biochemical attributes of the grapes. These results are explained by the higher capacity for water and nutrient absorption, the vigor provided by the rootstocks, and the better light distribution afforded by the high espalier. The integration of these practices can help balance the limitations of each variety, meeting the standards required for juices and wines, and resulting in higher quality products.

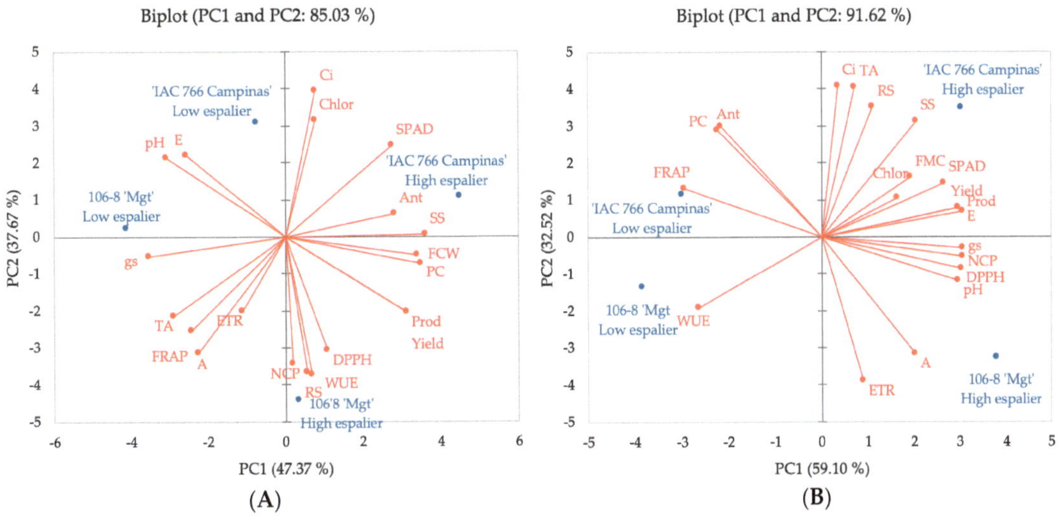

Figure 2. Principal component analysis of hybrid vines on different rootstocks and training systems. (**A**) 'IAC 138-22 Máximo' and (**B**) 'BRS Violeta'. Notes: internal carbon concentration (Ci), total chlorophyll (Chlor), anthocyanins (Ant), soluble solids (SS), fresh cluster weight (FCW), phenolic compounds (PC), productivity (Prod), water-use efficiency (WUE), reducing sugars (RS), number of clusters per plant (NCP), electron transport rate (ETR), assimilation rate (A), titratable acidity (TA), stomatal conductance (gs), transpiration rate (E).

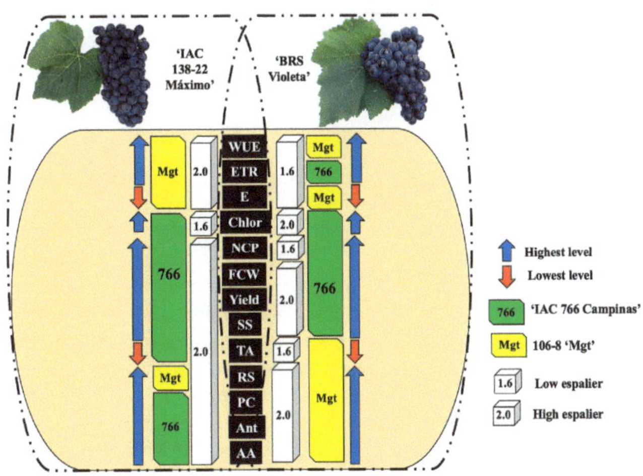

Figure 3. Graphical abstract of the two grapevine hybrids, 'IAC 138-22 Maximo' and 'BRS Violeta' and their interaction with the two trellis systems (low 1.6 m and high 2.0 m) and two rootstocks, 'IAC 766' (766) and '106-8 Mgt' (Mgt). Notes: water-use efficiency (WUE), electron transport rate (ETR), transpiration rate (E), total chlorophyll (Chlor), number of clusters per plant (NCP), fresh cluster weight (FCW), soluble solids (SS), titratable acidity (TA), reducing sugars (RS), phenolic compounds (PC), total monomeric anthocyanins (Ant), and antioxidant activity (AA).

In summary, the results indicate that the choice of training system and rootstock has a significant impact on the physiology, biochemistry, and final quality of the grapes from the hybrid vines 'IAC 138-22 Máximo' and 'BRS Violeta'. The combination of the high espalier with the 'IAC 766 Campinas' rootstock provides the best conditions for maximizing the photosynthetic efficiency and the synthesis of essential biochemical compounds, resulting in grapes of high nutritional and sensory quality. Therefore, this combination is recommended for the optimized production of these hybrids.

4. Discussion

The electron transport rate (ETR) is a fundamental indicator of photosynthetic performance, reflecting the plant's capacity to transport electrons through the electron transport chain in chloroplasts. This transport is crucial for ATP and NADPH production, which are utilized in the Calvin–Benson cycle for carbon fixation [37,38]. CO_2 enters the leaves by diffusing through the stomatal pores on the leaf surface, becoming available as a substrate for photosynthetic assimilation [39]. The sensitivity of vine leaves to solar exposure and water use by plants means that the canopy structure dimensions, geometric forms, and management practices are critical for achieving optimal physiological functioning [40]. Additionally, plants with higher chlorophyll content have a greater capacity to capture solar radiation, which, combined with optimized stomatal conductance (gs) and a higher transpiration rate (E), contributes to increased production of ATP and NADPH, essential for the Calvin–Benson cycle and the production of photoassimilates [37,38].

Physiologically, the improved leaf distribution with a high espalier maximizes solar light interception, which is essential for photosynthesis [14]. These results may be related to the interaction between the scion cultivar and the rootstock, as well as to the different vigor levels provided by the rootstocks to the scion cultivars, since the rootstock 'IAC 766 Campinas' is more vigorous than 106-8 'Mgt' [41–43]. The use of more vigorous rootstocks, such as 'IAC 766 Campinas', contributes to an increased capacity for water and nutrient absorption and translocation by the roots [44,45], which may result in a greater supply of resources for photosynthesis and plant metabolism. Biochemically, phenolic compounds and monomeric anthocyanins are important secondary metabolites present in grapes, playing crucial roles in the defense against environmental stresses and in the nutritional and sensory quality of the fruit. Phenolic compounds are known for their anti-inflammatory, anticancer, and cardioprotective properties [46,47]. Anthocyanins, in addition to providing the characteristic color to grapes, play an important role in the protection against chronic diseases due to their ability to neutralize free radicals and reduce oxidative stress in the body [48–51]. Flavonoids are derived from the shikimic and phenolic acid pathways [52].

The high espalier and 'IAC 766 Campinas' rootstock favor light exposure and canopy distribution, optimizing photosynthesis [14] and secondary metabolite production. The higher ETR indicates a more efficient transfer of electrons between the photosystems, leading to increased production of ATP and NADPH, which are essential for CO_2 assimilation in the Calvin cycle [37,38]. Increased stomatal conductance (gs) enhances gas exchange, increasing CO_2 uptake and facilitating a higher assimilation rate (A), which translates into greater photoassimilate production and plant growth [15,19]. The reduction in the internal carbon concentration (Ci) suggests that the available CO_2 is rapidly fixed during photosynthesis, preventing CO_2 accumulation in leaf cells and potentially reducing oxidative stress [53]. The SPAD index is an indirect indicator of chlorophyll content; high SPAD values are associated with greater photosynthetic capacity of the plant [54]. A higher assimilation rate (A) indicates a greater photosynthetic rate, resulting in increased carbohydrate production and plant growth [54]. Water-use efficiency (WUE), indicates the plant's capacity to assimilate a greater amount of carbon dioxide with less water lost through transpiration (E) [55]. An optimization in the plant's water balance allows efficient CO_2 assimilation, without excessive water loss through transpiration [55]. The high espalier configuration allows for improved light distribution over the canopy, enhancing light capture and CO_2 assimilation (A) [14].

'IAC 766 Campinas' is recognized by its strong vigor, translating into greater water and nutrient absorption capacity from the soil [41–43,56]. Higher photosynthetic efficiency is directly related to increased sugar production, which constitutes the main components of soluble solids [57].

The greater the affinity of the rootstock for potassium, the higher the pH of the grapes [58]. 'IAC 766 Campinas' rootstock has a higher affinity for this nutrient. The effect of rootstocks on the physicochemical composition of grapes may also be related to other factors such as vigor, water and nutrient absorption capacity, disease resistance, and interaction with the scion. These factors directly influence both primary and secondary plant metabolites and, consequently, grape quality [59–61].

Anthocyanin stability affects the color intensity of grapes, grape juice, and wine [62]. Anthocyanins are pigments responsible for grape coloration and play a significant role in antioxidant activity [48–51]. Greater light exposure with a high espalier induces the expression of genes related to anthocyanin biosynthesis, resulting in higher concentrations of these pigments [63]. Phenolic compounds are essential in defending plants against environmental stress, while also contributing significantly to the grapes' nutritional and sensory qualities [63]. The more vigorous 'IAC 766 Campinas' rootstock may enhance nutrient absorption, which in turn could lead to increased synthesis of these compounds [59–61]. The phenolic profile of grapes has garnered significant interest from researchers focused on wine and juice production [4,64–66]. These compounds, including anthocyanins, are renowned for their health benefits, especially in the prevention of chronic diseases. They act by neutralizing free radicals, namely unstable molecules that can damage cells and are linked to cardiovascular diseases, cancer, and neurodegenerative disorders [67]. Furthermore, the antioxidant properties of these compounds protect plant cells from damage by reactive oxygen species (ROS), enhancing the resistance to abiotic stress and promoting overall plant health [53]. Beyond mitigating oxidative stress, these compounds also improve immune function, reduce inflammation, and safeguard DNA from mutations [63].

The combination of the 'IAC 766 Campinas' rootstock with a high espalier is the most effective strategy for maximizing productivity, nutritional, and sensory quality of 'IAC 138-22 Máximo' and 'BRS Violeta' grapes. This approach simultaneously enhances the photosynthetic efficiency and the concentration of bioactive compounds, making it the ideal choice for viticulturists and the processing industry.

Author Contributions: F.J.D.N., J.D.R. and M.A.T. planned and designed the experiment. F.J.D.N. and C.V.B. performed the plant physiological analyses, chemical, biochemical and enzyme analyses. F.J.D.N., M.F.M. and J.L.H. performed data analyses. F.J.D.N., M.d.S.S. and M.L. created the tables and figures. F.J.D.N., J.D.R., E.O.O., G.P.P.L. and M.d.S.S. wrote and revised the manuscript. All authors have read and agreed to the published version of the manuscript.

Funding: This research was funded by the Fundação de Amparo à Pesquisa do Estado de São Paulo (Research Support Foundation of the State of São Paulo), FAPESP process no. 2016/07510-2 through granting the scholarship for the first author, several programs that supported the study in setting up the experimental area (FAPESP process no. 2011/03440-6, 2013/8915-5 and 2015/16440-5) and the CNPq for the Research Productivity Scholarship (process no. 307377/2021-0).

Data Availability Statement: The original contributions presented in the study are included in the article material, further inquiries can be directed to the corresponding author.

Conflicts of Interest: The authors declare no conflicts of interest.

References

1. Camargo, U.; Tonietto, J.; Hoffmann, A. Advances in grape culture in Brazil. *Rev. Bras. Frutic.* **2011**, *33*, 144–149. [CrossRef]
2. IBGE—Instituto Brasileiro de Geografia e Estatística. Fruticultura: Área Plantada e Quantidade Produzida. 2024. Available online: http://www.ibge.gov.br (accessed on 15 August 2024).
3. Burin, V.M.; Ferreira-Lima, N.; Panceri, C.P.; Bordignon-Luiz, M.T. Bioactive compounds and antioxidant activity of *Vitis vinifera* and *Vitis labrusca* grapes: Evaluation of different extraction methods. *Microchem. J.* **2014**, *114*, 155–163. [CrossRef]

4. Lago-Vanzela, E.S.; Procópio, D.P.; Fontes, E.A.F.; Ramos, A.M.; Stringheta, P.C.; Da-Silva, R.; Castillo-Muñoz, N.; Hermosín Gutiérrez, I. Aging of red wines made from hybrid grape cv. BRS Violeta: Effects of accelerated aging conditions on phenolic composition, color and antioxidant activity. *Food Res. Int.* **2014**, *56*, 182–189. [CrossRef]
5. Santos, J.A.; Fraga, H.; Malheiro, A.C.; Moutinho-Pereira, J.; Dinis, L.-T.; Correia, C.; Moriondo, M.; Leolini, L.; Dibari, C. Costafreda-Aumedes, S.; et al. A Review of the Potential Climate Change Impacts and Adaptation Options for European Viticulture. *Appl. Sci.* **2020**, *10*, 3092. [CrossRef]
6. Silva, M.J.R.; Paiva, A.P.M.; Pimentel, A.; Sánchez, C.A.P.C.; Callili, D.; Moura, M.; Leonel, S.; Tecchio, M.A. Yield performance of new juice grape varieties grafted onto different rootstocks under tropical conditions. *Sci. Hortic.* **2018**, *241*, 194–200. [CrossRef]
7. Tecchio, M.A.; Teixeira LA, J.; Terra, M.M.; Moura, M.F.; Paioli-Pires, E.J. Extração de nutrientes pela videira 'Niagara Rosada enxertada em diferentes porta-enxertos. *Rev. Bras. Frutic.* **2011**, *33*, 736–742. [CrossRef]
8. Rizk-Alla, M.S.; Sabry, G.H.; Abd-El-Wahab, M.A. Influence of Some Rootstocks on the Performance of Red Globe Grape Cultivar *J. Am. Sci.* **2011**, *7*, 71–81.
9. Tecchio, M.A.; da Silva, M.J.R.; Sanchez, C.A.P.C.; Callili, D.; Vedoato, B.T.F.; Hernandes, J.L.; Moura, M.F. Yield performance and quality of wine grapes (*Vitis vinifera*) grafted onto different rootstocks under subtropical conditions. *Bragantia* **2022**, *81*, 1 [CrossRef]
10. Silva, M.J.R.; Paiva, A.P.M.; de Souza, J.F.; Padilha, C.V.d.S.; Basílio, L.S.P.; Lima, M.d.S.; Pereira, G.E.; Corrêa, L.C.; Vianello, F. Lima, G.P.P.; et al. Phytochemical profile of Brazilian grapes (*Vitis labrusca* and hybrids) grown on different rootstocks. *PLoS ONE* **2022**, *17*, e0275489. [CrossRef]
11. Sánchez, C.A.P.C.; Tecchio, M.A.; Callili, D.; da Silva, M.J.R.; Basílio, L.S.P.; Leonel, S.; Alonso, J.C.; Lima, G.P.P. Productivity and Physicochemical Properties of the BRS Isis Grape on Various Rootstocks under Subtropical Climatic Conditions. *Agriculture* **2023**, *13*, 2113. [CrossRef]
12. Simonetti, L.M.; Sousa, M.C.; Moura, M.F.; Nunes, J.G.d.S.; Diamante, M.S.; da Silva, M.B.; da Silva, M.J.R.; Callili, D.; Lima, G.P.P.; Tecchio, M.A. The influence of different training systems and rootstocks on Sauvignon Blanc grapes. *Bragantia* **2021**, *80*, e2021 [CrossRef]
13. Pimentel Junior, A.; Domingues, F.J.; Basílio, L.S.P.; Monteiro, G.C.; Lima GP, P.; Tecchio, M.A. Training systems improved agronomic characteristics and quality of 'Niagara Rosada' table grapes. *Bragantia* **2023**, *82*, e20220252. [CrossRef]
14. Domingues Neto, F.J.; Pimentel Junior, A.; Modesto, L.R.; Moura, M.F.; Putti, F.F.; Boaro, C.S.F.; Ono, E.O.; Rodrigues, J.D.; Tecchio, M.A. Photosynthesis, Biochemical and Yield Performance of Grapevine Hybrids in Two Rootstock and Trellis Height. *Horticulturae* **2023**, *9*, 596. [CrossRef]
15. Del Zozzo, F.; Magnanini, E.; Poni, S. Physiological efficiency of grapevine canopies having varying geometries: Seasonal and diurnal whole canopy gas exchange assessment under well-watered and water deficit conditions. *Environ. Exp. Bot.* **2024**, *221*, 105716. [CrossRef]
16. Wu, Y.; Qiu, T.; Shen, Z.; Wu, Y.; Lu, D.; He, J. Effects of shading on leaf physiology and morphology in the 'Yinhong' grape plants. *Rev. Bras. Frutic.* **2018**, *40*, e-037. [CrossRef]
17. Miccichè, D.; de Rosas, M.I.; Ferro, M.V.; Di Lorenzo, R.; Puccio, S.; Pisciotta, A. Effects of artificial canopy shading on vegetative growth and ripening processes of cv. Nero d'Avola (*Vitis vinifera* L.). *Front. Plant Sci.* **2023**, *14*, 1210574. [CrossRef]
18. Martínez-Lüscher, J.; Brillante, L.; Kurtural, S.K. Flavonol Profile Is a Reliable Indicator to Assess Canopy Architecture and the Exposure of Red Wine Grapes to Solar Radiation. *Front. Plant Sci.* **2019**, *10*, 10. [CrossRef]
19. Yu, R.; Torres, N.; Tanner, J.D.; Kacur, S.M.; Marigliano, L.E.; Zumkeller, M.; Gilmer, J.C.; Gilmer, J.C.; Gilmer, J.C. Adapting wine grape production to climate change through canopy architecture manipulation and irrigation in warm climates. *Front. Plant Sci.* **2022**, *3*, 1015574. [CrossRef]
20. Bavougian, C.M.; Read, P.E.; Walter-Shea, E. Training System Effects on Sunlight Penetration, Canopy Structure, Yield, and Fruit Characteristics of 'Frontenac' Grapevine (*Vitis* spp.). *Int. J. Fruit Sci.* **2012**, *12*, 402–409. [CrossRef]
21. Würz, D.A.; Marcon Filho, J.L.; Allebrandt, R.; Bem, B.D.; Rufato, L.; Kretzschmar, A.A. Agronomic performance of 'Cabernet Sauvignon' grapevine in different training systems in high altitude regions of Santa Catarina State, Brazil. *Rev. Ciências Agroveterinárias* **2019**, *18*, 73–80.
22. Machado, T.O.X.; Portugal, I.; Kodel, H.A.C.; Droppa-Almeida, D.; Lima, M.S.; Fathi, F.; Oliveira, M.B.P.P.; Albuquerque-Júnior, R.L.C.; Dariva, C.; Souto, E.B. Therapeutic potential of grape pomace extracts: A review of scientific evidence. *Food Biosci.* **2024**, *60*, e104210. [CrossRef]
23. Wang, H.; Cao, G.; Prior, R.L. Oxygen radical absorbing capacity of anthocyanins. *J. Agric. Food Chem.* **1997**, *45*, 304–309. [CrossRef]
24. Sanchez-Moreno, C. Compuestos polifenolicos: Efectos fisiologicos. Actividad antioxidante. *Alimentaria* **2002**, *329*, 29–40.
25. Rustioni, L.; Rocchi, L.; Failla, O. Effect of anthocyanin absence on white berry grape (*Vitis vinifera* L.). *Vitis* **2015**, *54*, 239–242.
26. EMBRAPA. Sistema Brasileiro de Classificação do Solo–Brasília: EMBRAPA, Produção de Informação: 5. ed., Rev. e Ampl. 2018, p. 356. Available online: https://www.agroapi.cnptia.embrapa.br/portal/assets/docs/SiBCS-2018-ISBN-9788570358004.pdf (accessed on 19 July 2024).
27. Von Caemmerer, S.; Farquhar, G.D. Some relationships between the biochemistry of photosynthesis and the gas exchange of leaves. *Planta* **1981**, *153*, 376–387. [CrossRef]

28. Klughammer, C.; Schreiber, U. Complementary PS II quantum yields calculated from simple fluorescence parameters measured by PAM fluorometry and the Saturation Pulse method. *PAM Appl. Notes* **2008**, *1*, 27–35.
29. Sims, D.A.; Gamon, J.A. Relationships between leaf pigment content and spectral reflectance across a wide range of species, leaf structures and developmental stages. *Remote Sens. Environ.* **2002**, *81*, 337–354. [CrossRef]
30. Nelson, N.A. A photometric adaptation of Somogy method for the determination of glucose. *J. Biol. Chem.* **1944**, *135*, 136–175.
31. Singleton, V.L.; Rossi Junior, J.A. Colorimetry of total phenolics with phosphomolybidic-phosphotungstic acid reagents. *Am. J. Enol. Vitic.* **1965**, *16*, 144–158. [CrossRef]
32. Giusti, M.M.; Wrolstad, R.E. Characterization and measurement of anthocyanins by UV–Visible spectroscopy. *Curr. Protoc. Food Anal. Chem.* **2001**, *3*, 1–13. [CrossRef]
33. Brand-Williams, W.; Cuvelier, M.E.; Berset, C. Use of a free radical method to evaluate antioxidant activity. *Lebensm. Wiss. Technol.* **1995**, *28*, 25–30. [CrossRef]
34. Rossetto, M.R.M.; Vianello, F.; Rocha, S.D.; Lima, G.P.P. Antioxidant substances and pesticide in parts of beet organic and conventional manure. *Afr. J. Plant Sci.* **2009**, *3*, 245–253.
35. Benzie, I.F.F.; Strain, J.J. The ferric reducing ability of plasma (FRAP) as a measure of "antioxidant power": The FRAP assay. *Anal. Biochem.* **1996**, *239*, 70–76. [CrossRef] [PubMed]
36. Ferreira, D.F. Sisvar: A computer statistical analysis system. *Ciência Agrotecnologia* **2011**, *35*, 1039–1042. [CrossRef]
37. Taiz, L.; Zeiger, E.; Møller, I.M.; Murphy, A. *Fisiologia e Desenvolvimento Vegetal*; Artmed: Porto Alegre, Brazil, 2017; 888p.
38. Douthe, C.; Gago, J.; Ribas-Carbó, M.; Núñez, R.; Pedrol, N.; Flexas, J. Measuring photosynthesis and respiration with infrared gas analysers. In *Advances in Plant Ecophysiology Techniques*; Sánchez-Moreiras, A.M., Reigosa, M.J., Eds.; Springer: Cham, Switzerland, 2018; pp. 51–75.
39. Tominaga, J.; Shimada, H.; Kawamitsu, Y. Direct measurement of intercellular CO_2 concentration in a gas-exchange system resolves overestimation using the standard method. *J. Exp. Bot.* **2018**, *69*, 1981–1991. [CrossRef]
40. Hunter, J.K.; Tarricone, L.; Volschenk, C.; Giacalone, C.; Melo, M.S.; Zorer, R. Grapevine physiological response to row orientation-induced spatial radiation and microclimate changes. *OENO One* **2020**, *54*, 411–433. [CrossRef]
41. Santos Neto, J.R.A. *A Cultura da Videira*; Instituto Agronômico: Campinas, Brazil, 1973; p. 108.
42. Pommer, C.V.; Passos, I.R.S.; Terra, M.M.; Pires, E.J.P. *Variedades de Videira Para o Estado de São Paulo*; Instituto Agronômico, (IAC), Boletim Técnico: Campinas, Brazil, 1997; Volume 166, p. 59.
43. Simão, S. *Tratado de Fruticulture*; FEALQ: Piracicaba, Brazil, 1998; 760p.
44. Alvarenga, A.A.; Regina, M.A.; Fráguas, J.C.; Silva, A.L.; Souza, C.M.; Cançado GM, A.; Freitas, G.F. Indicação de porta-enxertos de videiras para o sul de Minas Gerais. In *Viticultura e Enologia: Atualizando Conceitos*; Epamig: Caldas, Brazil, 2002; pp. 243–256.
45. Pauletto, D.; Mourão Filho FD, A.A.; Kluge, R.A.; Scarpare Filho, J.A. Produção e vigor da videira 'Niágara Rosada' relacionados com o porta-enxerto. *Pesqui. Agropecuária Bras.* **2001**, *36*, 115–121. [CrossRef]
46. Decendit, A.; Mamani-Matsuda, M.; Aumont, V.; Waffo-Teguo, P.; Moynet, D.; Boniface, K.; Richard, E.; Krisa, S.; Rambert, J.; Mérillon, J.-M.; et al. Malvidin-3-O-β glucoside, major grape anthocyanin, inhibits human macrophage-derived inflammatory mediators and decreases clinical scores in arthritic rats. *Biochem. Pharmacol.* **2013**, *86*, 1461–1467. [CrossRef] [PubMed]
47. Granato, D.; Carrapeiro, M.d.M.; Fogliano, V.; van Ruth, S.M. Effects of geographical origin, varietal and farming system on the chemical composition and functional properties of purple grape juices: A review. *Trends Food Sci. Technol.* **2016**, *52*, 31–48. [CrossRef]
48. Abe, L.T.; Mota, R.V.; Lajolo, F.M.; Genovese, M.I. Compostos fenólicos e capacidade antioxidante de cultivares de uvas *Vitis labrusca* L. e *Vitis vinifera* L. *Food Sci. Technol.* **2007**, *27*, 394–400. [CrossRef]
49. Ferrari, D.; Speciale, A.; Cristani, M.; Fratantonio, D.; Molonia, M.S.; Ranaldi, G.; Saija, A.; Cimino, F. Cyanidin-3-O-glucoside inhibits NF-kB signalling in intestinal epithelial cells exposed to TNF-kB and exerts protective effects via Nrf2 pathway activation. *Toxicol. Lett.* **2016**, *264*, 51–58. [CrossRef]
50. Meng, X.; Maliakal, P.; Lu, H.; Lee, M.J.; Yang, C.S. Urinary and plasma levels of resveratrol and quercetin in humans, mice, and rats after ingestion of pure compounds and grape juice. *J. Agric. Food Chem.* **2004**, *52*, 935–942. [CrossRef]
51. Dani, C.; Oliboni, L.S.; Pasquali, M.A.; Oliveira, M.R.; Umezu, F.M.; Salvador, M.; Moreira, J.C.; Henriques, J.A. Intake of purple grape juice as a hepatoprotective agent in wistar rats. *J. Med. Food* **2008**, *11*, 127–132. [CrossRef] [PubMed]
52. Kumar, G.A.; Kumar, S.; Bhardwaj, R.; Swapnil, P.; Meena, M.; Seth, C.S.; Yadav, A. Recent advancements in multifaceted roles of flavonoids in plant–rhizomicrobiome interactions. *Front. Plant Sci.* **2024**, *14*, 1297706. [CrossRef] [PubMed]
53. Fu, J.; Li, L.; Wang, S.; Yu, N.; Shan, H.; Shi, Z.; Li, F.; Zhong, X. Effect of gibberellic acid on photosynthesis and oxidative stress response in maize under weak light conditions. *Front. Plant Sci.* **2023**, *14*, 1128780. [CrossRef]
54. Chaimala, A.; Jogloy, S.; Vorasoot, N.; CHolbrook, C.; Kvien, C.K.; Laohasiriwong, S. The variation of relative water content, SPAD chlorophyll meter reading, stomatal conductance, leaf area, and specific leaf area of Jerusalem artichoke genotypes under different durations of terminal drought in tropical region. *J. Agron. Crop Sci.* **2023**, *209*, 12–26. [CrossRef]
55. Alotaibi, M.; El-Hendawy, S.; Mohammed, N.; Alsamin, B.; Refay, Y. Appropriate application methods for salicylic acid and plant nutrients combinations to promote morpho-physiological traits, production, and water use efficiency of wheat under normal and deficit irrigation in an arid climate. *Plants* **2023**, *12*, 1368. [CrossRef]
56. Hartmann, H.T.; Kester, D.E. *Propagación de Plantas: Principios y Practicas*; Continental: Colonia San Juan Tlihuaca, México, 1990; p. 760.

57. Baldassi, C.; Berim, A.; Roeder, S.; Losciale, P.; Serra, S.; Gang, D.R.; Musacchi, S. Rootstock and Crop Load Effects on 'Honeycrisp'Photosynthetic Performance and Carbohydrate Accumulation. *Plants* **2023**, *12*, 4035. [CrossRef] [PubMed]
58. Kodur, S.; Tisdall, J.M.; Clingeleffer, P.R.; Walker, R.R. Regulation of berry quality parameters in 'Shiraz' grapevines through rootstocks (*Vitis*). *Vitis* **2013**, *52*, 125–128.
59. Dias, F.A.N.; da Mota, R.V.; Fávero, A.C.; Purgatto, E.; Shiga, T.M.; de Souza, C.R.; Pimentel, R.M.d.A.; Regina, M.d.A. Videira 'Syrah' sobre diferentes portaenxertos em ciclo de inverno no sul de Minas Gerais. *Pesqui. Agropecuária Bras.* **2012**, *47*, 208–215. [CrossRef]
60. Lee, J.; Steenwerth, K.L. 'Cabernet Sauvignon' grape anthocyanin increased by soil conservation practices. *Sci. Hortic.* **2013**, *159*, 128–133. [CrossRef]
61. Tecchio, M.A.; Moura, M.F.; Teixeira, L.A.J.; Pires, E.J.P.; Leonel, S. Influence of rootstocks and pruning times on yield and on nutrient content and extraction in 'Niagara Rosada' grapevine. *Pesqui. Agropecuária Bras.* **2014**, *49*, 340–348. [CrossRef]
62. Yamamoto, L.Y.; de Assis, A.M.; Roberto, S.R.; Bovolenta, Y.R.; Nixdorf, S.L.; García-Romero, E.; Gómez-Alonso, S.; Hermosín-Gutiérrez, I. Application of abscisic acid (*S* -ABA) to cv. Isabel grapes (*Vitis vinifera* × *Vitis labrusca*) for color improvement: Effects on color, phenolic composition and antioxidant capacity of their grape juice. *Food Res. Int.* **2015**, *77*, 572–583. [CrossRef]
63. Mian, G.; Colautti, A.; Belfiore, N.; Marcuzzo, P.; Tomasi, D.; Bell, L.; Celotti, E. Fertigation affects photosynthesis, modulation of secondary metabolism and sensory profiles of Vitis vinifera cv."Schioppettino" withered grapes and wines. *Sci. Hortic.* **2024**, *328*, 112954. [CrossRef]
64. Rebello, L.P.G.; Lago-Vanzela, E.S.; Barcia, M.T.; Ramos, A.M.; Stringheta, P.C.; Da-Silva, R.; Castillo-Muñoz, N.; Gómez-Alonso, S.; Hermosín-Gutiérrez, I. Phenolic composition of the berry parts of hybrid grape cultivar BRS Violeta (BRS Rubea × IAC 1398-21) using HPLC-DAD-ESI-MS/MS. *Food Res. Int.* **2013**, *54*, 354–366. [CrossRef]
65. Barcia, M.T.; Pertuzatti, P.B.; Gómez-Alonso, S.; Godoy, H.T.; Hermosín-Gutiérrez, I. Phenolic composition of grape and winemaking by-products of Brazilian hybrid cultivars BRS Violeta and BRS Lorena. *Food Chem.* **2014**, *159*, 95–105. [CrossRef]
66. Lima, M.d.S.; Silani, I.d.S.V.; Toaldo, I.M.; Corrêa, L.C.; Biasoto, A.C.T.; Pereira, G.E.; Bordignon-Luiz, M.T.; Ninow, J.L. Phenolic compounds, organic acids and antioxidant activity of grape juices produced from new Brazilian varieties planted in the Northeast Region of Brazil. *Food Chem.* **2014**, *161*, 94–103. [CrossRef]
67. Caiati, C.; Stanca, A.; Lepera, M.E. Free radicals and obesity-related chronic inflammation contrasted by antioxidants: A new perspective in coronary artery disease. *Metabolites* **2023**, *13*, 712. [CrossRef]

horticulturae

MDPI

Article

Structure and Trends in Climate Parameters of Wine-Growing Regions in Slovenia

Stanko Vršič *, Borut Pulko and Andrej Perko

Faculty of Agriculture and Life Sciences, University of Maribor, Pivola 10, 2331 Hoče, Slovenia; borut.pulko@um.si (B.P.); a.perko@um.si (A.P.)

* Correspondence: stanko.vrsic@um.si

Abstract: This study examined the structure and trends of climate parameters important for grape production from 1952 to 2022 in the wine-growing regions of Podravje, Posavje, and Primorska in Slovenia. Average and extreme temperature and precipitation data from six meteorological stations in three wine-growing regions were divided into annual and growing seasons. The results show that in the period 1991–2022, there was a warming in the growing season in all regions by 1.4–1.7 °C, except the southern part of Primorska (Koper station) 0.6 °C, compared to the reference period 1961–1990. The heat accumulation indices (GDDs and HI) increased significantly, which is mainly due to the increase in the maximum temperature in the growing season temperature (GST max) and the number of days with Tmax > 30 °C (NDT30). The NDT30 increased the most, by a factor of more than four. In the reference period (1961–1990), however, the trend in the number of hot days was even slightly negative. The mean seasonal temperature rose to around 17 °C in regions with a continental climate and to around 19 °C in the Mediterranean region, which could be reflected in the earlier ripening of the grapes. The trends show a decrease in total annual precipitation (AP) after 1991, but this was significant only at one inland location (Maribor), while the total precipitation during the growing season (GSP) decreased significantly at three locations (Maribor, Bilje, and Koper).

Keywords: climate change; grapevine; bioclimatic parameters; Slovenia

Citation: Vršič, S.; Pulko, B.; Perko, A. Structure and Trends in Climate Parameters of Wine-Growing Regions in Slovenia. *Horticulturae* **2024**, *10*, 854. https://doi.org/10.3390/horticulturae10080854

Academic Editors: Sarita Leonel, Sergio Ruffo Roberto and Simone Rodrigues da Silva

Received: 15 July 2024
Revised: 9 August 2024
Accepted: 10 August 2024
Published: 13 August 2024

1. Introduction

Climate is one of the most important limiting factors for agriculture production. Climate change is also challenging viticultural production everywhere, especially in regions with warm and dry climates. Frost and drought risk during the growing period are common problems in viticulture. Therefore, maintaining viticulture also requires adaptation to climate change, and the assessment of adaptation strategies needs to be more precise and multidisciplinary, tailored to local conditions [1].

For Western and Northern Europe, a temperature increase of 2.5 to 4.5 °C is predicted by the end of the 21st century and most climate models predict an increase in winter precipitation [2]. In recent years, warming trends have been observed in all seasons [3]. In the northern hemisphere, the warming will be stronger in the colder half of the year. It is expected that viticulture in many regions with cool climates will benefit from the higher temperatures in the growing season. Milder winters and warmer summers are expected, and extreme high temperatures will occur more frequently. The risk of low winter temperatures and unfavorable conditions during the flowering period is expected to be lower, while the risk of late frosts in spring will increase due to early budbreak [4–6].

Many researchers have studied the impact of climate change on the viability of viticulture production, examining changes in overwintering tolerance potential, frost incidence, growing season length, and heat and drought stress during the growing season [7–11]. The need to adapt to climate change is even greater for grapevines than for other crops, as the composition of berries, which is a key factor in fruit and wine quality, typicality, and market value, depends strongly on "terroir" (the particularities of the culture area) [12]. Spatial modeling research has shown that wine-growing regions will expand as parts of southern Europe become too hot to produce high-quality wines, and northern regions will acquire vineyard potential, as in the Middle Ages, from the 9th to the 13th century AD [13,14] However, despite these short-term benefits, the predicted rise in global temperatures over the next half century could require major changes in the wine industry. Slight changes in temperatures during the growing season could lead to shifts in varietal suitability in many regions [15] or require costly adaptation measures both in the vineyard and in the winery.

Unfavorable weather conditions could lead to varietal changes in some wine-growing regions. Due to climate change, the growing season has been extended, which also enables longer favorable conditions for the development of the two most persistent diseases (downy and powdery mildew). In Slovenia, there is usually greater pressure from downy mildew in the continental part (Podravje and Posavje) and powdery mildew in the Mediterranean (Primorska). The earlier appearance of downy mildew in the continental part may be a direct result of more favorable temperature conditions in May and June. In the worst-case climate scenario, two more fungicide sprays will be needed compared to the current management regimes, as reported by Salinari et al. in 2006 [16]. Therefore, from this point of view, sustainable varieties (PiWi, tolerant to downy and powdery mildew) are more adaptable to these conditions than traditional ones [17]. Adaptations to higher temperatures include also changes in plant material (e.g., rootstocks, varieties, and clones) and viticultural techniques (e.g., changes in trunk height, leaf area to fruit weight ratio, and timing of pruning) so that harvest dates fall within the optimal period of late September or early October in the northern hemisphere [18].

In some regions, projections show that an increase in precipitation can in turn affect soil development by increasing the amount of water flowing through the soil [18]. Climate change would therefore have a significant impact on soil development as it would lead to the loss of very fine particles of organic matter. The tendency towards more extreme weather events (more intense rainfall), as a consequence of climate change, can increase soil erosion, especially in vineyards on steep slopes [19–22].

The climatic factors are summarized in a series of climatic indices, of which the average temperature in the growing season (GSTavg), the Huglin Index (HI), and the growing degree days (GDDs) or Winkler Index are the most used [15]. The minimum heat requirement for the growth of vines is expressed as a value of the heat sum index, i.e., as growing degree days (GDDs) from April to October in the northern hemisphere, at a base temperature of 10 °C [15]. Becker [23] established a minimum GDDs of 1000 °C units, but later studies found a minimum value of 850 °C units [24,25]. After 1990, the developmental stages of the grapevine, such as budbreak, flowering, and ripening, occurred earlier on average than in the 1980s [26–29].

Slovenia is a very small wine-growing region in Central Europe with different climates (Mediterranean, Continental, and Pannonian climates). Despite the fact that Slovenia is not a globally important wine-growing country, most of the globally important grape varieties for quality wines are grown there. The aim of this research was to investigate the changes in temperature and precipitation as well as some bioclimatic indices based on which adaptation strategies to climate change can be recommended in the future, even for neighboring wine-growing regions with a similar climate.

2. Data and Methods

2.1. Study Area and Climate Data

For this study, the longest available data series from 6 meteorological stations (Maribor, Murska Sobota, Novo Mesto, Črnomelj, Bilje, and Koper) in three wine-growing regions (Podravje, Posavje, and Primorska) in Slovenia were used (Figure 1). The Podravje wine-growing region lies between the river Sava (SW) and the Hungarian border (NE). Geologically, the area is part of the former basin of the Pannonian Sea, which consists of folded and poorly interlocked marine sediments from the Neogene and has a Pannonian continental transitional climate [30]. The continental climate characteristics increase with increasing distance from the Alps. The vineyards (6000 ha) are predominantly planted with white grape varieties and are located on steep slopes with a gradient of 15–50% and at altitudes of 250 to 350 m [31]. The climate in the Posavje wine-growing region (SE) is also continental, with large seasonal temperature differences, cold winters, and moderately warm summers. The vineyards (3000 ha) are planted 50% with white and 50% with red grape varieties. In both wine-growing regions, half of the vineyards are on steep slopes with a gradient of 15 to 30%, and a quarter of the area has a gradient of more than 30% [31]. The Primorska (SW) wine-growing region lies along the Italian border, from the Adriatic to the Alps, and the general climate is sub-Mediterranean [30], characterized by an average annual minimum temperature of over 0 °C and a temperature in the warmest month of over 20 °C. Half of the vineyards are located on slopes with a gradient of up to 15%. The long-term average (1952–2022) of precipitation during the growing season (1 April–31 October) varies between 612 mm (Koper) and 870 mm (Bilje), and precipitation is very unevenly distributed throughout the year. The vineyards (6500 ha) are planted with 60% white and 40% red grape varieties [31].

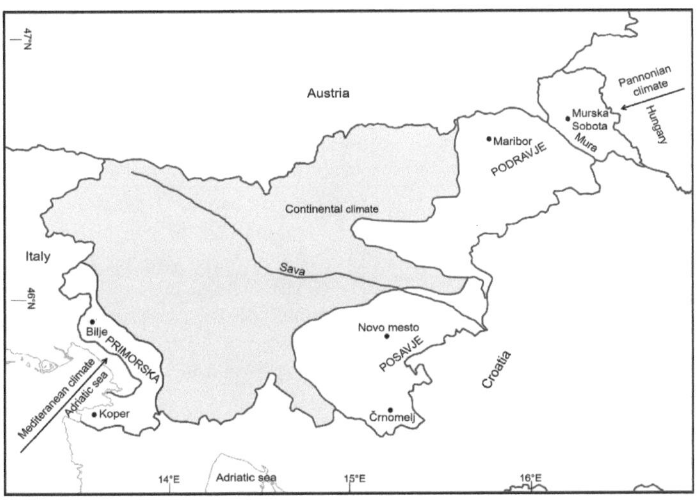

Figure 1. Map with study regions and climate stations (Maribor, Muska Sobota, Novo Mesto, Črnomelj, Bilje, and Koper) in the wine-growing regions of Podravje, Posavje, and Primorska in Slovenia.

The daily precipitation and temperature values (mean, maximum, and minimum) from six meteorological stations (1952–2022) were used for the analysis. The data were taken from the archives of the Slovenian Environment Agency (ARSO) [32]. All stations had sufficient records for a long-term analysis. The data provide a good reference for the general structure and trends of temperature and precipitation.

2.2. Climate Parameters and Bioclimatic Indices

An analysis of the observed climate was carried out for the periods 1952–2022, 1961–1990 (reference period for the 20th century with a minor change in climate parameters), and 1991–2022. The data from the individual stations were categorized according to growing seasons or important grapevine growing seasons and used to derive bioclimatic indices and extreme climate indices that are important for wine grape production (Table 1). For the growing season (April–October), precipitation and temperature (average, minimum, and maximum) of each station were summarized, as the averages of the growing season usually correlate significantly with wine varieties and type of wine production [15]. To assess the signs of heat stress, the number of days with temperatures above 30 °C was determined [9]. This temperature leads to premature ripening of the grapes (shorter growing season), lower total acidity, and lower aroma compounds [33].

Table 1. Analyzed bioclimatic parameters.

Parameter	Parameter Description
Tavg	Average annual temperature, °C
Tmax	Average annual maximum temperature, °C
Tmin	Average annual minimum temperature, °C
GSTavg	Average growing season temperature (April to October), °C
GSTmax	Average growing season maximum temperature (April–October), °C
GSTmin	Average growing season minimum temperature (April–October), °C
HI	Huglin Index (April to September), °C units
GDDs	Growing degree days (sum of temperature above 10 °C), °C units
TMJ	Average temperature in May and June, °C
NDTN20	Tropical nights: number of days with TN > 20 °C days
NDT25	Number of days with maximum temperature > 25 °C
NDT30	Number of days with maximum temperature > 30 °C
NDT35	Number of days with maximum temperature > 35 °C
NDF	Number of days with minimum temperature <0 °C (frost occurrence)
NDFF	Number of days between last and first frost (frost-free period length)
NDTN-2.5	Moderate cold days: number of days with TN < −2.5 °C days
NDTN-10	Extreme cold days: number of days with TN < −10 °C days
AP	Total annual precipitation, mm/m^2
GSP	Total growing season precipitation (April to October), mm/m^2

To obtain more information about the wine region and to determine general guidelines for the potential quality and style of the wine, the GDDs [34] and the Huglin Index (HI) [35] were calculated. These two bioclimatic indices make it possible to classify wine-growing regions according to the sum of the temperatures required for the development of the vines and the ripening of the grapes [36].

For the northern hemisphere, the Huglin Index [35] is calculated using the following formula:

$$HI = \sum_{01.04}^{30.09} d \cdot \left[\frac{(Tavg - 10) + (Tmax - 10)}{2}\right],$$

where Tavg is the daily mean air temperature (°C), Tmax is the daily maximum air temperature (°C), and d is the day length coefficient, ranging from 1.02 and 1.06 between the 40° and 50° of northern latitude. Baseline temperature = 10 °C. This index enables viticultural regions to be classified in terms of the sum of the temperatures required for the development of the vines and the ripening of the grapes. HI climatic ripening groups: very cold—HI-3 (HI ≤ 1500), cold—HI-2 (1500 < HI ≤ 1800), temperate—HI-1 (1800 < HI ≤ 2100), temperate-warm—HI+1 (2100 < HI ≤ 2400), warm—HI+2 (2400 < HI ≤ 3000), and very warm—HI+3 (3000 < HI) [35].

Growing degree days (GDD) [34] were calculated for 1 April to 31 October by summing the daily average temperatures (Tavg) above a base value of 10 °C, where values below 10 °C are set to zero:

$$\mathrm{GDD} = \sum_{01.04}^{30.10} \max[(\mathrm{Tavg} - 10)].$$

The GDD climatic ripening groups are as follows: Region I—very cold (≤1390), Region II—cold (1391–1670), Region III—warm (1671–1940), Region IV—hot (1941–2220), and Region V—very hot (≥2220).

Some indices for climate extremes were also calculated to determine changes in extreme temperatures. These indices are recommended by the WMO (World Meteorological Organization) and are currently being investigated by researchers [37]. The average temperatures in the period May–June (TMJ) were also calculated by Vršič et al., 2024 [38]. This parameter is important for predicting an increase in disease pressure (e.g., downy mildew), as more severe epidemics can be a direct result of more favorable temperature conditions in May and June [16]. Of the other extreme indices in Table 1, NDTN20, NDT25, NDT35, NDTN-2.5, and NDTN-10 were calculated for the first time for our wine-growing regions. NDTN20 and NDF indices were estimated annually as recommended by the ETCCDI (Expert Team on Climate Change Detection and Indices).

The variables were analyzed using descriptive statistics and trend analyses. As some of the parameters examined in this study were not normally distributed, a more stringent non-parametric Mann–Kendall trend test (MK test) with a significance level of 95% was used for all series [39]. The Mann–Kendall test, like other distribution-free or parametric tests, is very sensitive to an autocorrelation effect (persistence).

3. Results and Discussion

3.1. Climatic Structure in the Wine-Growing Regions of SLOVENIA

The general climate for the period 1952–2022 for the inland wine-growing regions of Podravje and Posavje is temperate continental, characterized by significant seasonal temperature fluctuations, cold winters, and moderately hot summers with an average annual temperature of 10.3 °C (5.7 to 15.5 °C) for Maribor, 9.9 °C (4.8 to 15.3 °C) for Murska Sobota, 10.2 °C (5.4 to 15.7 °C) for Novo Mesto, and 10.8 °C (5.6 to 16.4 °C) for Črnomelj, with annual precipitation of 998 mm, 801 mm, 1130 mm, and 1281 mm, respectively (Table 2). The Primorska climate is a sub-Mediterranean climate characterized by an average annual temperature of 12.6 °C (7.3 to 18.5 °C) for Bilje and 13.8 °C (9.9 to 18.1 °C) for Koper (near the Adriatic Sea) with a total annual precipitation of 1424 mm for Bilje and 995 mm for Koper.

For wine grape maturity potential, the location temperatures range between 15.5 and 15.8 °C (Podravje), 15.6 and 16.3 °C (Posavje), and 17.5 and 18.6 °C (Primorska) based on the long-term average temperatures of the growing season (GSTavg) (Table 2) [40]. The variability in temperature in the growing season (GSTavg and GSTmin) is similar at all locations, while GSTmax is more pronounced in the coastal location (Koper). The GSTmax temperatures are as follows: 21.5 °C for Maribor, 21.6 °C for Murska Sobota, 21.8 °C for Novo Mesto, 22.6 °C for Črnomelj, 24.0 °C for Bilje, and 23.2 °C for Koper; moreover, the GSTmin values are 10.6, 9.7, 10.0, 10.3, 11.6, and 14.2 °C, respectively.

The number of days with temperatures < 0 °C (NDF) was highest in Podravje (110 d in Murska Sobota) and the lowest in Primorska (29 d in Koper). The lowest number of days with moderately cold days (NDTN-2.5) and the number of days with extremely cold days (NDTN-10) was in the Podravje wine-growing region, followed by the Posavje, while the differences were not significant in the Primorska wine-growing region. The frost-free period (NDFF) was the longest for Primorska averaging 220 d (Bilje) to 264 d (Koper), followed by Podravje with 188-206 d, and Posavje with 193-197 d (Table 2). The number of days during the growing season with temperatures > 30 °C (NDT30) also follows the pattern from inland to the coast, with Podravje and Posavje having fewer days overall, around 13–14 days and 16–22 days, respectively, while Primorska has 22–31 days with maximum temperatures > 30 °C per year. The same pattern of increase is shown by the number of days with temperatures above 25 °C (NDT25), the days with a maximum temperature of more than 35 °C (NDT35), the number of tropical nights (NDTN20), and the average temperature in the period May to June (TMJ) (Table 2).

The values of growing degree days (GDDs) for Podravje, Posavje, and Primorska are between 1278 and 1325, between 1291 and 1420, and between 1631 and 1855 units, respectively (Table 2). These values place Podravje and Posavje in Winkler Region II (cool), which indicates a generally favorable climate for the production of quality wines [34]. The values for Primorska place it in Winkler Region II-III (cool to warm), which is also favorable for the production of quality wines but can be affected by excessive heat in this region, especially in the last two decades. In this region, the NDT30 value has doubled compared to the reference period 1961–1990 (Table 3).

Table 2. Mean and trend of bioclimatic parameters (standard and tested) for the 6 meteorological stations (in 3 wine-growing regions) Maribor and Murska Sobota (Podravje), Črnomelj and Novo Mesto (Posavje), and Bilje and Koper (Primorska) in Slovenia for the long-term period 1952–2022. Bold numbers indicate significant trends ($p \leq 0.05$).

Wine-Growing Region/Station	Podravje/Maribor					Posavje/Novo Mesto					Primorska/Bilje				
	Variable					Variable					Variable				
Parameters	Mean	SD	Trend yr^{-1}	MK-Test	p	Mean	SD	Trend yr^{-1}	MK-Test	p	Mean	SD	Trend yr^{-1}	MK-Test	p
Tavg	10.3	0.99	**0.037**	0.628	**0.001**	10.2	1.00	**0.051**	0.603	**0.001**	12.6	0.79	**0.017**	0.299	**0.001**
Tmax	15.5	1.14	**0.038**	0.542	**0.001**	15.7	1.19	**0.051**	0.440	**0.001**	18.5	0.82	0.021	0.367	0.060
Tmin	5.7	1.06	**0.042**	0.66	**0.001**	5.4	1.10	**0.053**	0.672	**0.001**	7.3	1.03	**0.021**	0.305	**0.001**
GSTavg	15.8	0.99	**0.037**	0.589	**0.001**	15.6	1.0	**0.039**	0.565	**0.001**	17.5	0.93	0.023	0.328	0.534
GSTmax	21.5	1.15	**0.038**	0.491	**0.001**	21.8	1.3	**0.034**	0.371	**0.001**	24.0	0.97	**0.024**	0.315	**0.001**
GSTmin	10.6	1.02	**0.041**	0.615	**0.001**	10.0	1.1	**0.045**	0.663	**0.001**	11.6	1.22	0.031	0.357	0.124
HI	1839	206	**7.03**	0.559	**0.001**	1849	218	**8.76**	0.462	**0.001**	2197	188	**6.18**	0.386	**0.020**
GDDs	1325	186	**6.88**	0.599	**0.001**	1291	192	**8.35**	0.573	**0.001**	1631	191	4.77	0.338	0.305
TMJ	17.0	1.28	**0.038**	0.48	**0.001**	16.73	**1.31**	**0.040**	0.478	**0.001**	18.26	**1.25**	**0.031**	0.365	**0.001**
NDTN20	1.55	2.54	**0.063**	0.444	**0.001**	0.45	1.04	**0.023**	0.412	**0.001**	3.04	3.75	**0.069**	0.297	**0.001**
NDT25	63.5	17.6	**0.63**	0.549	**0.001**	68.07	18.9	**0.52**	0.387	**0.001**	95.2	15.6	**0.29**	0.236	**0.004**
NDT30	13.3	11.8	**0.57**	0.502	**0.001**	16	14.2	**0.39**	0.390	**0.021**	31.0	15.6	**0.45**	0.383	**0.030**
NDT35	0.75	1.64	**0.025**	0.275	**0.003**	1.31	3.29	**0.055**	0.263	**0.005**	2.14	3.70	**0.10**	0.465	**0.001**
NDF	95	19	**−0.56**	−0.411	**0.001**	100	19	**−0.81**	−0.422	**0.002**	67	18.4	0.03	−0.008	0.927
NDFF	206	22	0.53	0.340	0.310	197	22	0.55	0.345	0.247	220	28.9	−0.20	−0.069	0.401
NDTN-2.5	54.9	16.1	**−0.49**	−0.411	**0.001**	54.51	16.1	**−0.44**	−0.394	**0.001**	36.2	14.7	0.045	0.0339	0.684
NDTN-10	8.2	7.77	**−0.22**	−0.447	**0.001**	9.06	7.76	**−0.20**	−0.448	**0.001**	1.39	2.39	0.014	−0.041	0.659
AP	998	150	**−2.88**	−0.252	**0.002**	1130	190	0.16	−0.058	0.481	1424	289	−2.78	−0.139	0.087
GSP	700	124	**−1.68**	−0.214	**0.008**	757	146	−0.18	−0.113	0.165	870	209	**−2.83**	−0.186	**0.022**
	Podravje/Murska Sobota					Posavje/Črnomelj					Primorska/Koper				
Tavg	9.9	0.99	**0.034**	0.565	**0.001**	10.8	0.97	**0.028**	0.484	**0.001**	13.8	0.64	**0.013**	0.259	**0.001**
Tmax	15.3	1.16	**0.036**	0.520	**0.001**	16.4	1.13	**0.030**	0.400	**0.010**	18.1	1.31	**0.040**	0.406	**0.006**
Tmin	4.8	1.05	**0.038**	0.599	**0.001**	5.6	0.92	**0.025**	0.405	**0.007**	9.9	0.91	−0.008	−0.123	0.131
GSTavg	15.5	1.0	**0.033**	0.511	**0.001**	16.3	1.0	**0.034**	0.452	**0.001**	18.6	0.86	**0.020**	0.283	**0.001**
GSTmax	21.6	1.2	**0.038**	0.446	**0.001**	22.6	1.2	**0.022**	0.263	**0.001**	23.2	1.61	**0.042**	0.404	**0.006**
GSTmin	9.7	1.0	**0.037**	0.576	**0.001**	10.3	1.0	**0.033**	0.466	**0.001**	14.2	0.93	−0.007	−0.106	0.192
HI	1831	213	**6.43**	0.478	**0.001**	1985	209	9.37	0.366	0.064	2227	236	**7.44**	0.412	**0.004**
GDDs	1278	186	**6.21**	0.532	**0.001**	1420	198	**5.80**	0.464	**0.001**	1855	180	**4.18**	0.277	**0.001**
TMJ	16.8	1.25	**0.037**	0.475	**0.001**	17.44	1.33	**0.035**	0.388	**0.001**	18.99	**1.23**	**0.025**	0.35	**0.001**
NDTN20	0.49	0.88	**0.021**	0.429	**0.001**	1.13	2.25	**0.061**	0.446	**0.001**	17.8	12.1	0.033	0.586	0.558
NDT25	64.8	17.4	**0.57**	0.495	**0.001**	79.3	17.5	**0.31**	0.243	**0.003**	84.2	25.1	**0.74**	0.376	**0.001**
NDT30	14.1	12.1	**0.37**	0.447	**0.001**	22.2	13.9	**0.35**	0.282	**0.001**	22.2	20.7	**0.70**	0.437	**0.001**
NDT35	0.83	2.04	**0.023**	0.209	**0.027**	1.8	3.38	**0.064**	0.253	**0.005**	1.37	3.29	**0.08**	0.413	**0.001**
NDF	110	18	**−0.43**	−0.374	**0.050**	100	16.7	−0.76	−0.138	0.092	29	17.6	**0.24**	0.225	**0.006**
NDFF	188	18	0.41	0.359	0.111	193	20.3	**0.41**	0.265	**0.001**	264	31	**−0.38**	−0.181	**0.039**
NDTN-2.5	64.9	15.5	**−0.364**	−0.313	**0.001**	59.6	13.4	−0.15	−0.123	0.135	9.75	8.67	0.04	0.083	0.313
NDTN-10	13.0	10.0	**−0.264**	−0.417	**0.001**	10.9	7.43	**−0.21**	−0.272	**0.001**	0.11	0.43	−0.003	−0.151	0.117
AP	801	112	−0.17	0.021	0.800	1281	184	0.59	0.051	0.532	995	185	−1.83	−0.133	0.101
GSP	574	94	0.28	0.037	0.655	803	165	0.62	0.057	0.487	612	152	**−1.85**	−0.208	**0.010**

Parameter abbreviations are in Table 1.

The average values of the Huglin Index, which is possibly more suitable than the Winkler Index for European regions [36], were for Podravje, Posavje, and Primorska between 1831 and 1839, between 1849 and 1985 (HI-1), and between 2197 and 2227 (HI+1), respectively (Table 2). These values assign Podravje and Posavje to the cool climate type, which is suitable for Chardonnay, Sauvignon Blanc, and Pinot Noir, for example, while the values for Primorska are assigned to the warm climate type according to Huglin and are more suitable for Cabernet Sauvignon and Merlot [35].

Growing season precipitation (GSP) values generally show that rainfall amounts decreased slightly throughout the period (1952–2022), although the trends were not significant for all locations. GSP differed significantly between stations, increasing from Murska Sobota (Podravje), 574 mm, towards Bilje (Primorska), 870 mm, per vintage (Table 2). The variability in the GSP in this long-term period shows a variation of 16 to 25% between years at all locations. The Murska Sobota location was drier (influence of the Pannonian climate) with a total GSP amount of 574 mm than Maribor, 700 mm, and both locations in Posavje (Novo Mesto, 757 mm, and Črnomelj, 803 mm), and then Bilje (Primorska), 870 mm (Table 3). A similar amount of GSP as in the eastern part of the Podravje wine-growing region (Murska Sobota) was also measured in the coastal area of the Primorska wine-growing region (Koper 613 mm) (Table 3). However, drier conditions with more frequent and longer dry spells were more likely, as higher temperatures can probably lead to a higher evapotranspiration rate, as also noted by Ramos et al. [41]. This is particularly pronounced in the Primorska wine-growing region, especially at the Bilje location, which has the highest average annual precipitation of all six locations at 1424 mm. However, precipitation at this location falls in the form of very intense short-term showers (due to the mixture of Mediterranean and Alpine climate) with increasingly intense dry periods, as the highest number of days with maximum temperatures > 30 °C (NDT30) is recorded here. During the growing season, the annual precipitation falls around 70%, 67–77%, and 60% in Podravje, Posavje, and Primorska, respectively.

Table 3. Averages of bioclimatic parameters for the six meteorological stations (Maribor, Murska Sobota, Novo Mesto, Črnomelj, Bilje, and Koper) in three wine-growing regions (Podravje *, Posavje **, Primorska ***) in Slovenia for the period 1952–2022, the reference period 1961–1990, the period 1991–2022, and by decades for the 1991–2022 period.

Periods/Parameters	GSTavg ± SD		GSP ± SD		GDDs ± SD		HI ± SD		T > 30 °C ± SD	
MARIBOR *										
1952–2022	15.8	± 0.99	700.5	± 124.3	1324.9	± 186.5	1839.0	± 206.6	13.2	± 11.8
1961–1990	15.2	± 0.59	725.4	± 116.5	1205.5	± 108.0	1704.5	± 120.0	5.8	± 3.8
1991–2022	16.6	± 0.80	669.2	± 128.2	1496.5	± 155.9	2017.1	± 199.8	21.9	± 12.7
1991–2000	16.2	± 0.75	738.9	± 142.1	1415.8	± 130.2	1914.0	± 163.4	13.5	± 9.9
2001–2010	16.7	± 0.48	692.9	± 97.4	1496.5	± 121.9	1996.9	± 169.2	21.8	± 13.2
2011–2022	17.0	± 0.57	591.3	± 128.8	1540.5	± 106.4	2119.7	± 145.5	28.5	± 8.9
MURSKA SOBOTA *										
1952–2022	15.5	± 1.00	574.5	± 94.1	1277.9	± 186.3	1831.2	± 212.8	14.1	± 12.1
1961–1990	14.8	± 0.65	577.9	± 100.3	1142.9	± 111.1	1686.3	± 125.4	6.3	± 4.6
1991–2022	16.3	± 0.87	578.2	± 93.6	1432.3	± 168.5	2002.9	± 201.6	23.0	± 12.6
1991–2000	15.9	± 0.75	571.4	± 138.7	1364.3	± 125.7	1914.7	± 164.8	17.3	± 12.6
2001–2010	16.3	± 0.47	581.4	± 88.7	1414.4	± 119.3	1985.3	± 164.8	23.9	± 12.4
2011–2022	16.8	± 0.59	581.3	± 74.4	1518.5	± 107.1	2091.1	± 151.7	27.0	± 10.5
NOVO MESTO **										
1952–2022	15.6	± 1.04	757.5	± 146.3	1291.2	± 193.8	1849.2	± 218.5	16.0	± 14.2
1961–1990	14.8	± 0.61	771.1	± 127.4	1146.4	± 109.2	1693.7	± 112.8	7.3	± 4.1
1991–2022	16.5	± 0.88	734.2	± 160.3	1450.2	± 161.4	2013.7	± 215.1	25.4	± 16.0
1991–2000	16.0	± 0.77	778.3	± 135.4	1374.4	± 127.2	1901.2	± 154.6	15.5	± 10.5
2001–2010	16.4	± 0.50	772.2	± 156.8	1443.4	± 122.8	1971.0	± 171.1	22.9	± 12.8
2011–2022	17.0	± 0.51	665.8	± 207.7	1538.9	± 90.0	2143.1	± 181.5	35.8	± 16.8
ČRNOMELJ **										
1952–2022	16.3	± 1.03	802.6	± 165.5	1419.6	± 208.2	1984.6	± 208.7	22.2	± 13.9
1961–1990	15.6	± 0.60	793.9	± 181.8	1279.0	± 111.2	1834.5	± 127.3	12.9	± 7.2
1991–2022	17.1	± 1.07	789.5	± 205.5	1579.5	± 208.2	2137.8	± 223.8	30.5	± 15.5
1991–2000	16.5	± 10.8	816.6	± 129.0	1467.3	± 191.1	2003.5	± 189.6	17.9	± 11.9
2001–2010	17.1	± 059	781.9	± 153.7	1581.1	± 137.8	2133.7	± 172.6	30.8	± 11.8
2011–2022	17.7	± 0.58	773.3	± 323.8	1696.1	± 115.2	2253.3	± 133.8	40.7	± 11.1
BILJE ***										
1952–2022	17.5	± 0.93	869.6	± 209.0	1631.3	± 190.8	2196.7	± 187.8	31.0	± 15.5
1961–1990	16.8	± 0.66	890.4	± 194.8	1479.0	± 133.9	2062.3	± 121.0	22.0	± 8.0
1991–2022	18.2	± 0.98	842.7	± 218.0	1753.2	± 199.0	2327.0	± 200.9	41.2	± 16.9
1991–2000	17.6	± 0.60	1026.9	± 220.2	1648.0	± 116.2	2190.6	± 123.3	29.6	± 9.8
2001–2010	18.1	± 0.47	758.4	± 219.4	1752.0	± 116.0	2313.9	± 162.3	38.6	± 14.6
2011–2022	18.7	± 0.64	759.3	± 195.5	1877.3	± 126.3	2451.5	± 118.3	53.1	± 14.6
KOPER ***										
1952–2022	18.6	± 0.86	612.7	± 152.4	1854.6	± 181.4	2227.4	± 236.3	22.2	± 20.7
1961–1990	18.3	± 0.60	653.8	± 165.7	1794.1	± 125.1	2068.3	± 138.3	8.2	± 7.9
1991–2022	18.9	± 0.99	573.5	± 155.6	1904.7	± 205.4	2399.6	± 260.2	38.2	± 23.3
1991–2000	18.2	± 0.62	598.8	± 126.3	1774.5	± 124.2	2225.7	± 144.2	22.8	± 13.7
2001–2010	18.6	± 0.50	574.7	± 171.8	1854.1	± 112.4	2343.1	± 144.4	33.7	± 14.8
2011–2022	19.9	± 1.07	551.4	± 147.7	2115.1	± 221.2	2591.7	± 204.3	54.8	± 19.1

GSTavg—average growing season temperature; GSP—total growing season precipitation; GDD—growing degree days (1 April to 31 October) °C units; HI—Huglin Index °C units; T > 30 °C—number of days with maximum temperature > 30 °C; and SD—Pearson's standard deviation.

3.2. Temperature Parameter Trends

The annual trends of the individual temperature parameters for 71 years (1952–2022) for the wine-growing regions in Slovenia are shown in Table 2. An increase in the average annual temperature (Tavg) for the period 1952–2022 ranged from 0.13 (Koper) to 0.51 °C (Novo Mesto) per decade (Table 2), and the average growing season temperature (GSTavg) ranged from 0.20 (Koper) to 0.39 °C (Novo Mesto) per decade (Table 2 and Figure 2). Over the periods studied, this corresponds to a change in GSTavg of 2.6 °C in Maribor, 2.3 °C in Murska Sobota, 2.8 °C in Novo Mesto, 2.4 °C in Črnomelj, 1.6 °C in Bilje, and 1.4 in Koper (Table 2). These changes can have a major impact on wine production. Many studies have confirmed that viticulture is one of the sectors most sensitive to climate change [42–45]. Similar trends have been observed in other world's wine-growing regions [15,46,47], and also in Central Europe. In Italy, the Venetian area experienced an increase in the average vegetation temperature (1964 to 2009) of up to 2.3 °C [40,48]. Lower warming (1.5 °C) was observed in moderate climate conditions of northern wine regions in Slovakia, which has not yet caused sufficient changes in the grapevine phenology to require serious adaptation measures [29]. Climate change led to the earlier development of phenological stages as reported by Ruml et al. (2016) [49] in Serbia and Prša et al. (2022) [50] and Omazić et al. (2024) [51] in Croatia. Some regions in Croatia are becoming less suitable for economically sustainable grape production [52]. In the Western part of the Carpathian basin (Hungary), climate change has several positive effects in the Sopron wine-growing region, this may result in the cultivation of more quality wine grapes and wines [53], which is also expected in other Hungarian regions [54]. The trend of earlier ripening of grapes was also found in northeastern Slovenia in the period 1980–2009 [55]. In Austria, based on temperature evolution, a doubling of the areas suitable for viticulture is predicted to occur by the 2050s [56].

The trends in minimum and maximum temperatures (annual and growing season) are similar at all locations, but the trends in minimum temperatures (Tmin and GSTmin) are slightly more pronounced, except at the Koper and Murska Sobota locations. This could be related to the lower humidity, as found in other wine-growing regions in the USA [57] and Europe [47]. These two locations have less rainfall than the other four. However, the average warming rates of the growing season for the six stations studied in Slovenia are determined by changes in maximum temperatures, with a significant increase in the number of days with a maximum temperature above 30 °C (NDT30), by an average of 3.5 (Črnomelj) to 7 (Koper) days per decade (Table 2). The temperature trends at the six studied locations have led to significant changes in the heat summation indices HI and GDD, whose values have increased on average from 6.2 (Bilje) to 9.4 (Črnomelj) and from 4.2 (Koper) to 8.3 (Novo Mesto) °C units per year, respectively, which for HI means from 440 to 667 and GDDs from 298 to 589 °C units in 71 years of the studied period. Heat accumulation has also increased in other European wine-growing regions, in Spain by 155–464 °C units [41] or 250–300 °C units in the last 30–50 years [47,58]. When comparing these data, two facts must be considered, namely the fact that our study period is almost two decades longer, and the fact that climate change trends are more pronounced in our area after 1990 [33], which is explained in more detail in the following paragraphs.

The trend in the number of days with temperatures < 0 °C (NDF) is decreasing in all locations in inland wine-growing regions. It is more pronounced in locations in Posavje (around 8 days per decade). The number of days between the last spring frost and the first fall frost (NDFF) is increasing (4.1 to 5.5 days per decade). At locations with a Mediterranean climate (Bilje, Koper), however, the trends are in the opposite direction, with the values for NDF increasing and for NDFF decreasing, although the NDFF trends for Bilje are not significant.

Other indices for temperature extremes (NDT25, NDT35, NDTN20, and TMJ) showed an increasing trend at all locations, with the exception of NDTN20 at the Koper location (Table 2). On average, the NDT25 increased from 2.9 d in Bilje to 7.4 d in Koper per decade and the NDT35 from 0.2 d in Murska Sobota to 1 d in Bilje per decade. The same trend can be observed in the number of days with tropical nights (NDTN20). The tendency towards an increase in the mean temperature in May and June (TMJ) was also significant for all locations ($p = 0.001$), namely the TMJ increased from 0.25 in Koper to 0.4 °C in Novo Mesto per decade (Table 2). More favorable temperature conditions in May and June may lead to higher disease pressure. In response to adaptation to future climate change, more attention will need to be paid to managing early downy mildew infections. Salinari et al. (2006) [16] reported that in response to adaptation to future climate change, more attention needs to be paid to the management of early powdery mildew infections. Their study found that under the most unfavorable climate scenario, two additional fungicide sprays are required compared to current management regimes. This could also be applied to neighboring wine-growing areas in the region.

A more detailed analysis of the individual periods within the long-term study period 1952–2022 shows even clearer changes. Minor changes in bioclimatic parameters were observed for the reference period (1961–1990) (Supplementary Figures S1–S3). The growing season average temperatures (GSTavg) were close to 15.0 °C for all stations in Podravje and Posavje, namely 14.8 °C for Murska Sobota and Novo Mesto, 15.2 °C for Maribor, and 15.6 °C for Črnomelj. The GSTavg values were higher in Primorska, 16.8 °C in Bilje, and 18.3 °C in Koper. The changes between 1991 and 2022 generally appear to be the most dramatic. The GSTavg values increased for all six stations and amounted to 16.6, 16.3, 16.5, 17.1, 18.2, and 18.9 °C (Table 3), with GSTavg trends for this period being 0.034, 0.040, 0.048, 0.057, 0.056, and 0.084 °C per year for Maribor, Murska Sobota, Novo Mesto, Črnomelj, Bilje, and Koper, respectively (Supplementary Figures S1–S3). The warming was due to the changes in GSTmin and GSTmax at all locations, but the trends in GSTmax were higher, except at the Črnomelj station, where the trend in GSTmin was more pronounced than in GSTmax (Supplementary Figures S1–S3).

The number of days with maximum temperatures > 30 °C (NDT30) increased at all locations in the period 1991–2022. Compared to the reference period 1961–1990, the NDT30 increased from 3 to 15 days per decade (Supplementary Figures S1–S3). The NDT30 for Podravje, Posavje, and Primorska was 22 to 23, 25 to 30, and 38 to 41 days, respectively (Table 3), and was two to four times higher compared to the reference period (2.5 to 6.6 times higher after 2010). In the period 1991-2022, the NDT30 increased in the Primorska wine-growing region by 1.1 days per year in Bilje and by 1.5 days per year in Koer. The NDT30 increased from 30 and 23 days in the first decade of this period to 53 and 55 days after 2010 in Bilje and Koper, respectively. A similar trend can also be observed at the Črnomelj location (1 day per year). At the other three meteorological stations, Novo Mesto, Maribor, and Murska Sobota, the trend was less pronounced; in Murska Sobota it amounted to 0.3 days per year. In the reference period (1961–1990), the NDT30 trend was even slightly negative.

If the warming trend continues over the next 30 years at the same rate as it has since the 1990s, it is to be expected that the wine-growing regions of Podravje and Posavje will also move completely into the warm climate group. Daily maximum temperatures of 30 °C are critical for optimal grapevine development and can lead to plant stress, a decrease in photosynthesis, a greater lack of water, premature ripening of the grapes, and drying of the berries, even in early ripening varieties such as Bouvier in Slovenia [33]. However, a few days with temperatures above 30 °C during the ripening period can be beneficial [9,48], especially for late-ripening varieties [33]. At the Bilje location, there were already an average of 22 hot days in the reference period, which almost doubled in the period 1991–2022. At the Koper location, there were only eight such days during the reference period, and the number of NDT30 days was five times higher in the following thirty years (1991–2022).

The total warming of average temperatures in the growing season (GSTavg) was between 1.4 and 1.7 °C for all locations, except for the coastal location (Koper), where the warming was only 0.6 °C, over the respective periods (1991–2022) compared to the reference period (1961–1990). In addition, the GSTavg warming after 2010 was 1.8–2.0 °C, 2.1–2.2 °C, and 1.6–1.9 °C in Podravje, Posavje, and Primorska (Table 3). Similar results were also found in other European wine-growing regions [47] with an average warming of the growing seasons of 1.7 °C in the last 30–50 years.

The warming trends in the period between 1991 and 2022 are also confirmed by the increase in the heat sum indices. The growing degree days (GDDs), also known as the Winkler Index (WI) and Huglin Index (HI), are often used to assess the climatic suitability for specific grape varieties and/or wine styles and are variations of degree days (heat sum) or heat accumulation (Table 1). The trends for both parameters show significant changes at each location. However, changes were greater in relative terms, with HI giving more weight to maximum temperatures (Table 1) and GSTmax increasing more at all locations. The GDDs increased by 54, 67, 80, 103, 107, and 170 °C units, while HI increased by 82, 75, 106, 113, 121, and 183 °C units per decade for Maribor, Murska Sobota, Novo Mesto, Črnomelj, Bilje, and Koper (Supplementary Figures S1–S3). The average values of these two indices in this period were around 300 °C units higher for the GDDs (except Koper 110) and from 265 to 331 °C units higher for HI than in the reference period (Table 3). Heat accumulation has also increased by 250–300 °C units in other European wine-growing regions over the last 30–50 years [47], and increased heat accumulation (WI and HI) in inland Spanish wine-growing regions has been reported, but not in coastal regions.

HI values of 1700–1900 °C units [35] indicate that the Podravje and Posavje wine-growing region is suitable for medium-late varieties such as Chardonnay, Sauvignon Blanc, etc. In the reference period (1961–1990), the HI value exceeded the value of 1900 °C units at both stations in Podravje and in Novo Mesto in Posavje only once (1983) and in Črnomelj ten times, while in the period 1991–2021, the HI value exceeded this value in more than two-thirds of the years at all stations in Podravje and Posavje (in Črnomelj, the value of 1900 °C units was not exceeded in only three years in the 1991–2021 period). In addition, the HI also exceeded the value of 2100 °C units in this period, which was most pronounced after 2010, e.g., in Črnomelj in half of the years. In general, in the period 1991–2022, the values of the GDDs increased by 23–26% and HI by 16–18%, while after 2010, even the GDDs increased by 27–34% and HI by 23–26% compared to the reference period (1961–1990) (Table 3).

Based on the classification of wine-growing regions into climate maturity groups [24] and the increase in GSTavg in the last decade of the observation period, it can be concluded that this wine-growing region is suitable for growing some grapevine varieties from the warm climate maturity group [34]. Initial results from the process of Merlot introduction in these two regions (not published) confirm this prediction.

In the Primorska wine-growing region, the HI value did not exceed the value of 1900 °C units (first half of HI-1 ripening group) in the reference period (1961–1990) in only four years at either station. In Bilje and Koper, it averaged between 2062 and 2068 °C units (second half of HI-1), while the HI value exceeded the value of 2300 °C units (second half of HI+1) in the period 1991–2022. The HI value for Bilje was 2327 and for Koper 2400 °C (on the border of HI+2) units, which shows that this region belongs to the moderately warm ripening group for grapes, and based on the values in the last decade (2011–2022), Bilje with an average of 2451 °C units (HI+2) and Koper with 2591 °C units (HI+2), these locations are already in the warm ripening group [59] according to Jones et al. [25]. The result of this warming may be that some areas that are still suitable for growing certain varieties will no longer exist in the future, or it will no longer be possible to produce premium wines on them as reported by White et al. 2006 [60]. As Tate 2001 [61] also noted, pests and diseases that are currently restricted by the winter cold will extend their range northwards.

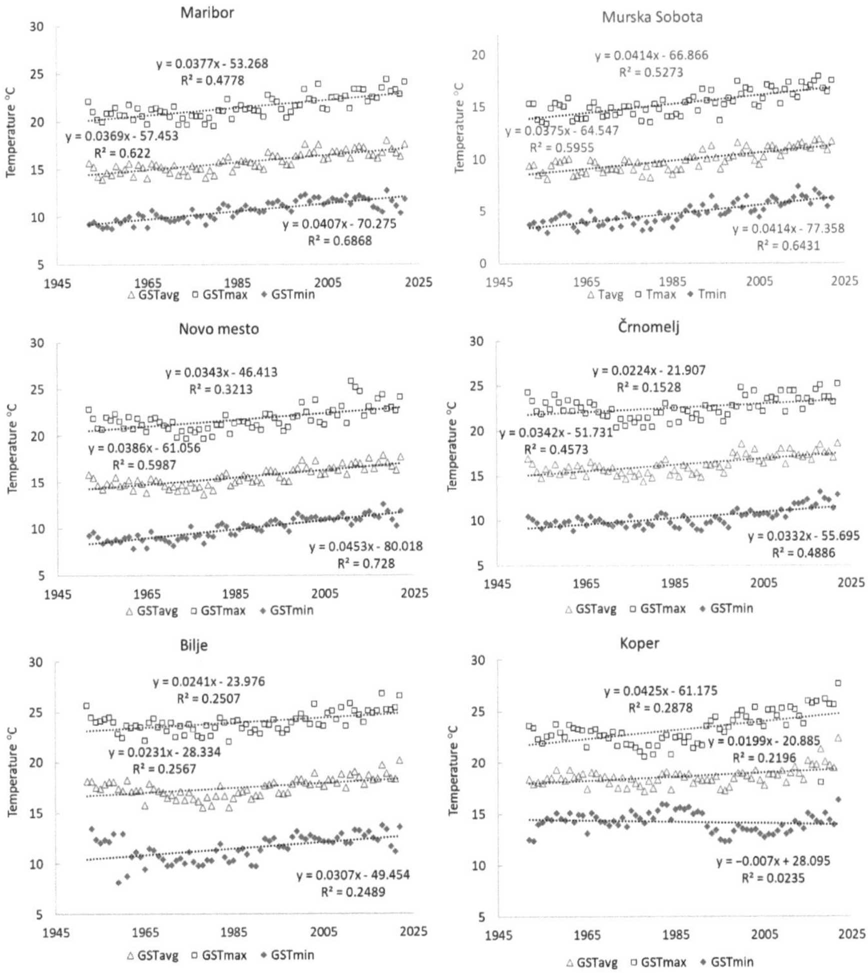

Figure 2. Temperature trends (average, maximum, and minimum) during the growing season (1 April to 31 October) in Maribor, Murska Sobota (Podravje), Novo Mesto, Črnomelj (Posavje), Bilje, and Koper (Primorska) in Slovenia for the long-term period 1952–2022.

3.3. Precipitation Parameters Trends

The high inter- and intra-annual precipitation variability has weakened many of the trends in precipitation parameters (Figure 3). Annual precipitation (AP) did not change significantly at any of the locations in the long-term period (1952–2022), only the trend for Maribor (Podravje) showed a significant decrease in precipitation (-3 L m^{2}yr^{-1}). The total growing season precipitation (GSP) showed significantly decreasing trends in Maribor (-1.7 mm m^{2}yr^{-1}), Bilje (-2.8 mm m^{2}yr^{-1}), and Koper (-1.8 mm m^{2}yr^{-1}) (Table 2). The inter-annual variability in the climate makes it difficult to assess tendencies in precipitation distribution patterns and possible effects of climate change. Nevertheless, some recent studies indicate significant changes in extreme events, such as more frequent and more extreme droughts, an increase in precipitation in the cold season, and drying out in the warm season [62–64]. It is therefore to be expected that the frequency, intensity, and distribution of precipitation will change due to the increased speed of the water cycle, which will probably also have an impact on the water supply in agriculture. In Europe, decreasing precipitation trends or changes in the seasonality of precipitation have been observed for large parts of the Mediterranean region [63,65]. This also confirms our results for the growing season in the Mediterranean region of Primorska (both locations). In the continental part of Europe, these trends were exceptionally absent [47]. Many of these studies were conducted two decades ago, and in many places, the precipitation pattern has changed. The trends in annual and growing season precipitation for the Maribor location in the continental wine-growing region of Podravje confirmed our assumption.

The precipitation pattern is spreading from the east towards Maribor (the eastern part of the Alps), especially after 2010, and is becoming more and more similar to the Murska Sobota location (on the edge of the Pannonian climate) (Figure 3). The long-term average (1952–2022) of precipitation in the growing season (April–October) for Maribor is 700 mm m^2yr^{-1} and decreases to 591 mm m^2yr^{-1} after 2010. The average amount of precipitation after 2010 is 134 mm m^2yr^{-1} lower (−18%), compared to the 725 mm m^2yr^{-1} in the reference period 1961–1999 (Table 3). A similar pattern of decline in GSP after 2010 as in Maribor is also observed at both locations in Primorska, namely 131 mm m^2yr^{-1} (Bilje) and 103 mm m^2yr^{-1} (Koper). These decreases could be critical, as the vines should have sufficient access to soil moisture from flowering to véraison and should not be exposed to high drought stress.

The most stable amount of precipitation remains at the Črnomelj location with 773 mm m^2yr^{-1}. The least changed amount of precipitation is recorded at the Murska Sobota location, which amounts to 575 mm over the entire long-term period (1952–2022) (Table 3). The highest precipitation amounts were recorded in Bilje (inland part of Primorska), where large-scale erosion events frequently occur. This is due to both the high rainfall and soil management system (soil tillage) in Primorska (in Podravje and Posavje the soil is green-covered), and the increasing extremes are likely to have additional negative impacts. Jones et al. [47] found that precipitation levels have not changed significantly in other European wine regions. However, higher temperatures lead to higher evapotranspiration. Moisture deficiency in the berry growth phase, especially in Koper (Primorska) and Murska Sobota (Podravje), could reduce cell division and lead to significant dehydration and sunburn, which in combination can lead to a reduction in berry size and yield, as also noted by Peacock [66]. In Murska Sobota, the amount of precipitation in the vegetation was already at a similar level in the reference period and in the last three decades as in the last decade in Maribor, which could indicate to a certain extent that the influence of the Pannonian climate on Maribor has become increasingly noticeable in the last decade.

Even though the trends of decreasing precipitation for Novo Mesto are not significant, on average 14% less precipitation was recorded after 2010 than in the reference period. In Črnomelj, the amount of precipitation in the vegetation was consistently at a similar level (over 770 mm m^2yr^{-1}), although the distribution of precipitation and the intensity of individual weather phenomena must be taken into account. The average amount of precipitation in the coastal area (Koper) is only two-thirds of the amount of precipitation in Bilje. The distribution of precipitation in Primorska is the least favorable of all wine-growing regions. The distribution of precipitation and the intensity of individual weather phenomena must also be evaluated.

As far as the future climate is concerned, an average warming of 2.0 °C is predicted for most of the world's best wine-growing regions by 2050 [67–69]. It is predicted that parts of southern Europe will become too hot to produce high-quality wines and that northern regions will acquire vineyard potential. Further warming could lead to grape varieties exceeding their climatic optimum, making current wine styles more difficult to produce.

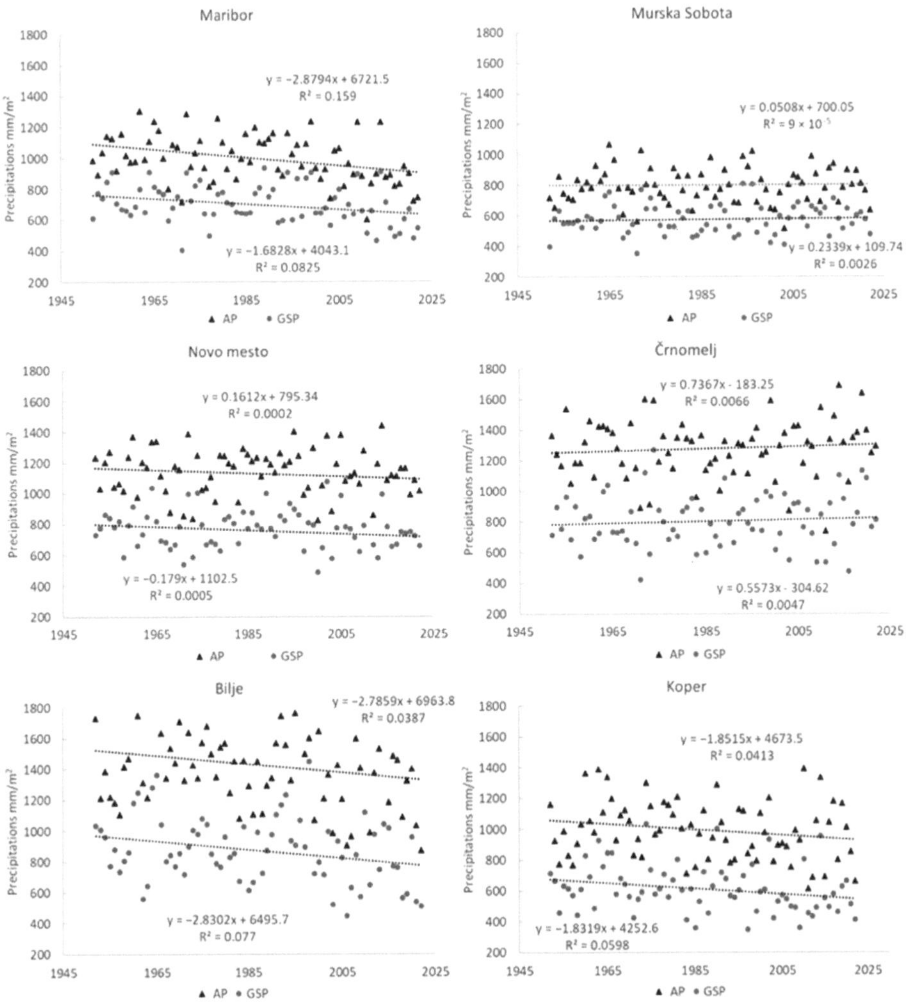

Figure 3. Precipitation trends (AP—annual and GSP—growing season, from 1 April to 31 October) in Maribor, Murska Sobota (Podravje), Novo Mesto, Črnomelj (Posavje), Bilje, and Koper (Primorska) in Slovenia for the long-term period 1952–2022.

4. Conclusions

The general conclusion of this study is that the greatest effects of climate change were felt in all three wine-growing regions after 1990. The greatest increase was in the number of hot days with a temperature above 30 °C (NDT30), which also had the greatest impact on other bioclimatic parameters, e.g., the average air temperature in the growing season, the sum of effective temperatures (GDDs) and the Huglin Index. In the period 1991–2022, the average growing season temperature (GSTavg) increased by about 1.5 °C or more in all wine-growing regions compared to the reference period, except in the coastal region of Koper, where the GSTavg temperature is 0.6 °C higher. In regions with a continental climate, the GSTavg temperature rose to around 17 °C and in the Mediterranean region to around 19 °C, which may be reflected in the earlier ripening of the grapes. If the warming trend continues in the next 30 years in a similar way as it has since the 1990s, it is expected that the wine-growing regions of Podravje and Posavje will completely transition to the warm climate ripening group. Another very important bioclimatic parameter at higher temperatures is the amount of precipitation during the growth of the vines. The total amount of precipitation in the growing season (GSP) shows a downward trend in all three wine-growing regions. The total amount of precipitation has decreased significantly in three locations, both in the maritime locations (Bilje and Koper) and in the inland locations (Maribor). As far as recent precipitation is concerned, it appears to be increasingly unevenly distributed, not only in the vegetation areas but also in the wine-growing regions. Therefore, the network of meteorological stations is even more important for the monitoring of weather phenomena and for the adoption and implementation of technological measures in the vineyards. For the network of meteorological stations to function well, its financing should continue to be the responsibility of the state and the municipalities.

Supplementary Materials: The following supporting information can be downloaded at: https://www.mdpi.com/article/10.3390/horticulturae10080854/s1, Figure S1: Trends in the bioclimatic parameters fort the meteorological stations Maribor and Murska Sobota (wine-growing region Podravje) in Slovenia for the periods 1961–1990 and 1991–2022; Figure S2: Trends in the bioclimatic parameters fort the meteorological stations Novo mesto and Črnomelj (wine–growing region Posavje) in Slovenia for the periods 1961–1990 and 1991–2022; Figure S3: Trends in the bioclimatic parameters fort the meteorological stations Bilje and Koper (wine–growing region Primorska) in Slovenia for the periods 1961–1990 and 1991–2022.

Author Contributions: Conceptualization, S.V.; methodology, S.V. and B.P.; software, A.P.; validation S.V. and A.P.; formal analysis, S.V. and A.P.; investigation, S.V.; resources, S.V.; data curation, S.V., B.P. and A.P.; writing—original draft preparation, S.V.; writing—review and editing, S.V., B.P. and A.P.; visualization, S.V.; supervision, S.V. and B.P.; project administration, S.V.; funding acquisition, S.V. All authors have read and agreed to the published version of the manuscript.

Funding: This research was funded by the Slovenian National Research Agency and the Ministry of Agriculture Forestry and Food of the Republic of Slovenia (grant no. CRP V4-2202) via the project "Guidelines for adapting grape and wine production to climate change and market demands".

Data Availability Statement: The raw data supporting the conclusions of this article will be made available by the authors on request.

Acknowledgments: We thank the Slovenian Environment Agency for providing the climate data.

Conflicts of Interest: The authors declare no conflicts of interest. The funders had no role in the design of this study; in the collection, analyses, or interpretation of data, in the writing of the manuscript; or in the decision to publish the results.

References

1. Naulleau, A.; Gary, C.; Prévot, L.; Hossard, H. Strategies for adaptation to climate change in grapevine production—A systematic review. *Front. Plant Sci.* **2021**, *11*, 607859. [CrossRef]
2. Schultz, H.R.; Lebon, E. Modelling the effect of climate change on grapevine water relations.VII International Symposium on Grapevine Physiology and Biotechnology. *Acta Hortic.* **2005**, *689*, 71–78. [CrossRef]
3. Branković, Č.; Srnec, L.; Patarčić, M. An Assessment of Global and Regional Climate Change Based on the EH5OM Climate Model Ensemble. *Clim. Chang.* **2010**, *98*, 21–49. [CrossRef]
4. Mosedale, J.R.; Wilson, R.J.; Maclean, I.M.D. Climate change and crop exposure to adverse weather: Changes to frost risk and grapevine flowering conditions. *PLoS ONE* **2015**, *10*, e0141218. [CrossRef]
5. Venios, X.; Korkas, E.; Nisiotou, A.; Banilas, G. Grapevine Responses to Heat Stress and Global Warming. *Plants* **2020**, *9*, 1754. [CrossRef]
6. Fraga, H. Climate change: A new challenge for the winemaking sector. *Agronomy* **2020**, *10*, 1465. [CrossRef]
7. Carter, T.R.; Parry, M.L.; Porter, J.H. Climatic Change and Future Agroclimatic Potential in Europe. *Int. J. Climatol.* **1991**, *11*, 251–269. [CrossRef]
8. Menzel, A.; Fabian, P. Growing season extended in Europe. *Nature* **1999**, *397*, 659. [CrossRef]
9. Jones, G.V.; Davis, R.E. Climate Influences on Grapevine Phenology, Grape Composition, and Wine Production and Quality for Bordeaux, France. *Am. J. Enol. Vitic.* **2000**, *51*, 249–261. [CrossRef]
10. Keller, M. Climate Change Impacts on Vineyards in Warm and Dry Areas: Challenges and Opportunities. *Am. J. Enol. Vitic.* **2023**, *74*, 0740033. [CrossRef]

11. Lobell, D.B.; Field, C.B.; Cahill, K.N.; Bonfils, C. Impacts of future climate change on Californian perennial crop yields: Model projections with climate and crop uncertainties. *Agric. For. Meteorol.* **2006**, *141*, 208–218. [CrossRef]
12. Delrot, S.; Grimplet, J.; Carbonell-Bejerano, P.; Schwandner, A.; Bert, P.F.; Bavaresco, L.; Costa, L.D.; Di Gaspero, G.; Duchêne, E.; Hausmann, L.; et al. Genetic and genomic approaches for adaptation of grapevine to climate change. In *Genomic Designing of Climate-Smart Fruit Crops*; Kole, C., Ed.; Springer: Cham, Switzerland, 2020; pp. 157–270. [CrossRef]
13. Kenny, G.J.; Harrison, P.A. The effects of climate variability and change on grape suitability in Europe. *J. Wine Res.* **1992**, *3*, 163–183. [CrossRef]
14. Laget, F.; Tondut, J.; Deloire, A.; Kelly, M.T. Climate trends in a specific Mediterranean viticultural area between 1950 and 2006. *J. Int. Sci. Vigne Vin.* **2008**, *42*, 113–123. [CrossRef]
15. Jones, G.V.; White, M.A.; Cooper, O.R.; Storchmann, K. Climate change and global wine quality. *Clim. Chang.* **2005**, *73*, 319–343. [CrossRef]
16. Salinari, F.; Giosuè, S.; Tubiello, F.N.; Rettori, A.; Rossi, V.; Spanna, F.; Rosenzweig, C.; Gullino, M.L. Downy mildew (*Plasmopara viticola*) epidemics on grapevine under climate change. *Glob. Chang. Biol.* **2006**, *12*, 1299–1307.
17. Töpfer, R.; Trapp, O. A cool climate perspective on grapevine breeding: Climate change and sustainability are driving forces for changing varieties in a traditional market. *Theor. Appl. Gen.* **2022**, *135*, 3947–3960. [CrossRef]
18. van Leeuwen, C.; Destrac-Irvine, A.; Dubernet, M.; Duchêne, E.; Gowdy, M.; Marguerit, E.; Pieri, P.; Parker, A.; de Rességuier, L.; Ollat, O. An Update on the Impact of Climate Change in Viticulture and Potential Adaptations. *Agronomy* **2019**, *9*, 514. [CrossRef]
19. Montagne, D.; Cornu, S. Do we need to include soil evolution module in models for prediction of future climate change? *Clim. Chang.* **2010**, *98*, 75–86. [CrossRef]
20. Prior, B. Inventory management to adapt to climate change. *Das Dtsch. Weinmagazin* **2007**, *10*, 22–27.
21. Vršič, S.; Ivančič, A.; Pulko, B.; Valdhuber, J. Effect of soil management systems on erosion and nutrition loss in vineyards on steep slopes. *J. Environ. Biol.* **2011**, *32*, 289–294.
22. Vršič, S. Soil erosion and earthworm population responses to soil management systems in steep-slope vineyards. *Plant Soil Environ.* **2012**, *57*, 258–263. [CrossRef]
23. Becker, N. Site Selection for Viticulture in Cooler Climates Using Local Climatic Information. In Proceedings of the International Symposium on Cool Climate Viticulture and Enology, the Valley River Inn, Eugene, Oregon, 25–28 June 1984; Oregon State University Technical Publication 7628: Corvallis, OR, USA, 1985; pp. 20–34.
24. Kenny, G.J.; Shao, J. An assessment of a latitude–temperature index for predicting climate suitability for grapes in Europe. *J. Hortic. Sci.* **1992**, *67*, 239–246. [CrossRef]
25. Jones, G.V.; Duff, A.A.; Hall, A.; Myers, J. Spatial analysis of climate in winegrape growing regions in the western United States. *Am. J. Enol. Vitic.* **2010**, *61*, 313–326. [CrossRef]
26. Duchene, E.; Huard, F.; Dumas, V.; Schneider, C.; Merdinoglu, D. The challenge of adapting grapevine varieties to climate change. *Clim. Res.* **2010**, *41*, 193–204. [CrossRef]
27. Jorquera-Fontena, E.; Orrego-Verdugo, R. Impact of global warming on the phenology of a variety of grapevine grown in Southern Chile. *Agrociencia* **2010**, *44*, 427–435.
28. Webb, L.B.; Whetton, P.H.; Barlow, E.W.R. Observed trends in winegrape maturity in Australia. *Glob. Change Biol.* **2011**, *17*, 2707–2719. [CrossRef]
29. Bernáth, S.; Paulen, O.; Šiška, B.; Kusá, Z.; Tóth, F. Influence of Climate Warming on Grapevine (*Vitis vinifera* L.) Phenology in Conditions of Central Europe (Slovakia). *Plants* **2021**, *10*, 1020. [CrossRef]
30. Ogrin, D.; Repe, B.; Štaut, L.; Svetlin, D.; Ogrin, M. Climate classification of Slovenia based on data from the period 1991–2020. *Dela* **2023**, *59*, 5–89. [CrossRef]
31. Vršič, S.; Pulko, B.; Perko, A. Climate change trends in the wine-growing regions of Slovenia. In Proceedings of the 6th Slovenian Viticulture and Wine Congress, Ptuj, Slovenia, 21–22 April 2023; Vršič, S., Ed.; Agricultural Forestry Institute Ptuj: Ptuj, Slovenia, 2023; pp. 97–111.
32. ARSO (Slovenian Environment Agency). *The Daily Precipitation and Temperature Values for the Period from 1952 to 2022*; Slovenian Environment Agency: Ljubljana, Slovenia, 2024.
33. Vršič, S.; Vodovnik–Plevnik, T. Reactions of vines varieties to climate changes in NE Slovenia. *Plant Soil Environ.* **2012**, *58*, 34–41. [CrossRef]
34. Winkler, A.J.; Cook, J.A.; Kliewer, W.M.; Lider, L.A. *General Viticulture*; University of California Press: Berkeley, CA, USA, 1974.
35. Huglin, P. Nouveau mode d'évaluation des possibilités héliothermiques d'un milieu viticole. In *Proc Symp Int sur L'ecologie de la Vigne*; Ministére de l'Agriculture et de l'Industrie Alimentaire: Contança, Romania, 1978; pp. 89–98.
36. Blanco-Ward, D.; Garcia-Queijeiro, J.M.; Jones, G.V. Spatial climate variability and viticulture in the Miño River Valley of Spain. *Vitis* **2007**, *46*, 63–70.
37. Zhang, X.; Alexander, L.; Hegerl, G.C.; Jones, P.; Tank, A.K.; Peterson, T.C.; Trewin, B.; Zwiers, F.W. Indices for monitoring changes in extremes based on daily temperature and precipitation data. *Wiley Interdiscip. Rev. Clim. Chang.* **2011**, *2*, 851–870. [CrossRef]
38. Vršič, S.; Pulko, B.; Vodovnik-Plevnik, T.; Perko, A. The Impact of Climatic Warming on Earlier Wine-Grape Ripening in Northeastern Slovenia. *Horticulturae* **2024**, *10*, 611. [CrossRef]

39. Hirsch, R.M.; Alexander, R.B.; Smith, R.A. Selection of methods for the detection and estimation of trends in water quality. *Water Resour. Res.* **1991**, *27*, 803–813. [CrossRef]
40. Jones, G. Climate change and wine: Observations, impacts and future implications. *Wine Ind. J.* **2006**, *21*, 21–26.
41. Ramos, M.C.; Jones, G.V.; Martínez-Casasnovas, J.A. Structure and trends in climate parameters affecting winegrape production in northeast Spain. *Clim. Res.* **2008**, *38*, 1–15. [CrossRef]
42. Machar, I.; Vlckova, V.; Bucek, A.; Vrublova, K.; Filippovova, J.; Brus, J. Environmental Modelling of Climate Change Impact on Grapevines: Case Study from the Czech Republic. *Pol. J. Environ. Stud.* **2017**, *26*, 1927–1933. [CrossRef]
43. Carroquino, J.; Garcia-Casarejos, N.; Gargallo, P. Classification of Spanish wineries according to their adoption of measures against climate change. *J. Clean. Prod.* **2020**, *244*, 142–155. [CrossRef]
44. Olesen, J.E.; Bindi, M. Consequences of climate change for european agricultural productivity, land use and policy. *Eur. J. Agron.* **2002**, *16*, 239–262. [CrossRef]
45. Keller, M. Managing grapevines to optimise fruit development in a challenging environment: A climate change primer for viticulturists. *Aust. J. Grape Wine Res.* **2010**, *16*, 56–69. [CrossRef]
46. Mozell, M.R.; Thach, L. The impact of climate change on the global wine industry. *Wine Econ. Policy* **2014**, *3*, 81–89. [CrossRef]
47. Jones, G.V.; Duchêne, E.; Tomasi, D.; Yuste, J.; Braslavska, O.; Schultz, H.; Martinez, C.; Boso, S.; Langellier, F.; Perucho, C.; et al. *Changes in European Winegrape Phenology and Relationships with Climate*; Proceedings GESCO: Geisenheim, Germany, 2005; pp. 55–61.
48. Tomasi, D.; Jones, G.V.; Giust, M.; Lovat, L.; Gaiotti, F. Grapevine phenology and climate change: Relationships and trends in the Veneto region of Italy for 1964–2009. *Am. J. Enol. Vitic.* **2011**, *62*, 329–339. [CrossRef]
49. Ruml, M.; Korać, N.; Vujadinović, M.; Vuković, A.; Ivanišević, D. Response of grapevine phenology to recent temperature change and variability in the wine-producing area of Sremski Karlovci, Serbia. *J. Agric. Sci.* **2016**, *154*, 186–206. [CrossRef]
50. Prša., I.; Rakić, V.; Rašić, D.; Vučetić, V.; Telišman Prtenjak, M.; Omazić, B.; Blašković, L.; Karoglan, M.; Preiner, D.; Drenjančević, M.; et al. Influence of Weather and Climatic Conditions on the Viticultural Production in Croatia. In Proceedings of the Terclim 2022, XIVth International Terroir Congress 2nd ClimWine Symposium, Bordeaux, France, 3–8 July 2022.
51. Omazić, B.; Telišman Prtenjak, M.; Kvakić, M.; Meštrić, J.; Bubola, M.; Prša, I.; Karoglan, M. Changes in grapevine budburst and harvest dates in Croatia under current and future climate conditions. EGU24-4051. In Proceedings of the EGU General Assembly 2024, Vienna, Austria, 14–19 April 2024.
52. Omazić, B.; Telišman Prtenjak, M.; Prša, I.; Belušić Vozila, A.; Vučetić, V.; Karoglan, M.; Karoglan Kontić, J.; Prša, Ž.; Anić, M.; Šimon, S.; et al. Climate change impacts on viticulture in Croatia: Viticultural zoning and future potential. *Int. J. Climatol.* **2020**, *40*, 5634–5655. [CrossRef]
53. Kovacs, E.; Puskas, J.; Pozsgai, A. Positive Effects of Climate Change on the Field of Sopron Wine-Growing Region in Hungary. In *Perspectives on Atmospheric Sciences*; Karacostas, T., Bais, A., Nastos, P., Eds.; Springer Atmospheric Sciences: Cham, Switzerland, 2017.
54. Gaal, M.; Moriondo, M.; Bindi, M. Modelling the impact of climate change on the Hungarian wine regionsusing random forest. *Appl. Ecol. Environ. Res.* **2012**, *10*, 121–140. [CrossRef]
55. Vršič, S.; Šuštar, V.; Pulko, B.; Kraner-Šumenjak, T. Trends in climate parameters affecting winegrape ripening in northeastern Slovenia. *Clim. Res.* **2014**, *58*, 257–266. [CrossRef]
56. Eitzinger, J.; Kubu, G.; Formayer, H.; Gerersdorfer, T. Climatic wine growing potential under future climate scenarious in Austria. In Proceedings of the Sustainable Development and Bioclimate: Reviewed Conference Proceedings, Vienna, Austria, 5–8 October 2009; pp. 146–147.
57. Jones, G.V. Climate change in the western United States grape growing regions. *Acta Hortic.* **2005**, *689*, 41–60. [CrossRef]
58. Duchêne, E.; Schneider, C. Grapevine and climatic changes: A glance at the situation in Alsace. *Agron. Sustain. Dev.* **2005**, *25*, 93–99. [CrossRef]
59. Huglin, P. *Biologie et Écologie de la Vigne*; Lavoisier: Paris, France, 1986; 372p.
60. White, M.A.; Diffenbaugh, N.S.; Jones, G.V.; Pal, J.S.; Giorgi, F. Extreme heat reduces and shifts United States premium wine production in the 21st century. *Proc. Natl. Acad. Sci. USA* **2006**, *103*, 11217–11222. [CrossRef] [PubMed]
61. Tate, A.B. Global warming's impact on vine. *J. Wine Res.* **2001**, *12*, 95–109. [CrossRef]
62. Ramos, M.C. Rainfall distribution patterns and their change over time in a Mediterranean area. *Theor. Appl. Climatol.* **2001**, *69*, 163–170. [CrossRef]
63. Sumner, G.; Homar, V.; Ramis, C. Precipitation seasonality in eastern and southern coastal Spain. *Int. J. Climatol.* **2001**, *21*, 219–247. [CrossRef]
64. Brunetti, M.; Maugeru, M.; Nanni, T.; Navarra, A. Droughts and extreme events in regional daily Italian precipitation series. *Int. J. Climatol.* **2002**, *22*, 543–558. [CrossRef]
65. Lana, X.; Serra, C.; Burgueno, A. Trends affecting pluvio- metric indices at the Fabra observatory (Barcelona, NE Spain) from 1917 to 1999. *Int. J. Climatol.* **2003**, *23*, 315–332. [CrossRef]
66. Peacock, B. Water Management for Grapevines. University of California, Tulare County, Pub. IG1-2005, 95. Available online: https://ucanr.edu/sites/Tulare_County/files/82035.pdf (accessed on 12 July 2024).
67. Lebon, E. Changements climatiques: Quelles conséquences prévisibles sur la viticulture? In *6émes Rencontres Rhodaniennes*; Institut Rhodanien: Orange, France, 2002; pp. 31–36.

68. Stock, M. Klimaveränderungen fordern die Winzer—Bereitschaft zur Anpassung ist erforderlich. *Geisenheimer Berichte* **2005**, *57*, 29–48.
69. Webb, L.B.; Whetton, P.H.; Barlow, E.W.R. Modelled impact of future climate change on the phenology of winegrapes in Australia. *Aust. J. Grape Wine Res.* **2007**, *13*, 165–175. [CrossRef]

horticulturae

MDPI

Article

Long-Term Evolution of the Climatic Factors and Its Influence on Grape Quality in Northeastern Romania

Roxana Mihaela Filimon [1,†], Claudiu Ioan Bunea [2,†], Răzvan Vasile Filimon [1,*], Florin Dumitru Bora [3] and Doina Damian [1]

1 Research Development Station for Viticulture and Winemaking Iasi, 48 Mihail Sadoveanu Alley, 700490 Iasi, Romania; roxanacotovanu@yahoo.com (R.M.F.); doinadamian@yahoo.com (D.D.)
2 Department of Viticulture and Oenology, Faculty of Horticulture and Business in Rural Development, University of Agricultural Sciences and Veterinary Medicine Cluj-Napoca, 3-5 Mănăștur Street, 400372 Cluj-Napoca, Romania; claudiu.bunea@usamvcluj.ro (C.I.B.)
3 Department of Viticulture and Oenology, Advanced Horticultural Research Institute of Transylvania, Faculty of Horticulture and Business in Rural Development, University of Agricultural Sciences and Veterinary Medicine Cluj-Napoca, 3-5 Mănăștur Street, 400372 Cluj-Napoca, Romania; boraflorindumitru@gmail.com (F.D.B.)
* Correspondence: razvan_f80@yahoo.com (R.V.F.)
† These authors contributed equally to this work.

Abstract: Climate change is currently the greatest threat to the environment as we know it today. The present study aimed to highlight the changes in the main climatic elements during the last five decades (1971–2020) in northeastern Romania (Copou-Iaşi wine-growing center) and their impact on grape quality, as part of precision viticulture strategies and efficient management of grapevine plantations. Data analysis revealed a constant and significant increase in the average air temperature in the last 50 years (+1.70 °C), more pronounced in the last 10 years (+0.61 °C), with a number of days with extreme temperatures (>30 °C) of over 3.5-fold higher, in parallel with a fluctuating precipitation regime. The increase in average temperatures in the last 40 years was highly correlated with the advancement of the grape harvest date (up to 12 days), a significant increase in *Vitis vinifera* L. white grape sugar concentration (+15–25 g/L), and a drastic decrease in total acidity (−2.0–3.5 g/L tartaric acid). The significant increase in the values of the bioclimatic indices require the reclassification of the wine-growing area in higher classes of favorability, raising the opportunity to grow cultivars that are more suited to warmer climates, ensuring the efficiency of the plantation, and meeting current consumer expectations.

Keywords: bioclimatic indices; climate change; climate suitability; precision viticulture; *Vitis vinifera* L.

Citation: Filimon, R.M.; Bunea, C.I.; Filimon, R.V.; Bora, F.D.; Damian, D. Long-Term Evolution of the Climatic Factors and Its Influence on Grape Quality in Northeastern Romania. *Horticulturae* **2024**, *10*, 705. https://doi.org/10.3390/horticulturae10070705

Academic Editors: Sarita Leonel, Sergio Ruffo Roberto and Simone Rodrigues da Silva

Received: 3 June 2024
Revised: 27 June 2024
Accepted: 2 July 2024
Published: 3 July 2024

1. Introduction

Viticulture and wine production are ancient occupations that have accompanied humans for thousands of years. Currently, the world area under grapevines is estimated to be 7.2 mha, with the European Union (EU) being the world's largest wine producer (144.5 mhL) and consumer (107 mhL; 48% of the world total volume) [1]. The latest international data highlight a decrease in the areas planted with grapevine in the interval 2001–2023, but especially the gradual decrease in wine production. In 2023, world wine production was estimated to be 237 mhL (−9.6% compared to 2022), a value that represents the smallest volume of wine recorded in the last 60 years [1]. Moreover, the OIV report mentioned that Italy, Spain, and Greece registered significant decreases in wine production compared to 2022 due to unfavorable weather conditions (extreme temperatures, along with flood and hail damage) that led to the appearance of downy mildew and severe droughts during the growing season [1]. These fluctuations are mainly attributed to climate changes, indicating a high variability and a lack of predictability in terms of grape yield and wine production.

In Romania, viticulture is a traditional occupation that appeared and evolved due to the favorable climate. Currently, Romania is one of the main wine-producing countries, occupying the 10th place in the world in terms of the area planted with grapevine. This represents over 187,000 ha, resulting, in 2023, in relatively large wine production of 4.6 mhl (+21.2% compared to 2022) [1]. In recent decades, the reality of climate change has been accepted by the majority of the scientific community [2]. As is often said, climate change is currently the greatest threat to the environment as we know it today, and we still do not know how things will evolve. Fortunately, grapevines managed to adapt to different climatic conditions, growing on six out of seven continents, in a large diversity of climates (mainly in the temperate climate of the northern hemisphere) [3]. However, climate plays a vital role in grapevine growing, influencing plant physiology, grape yield, and physico-chemical characteristics [3,4]. Each cultivar responds differently to climate change, being proof of grapevine plasticity and resilience to climate change [5].

The gradual warming of the climate affects the evolution of natural factors in the viticultural ecosystems; summers become hotter and drier, with longer autumns, shorter and excessive winters, more frequent periods of drought, and wet periods with excessive rain. Although it can adapt in time, at this point, the consequences for the grapevine are inevitable: compression of grapevine phenology, plants less prepared for the cold season and more sensitive to extreme temperatures, forced ripening of grapes, and advancement of harvest dates, the grapes being poor in their natural components [6–8]. As a result of global warming, the maintenance of grapevines on the perimeter of the traditional wine-growing regions remains a challenge for the producers, requiring the application of rigorous measures to protect or adapt the plants.

Climate change modifies the viticultural potential of vineyards, their specific wine styles, and even their limits [4]. In recent decades, several studies have highlighted the impact of climate change on viticulture [4,9–11], or how to adapt viticultural or winemaking practices under new climate conditions [12–14]. Most studies include the analysis of the risk factors under controlled conditions, and the design of favorability maps based on bioclimatic indices or statistical modeling, without the estimation of the economic impact and the real possibility of implementation of the proposed measures.

Jones et al. [9] suggested that growing season temperature indices can be effectively used to define vineyards' climate, the impact of climate change being highly variable. In a certain vineyard, climatic conditions vary from one year to the other, inducing the "vintage effect", respectively, annual changes in yield, quality, and wine typicity [12]. Wine-growing regions are characterized by a particular climate, which determines the use of specific cultivars and viticultural practices [3]. In the same context, the French term "terroir" is used to describe the environmental factors and farming practices that affect a crop's phenotype, providing distinctive characteristics to the final product [15,16]. Climate plays an important role in the terroir concept, but current climate changes modify the known conditions, leading to a change in the terroir perception.

Jones et al. [9] showed that the mean air temperatures in the growing season increased by 1.3 °C between 1950 and 1999 and 1.7 °C from 1950 to 2004 for the main viticultural areas of the world. Certainly, in some northern areas, climate change may be beneficial for viticulture, allowing grape maturation. Still, most of the world's vineyards will suffer due to warming and extreme climatic events, the producers being forced to find new cultivation strategies, including cultivars better adapted to the new conditions or relocation [3,17].

Among climate change-related effects are often mentioned advanced harvest data, increased grape sugar content (which leads to higher alcohol levels in wine), lower acidity, and modification of varietal aroma compounds [10]. Over time, the alcohol content of wine has evolved progressively, a trend related to climate change, respectively, to the gradual increase in air temperatures, which led to increasing sugar amounts in grapes [18,19]. In the general context of functional food consumption and healthy nutrition concept, a higher sugar content of grapes means wines with an increased alcohol content that are no longer in line with current consumer requirements [18,20,21]. High temperatures favor the degradation of grape acidity, modifying the sensory profile of the final wine [3,22]. Moreover, must and wine acidity corrections are more often necessary in recent years [23].

Global surface temperature was 1.09 °C higher in 2011–2020 than 1850–1900, with larger increases over land than over the ocean [24]. Temperatures in Europe have increased more than twice the global average over the past 30 years, the highest of any continent [25]. Moreover, over the 1991–2021 period, temperatures in Europe have warmed significantly, with about +0.5 °C per decade [25]. Recent scientific studies carried out in Romania indicate an expansion of the area of the suitability of grapevine culture by about 2.4 million ha, an increase in favorable altitude by approx. 180 m (to a maximum of 835 m), and a northward extension of the area of favorability by about 0.036° [4]. At the same time, the climatic suitability for the cultivation of white wine grape cultivars has decreased, the respective regions becoming more favorable for the cultivation of red wine cultivars that require higher temperatures and longer time intervals for the phenolic maturation of the grapes.

In the Romanian viticultural landscape, the wine-growing area of Moldavia occupies a very important place. Located in the northeast of Romania, the Iaşi wine-growing region is one of the oldest in the country and includes the Copou-Iaşi wine-growing center, located in the east-northeast of the Moldavian Plateau, in the area where the parallel of 47°10′ north latitude meets the meridian of 27°35′ east longitude. In the Copou-Iaşi wine-growing center, white wines represent the main direction of production, the largest areas being currently occupied by the cultivars Fetească albă, Fetească regală, Aligoté, Muscat Ottonel, Sauvignon blanc, and Chardonnay. In the east of Romania, the climate conditions are influenced by the Siberian anticyclone in the cold season and by the Asiatic cyclone during summer [26]. According to the Köppen–Geiger classification, the climate is currently characterized as humid continental, estimated by Beck et al. [27] to become a humid subtropical climate by 2100. Considering that climatic factors determine the wine-growing area and set the production directions, the current study aimed at highlighting the changes in the main climatic elements during the last five decades (1971–2020) in the vineyards of N-E Romania and their impact on sugar accumulation and total acidity reduction in grapes of some white wine cultivars (*Vitis vinifera* L.), as part of precision viticulture strategies and efficient management of actual grapevine plantations. In the context of the climate changes, it is necessary to continuously evaluate vineyards' climate suitability and promote cultivars that correspond to the new climatic conditions that are better adapted and will maintain high production and satisfy consumer demands.

2. Materials and Methods

2.1. Climate Data Collection

The study was carried out on the grapevine plantations of the Research and Development Station for Viticulture and Winemaking Iasi (Copou-Iaşi wine-growing center, Iasi wine-growing region, NE of Romania), being focused on the analysis of the evolution of climatic factors in the period 1971–2020 (50 years). To allow data interpretation, the analyzed period was divided into five decades: 1971–1980, 1981–1990, 1991–2000, 2001–2010, and 2011–2020. The meteorological data were recorded daily using an AgroExpert® weather station generation 1 (Metrilog Systems, Bucharest, Romania) located in the experimental field (47°12′18″ N, 27°32′03″ E; at 191 m altitude) (Figure 1), connected to a computer.

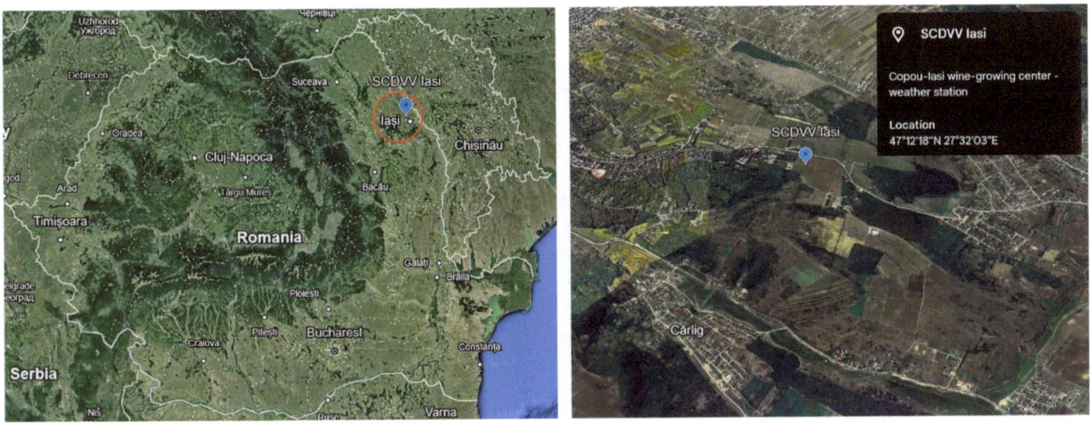

Figure 1. The location of the Copou-Iaşi wine-growing center (NE of Romania). Source: Google Earth [28].

2.2. Studied Parameters and Bioclimatic Indices

To highlight the climate changes, the following parameters were calculated: annual average temperature, growing season (April 1 to September 30) average temperature, annual precipitations, growing season precipitations, number of days with temperatures > 30 °C and with temperatures >35 °C, number of days with temperatures lower than −2 °C (spring frost) and −15 °C (winter frost), the sum of positive temperatures (Σt°g; the sum of average daily temperatures > 0 °C during the growing season (01.04–31.09)), the sum of active temperatures (Σt°a; the sum of growing season average temperatures >10 °C), and the sum of effective temperatures (Σt°u; the sum of differences between average daily temperatures > 10 °C and biological threshold for grapevine (10 °C)). To assess the heliothermic and water resources of the area, we calculated the hydrothermal coefficient (HC; the ratio between the growing season average precipitations and the sum of active temperatures, multiplied by 10) [29], De Martonne aridity index (IDM; the ratio between average annual precipitation and annual mean temperature plus 10) [30], actual heliothermal index (IHr; the result of multiplication of the growing season hours of real insolation and the Σt°u, multiplied by 10^6) [31], grapevine bioclimatic index (Ibcv; the ratio between the real insolation multiplied by Σt°a, and growing season precipitations multiplied by the number of days in the growing season with average temperatures >10 °C, divided by 10) [32], the oenoclimatic aptitude index (IAOe; the sum of real insolation and Σt°a, from which is subtracted the growing season average precipitations and 250) [33], the Huglin heliothermal index (HI; the sum of the growing season maximum temperature minus 10 and the growing season minimum temperature minus 10, divided by 2 and multiplied by the length of daylight hours coefficient—varying from 1.02 to 1.06 between 40° and 50° latitude) [34], the Winkler index (WI; the sum of daily average temperatures above 10 °C, from April 1 to October 31) [35], the cool night index (the average minimum temperatures of September, in the Northern Hemisphere) [36]. The values are shown as a 10-year average (decade average), with the multiannual average value (50 years) for each analyzed parameter also being presented.

2.3. Grapevine Cultivars and Growing Conditions

Two autochthonous *Vitis vinifera* L. cultivars for white wines (Fetească Albă and Fetească Regală), and two *Vitis vinifera* L. international cultivars for white wines (Aligoté and Muscat Ottonel), widely planted in most of the major wine-producing regions in Romania, were monitored for 40 years (1981–2020) in terms of sugar content and total acidity of grapes at ripening. All four cultivars are growing in the grapevine plantations of the Research and Development Station for Viticulture and Winemaking Iaşi; planting distances were 2.2 between rows and 1.2 m between plants, in the semi-high trunk system (80 cm), Guyot training system, with plants grafted on SO_4 rootstock (*Vitis berlandieri* × *Vitis riparia*). The soil is cambic chernozem with a clayey–loamy texture (pH 6.8). The elevation is 150–160 m, S-SW, 6–7% slope, and N-S row orientation. The main applied agrotechnical operations were specific to the standard grapevine growing technology (manual pruning, weeding, and harvesting). Grape maturity and harvest date were established according to Eichhorn–Lorenz phenological stages (EL 38—berries ripe for harvest) [37], by weekly determination of sugar and total acidity concentrations.

2.4. Chemical Determinations

Grape must sugar (g/L), equated according to the total soluble solids content (digital refractometer HI96801, Hanna Instruments, Cluj Napoca, Romania), and total acidity (g/L as tartaric acid) were determined using the standard methods specified in the OIV Compendium of International Methods of Wine and Must Analysis [38].

2.5. Statistical Procedures

The one-way analysis of variance (ANOVA) was used to determine the statistically significant differences between the groups of data (n = 10) in XLSTAT 2021.5.1 for Microsoft® Excel 2019. The method used to discriminate among the means was Tukey's test at 95% confidence level. Values noted with the same letter indicate statistically non-significant differences ($p > 0.05$). Regression analysis was performed to look for relationships between data (Microsoft® Excel; data analysis) (Pearson correlation). Principal Component Analysis and Agglomerative Hierarchical Clustering (Ward's method) were performed to investigate data group formation using the XLSTAT 2021.5.1 statistical software.

3. Results and Discussion

3.1. Climatic Elements

Climate is defined as the long-term pattern of weather conditions in a particular area [39]. Global warming gradually modifies the climate conditions, creating new patterns, increasing climate variability. According to Antón et al. [40], climate change is expected to increase the variability of weather conditions and the frequency of extreme events. However, temperature is considered the main climatic element and a good indicator of climate change because it has a significant impact on the entire ecosystem by guiding the life cycle of various organisms [41,42]. Temperature tolerances differ for various species and cultivars, but grapes are generally produced in areas with an annual average temperature between 10 and 20 °C, and growing season average temperatures between 13 and 21 °C, mostly in the mid-latitude regions of the continents [43,44]. In the Copou-Iasi wine-growing center, NE of Romania, in the period 1971–2020, the multiannual average temperature was calculated as 9.89 ± 0.98 °C. The lowest temperature recorded in the studied interval was −27.2 °C (December 1996), and the highest was 42.3 °C (July 2007). In the analyzed period, the average air temperature increased from one decade to another by at least 0.28 °C. Thus, the difference between the first analyzed decade (1971–1980) and the last decade (2011–2020) was +1.70 °C (Figure 2a). A significant increase in annual temperatures was highlighted starting with the year 2000, the period 1991–2000 being one of transition. If in the interval 1971–1990 were years when the average air temperature frequently reached 7–8 °C, in recent years, due to climate change, the lowest average air temperature is around 10 °C. According to the U.S. Environmental Protection Agency, worldwide, the decade 2012–2021 was the warmest since thermometer-based observations were made, confirming the similar trend of increasing air temperatures [45].

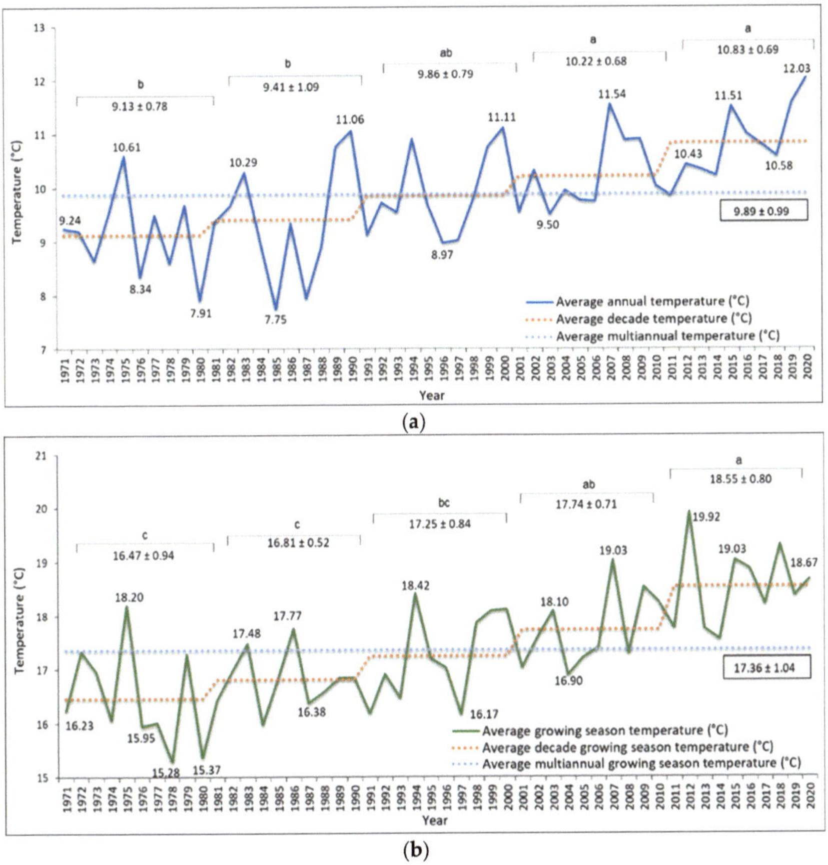

Figure 2. Evolution of annual average temperatures (**a**) and growing season average temperature (**b**) in the Copou-Iaşi wine-growing center, NE of Romania (1971–2020). Note: The mean values of the decades are presented as the average of the annual data (n = 10) with standard deviation (±). Values with the same letter are not statistically significant ($p > 0.05$) using Tukey's test.

Regarding the average temperatures in the growing season (April–September), recent studies conducted in most European countries indicated a general increasing trend [3]. In our study, a significant increase in air temperatures was observed during the growing season (Figure 2b). The multiannual average temperature in the growing season was 17.36 ± 1.04, the difference between the first analyzed decade and the 2011–2020 interval being +2.08 °C. The rise in air temperature during the growing season was more evident starting with the year 2000, the statistical differences being significant compared to the 1971–1990 interval. However, the most important increase in the growing season average temperature was observed in the last decade, over 0.81 °C.

Precipitation has important effects on the development of ecosystems. Rain and snow can affect the amount of water on the surface, while the timing of snowmelt influences the availability of groundwater for irrigation. As the temperatures rise, a more intense evaporation of water takes place, which subsequently generates more consistent precipitation; thus, theoretically, climate warming is expected to increase the amount of precipitation. According to US EPA [46], on average, since 1901, global precipitation has increased from decade to decade at an average rate of 1.016 mm (0.04 inches). On the other hand, grapevine is a species with moderate demands on water, the daily consumption of a grapevine stock being 0.2–1.5 L, which means approx. 8000 L/day/ha [47]. Considering losses through evaporation and soil infiltration, for the correct supply of grapevine, a rainfall volume of 300–350 mm is necessary during the growing season, respectively, 500–700 mm annual precipitation [43]. The IPCC group predicts that most vineyards will face an increase in drought intensity and a reduction in surface and groundwater resources in the future; the cultivars to be planted must be selected according to their drought tolerance [2]. In the Copou-Iaşi area, after a maximum value of precipitations reached in the interval 1991–2000 (613.15 ± 147.75 mm), the average values of the decades remained relatively constant, with a slight non-significant decrease (Figure 3a).

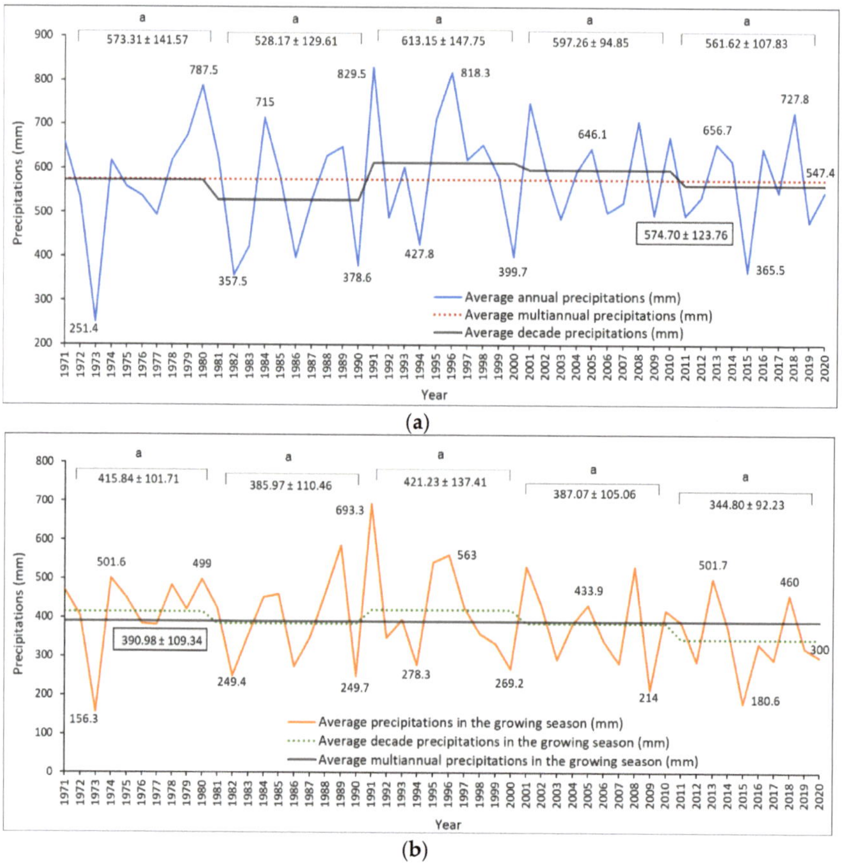

Figure 3. Changes in the annual average precipitation (**a**) and growing season average precipitation (April–September) (**b**) in the Copou-Iaşi wine-growing center, NE of Romania (1971–2020). Note: The mean values of the decades are presented as the average of the annual data (n = 10) with standard deviation (±). Values with the same letter are not statistically significant ($p > 0.05$) using Tukey's test.

The multiannual average of precipitation was 574.70 mm, with a high standard deviation (±123.76 mm) that indicates a large variability of the annual values. There were years (e.g., 1991, 1996) with high precipitation values (>800 mm) followed by dry years, in which the precipitation was up to two times lower (250–400 mm); in general, the conditions for efficient cultivation of grapevine were found in eight out of ten years.

Water deficit in the growing season affects photosynthesis and shoot growth and reduces berry size, while excessive water deficit stress can damage the leaves and stop grape ripening [12]. Also, excess water in soil causes a vigorous growth of shoots, sensitivity to fungal diseases, a delay in grape ripening, and lower rates of sugar accumulation [43,48,49]. In our study, the multiannual (1971–2020) average of growing season precipitation was 390.98 ± 109.34 mm, the decrease in the amount of precipitation being more obvious (−71 mm between the first and the last analyzed decades), but still statistically non-significant due to the very high data dispersion (Figure 3b).

Worldwide, over recent decades, no significant changes in the average precipitation regime during the growing season were reported [50]. However, Chen et al. [51], showed that the number of rainfall days decreased significantly in northeast China, increasing the daily rainfall amounts. In general, precipitation in the high latitudes of the northern hemisphere have increased, while rainfall in eastern Asia, Australia, and the Pacific region has declined, with rainfall variability increasing almost everywhere in the world [52]. Across Europe, a significant positive annual trend in precipitation was reported, with regional differences, however, with northern Europe becoming wetter and southern Europe becoming drier [53].

High-temperature variations directly affect photosynthesis and grapevine growth, the physiological processes being reduced when the temperature rises above an optimum limit [54]. The optimum photosynthetic temperature for grapevine is between 25 and 35 °C [55]. Above 35 °C, vegetation activity is impaired, and in some extreme cases, vineyards may suffer severe and irreversible damage [11,54]. Heat stress was evaluated based on the number of days with maximum temperatures above 30 and 35 °C. In the N-E of Romania, the parameter that increased most during the 1971–2020 period was the number of days with temperatures >30 °C. If in the first analyzed decade (1971–1980) the number of days with temperatures >30 °C was low (~9 days); during the following decades, there was a gradual and significant increase in the number of days with maximum temperatures >30 °C, exceeding an average of 30 days in the 2011–2020 interval (Figure 4a). If, until 2000, there were years when the temperatures did not exceed 30 °C (e.g., 1991), after the year 2000, the number of days with risky temperatures for grapevine was between 9 and 51, the heat stress on the plants increasing exponentially.

An upward trend was also registered in the case of the number of days with temperatures above 35 °C, the values being 10-fold higher in the decades after 2000, compared to the 1971–2000 interval (Figure 4b). Even if statistically the differences between the decades are non-significant due to the large fluctuations from year-to-year values, there is a clear trend towards an increase in the number of days with maximum temperatures above 35 °C, which seriously affects the normal development of plants, and, finally, the grape yield and quality. Prolonged exposure to extremely high temperatures (above 35 °C) can negatively affect plant development, with severe sunburns being reported, which increases the incidence of fungal infection [11,56]. A similar trend was reported by Bucur and Dejeu [57] in seven viticultural wine-growing regions in Romania, the number of days with temperatures above 30 and 35 °C gradually increasing significantly in the last 20 years.

The grapevine is a deciduous perennial fruit crop, and annually the canes and buds should withstand low temperatures during winter. However, grapevine has a limited resistance to winter frosts, varying depending on genetic factors (species and variety). The widely cultivated *Vitis vinifera* L. is a cold-sensitive species and cannot survive severe winter in regions with extremely low temperatures [58]. The most resistant to negative temperatures are the cultivars originating from temperate-continental climates, while the most sensitive cultivars come from the Mediterranean area [43,59]. Cold stress, including chilling (0 to 15 °C) and freezing (<0 °C) stresses, has an adverse effect on plant growth, development, and productivity, limiting grapevine geographical distribution [58,60].

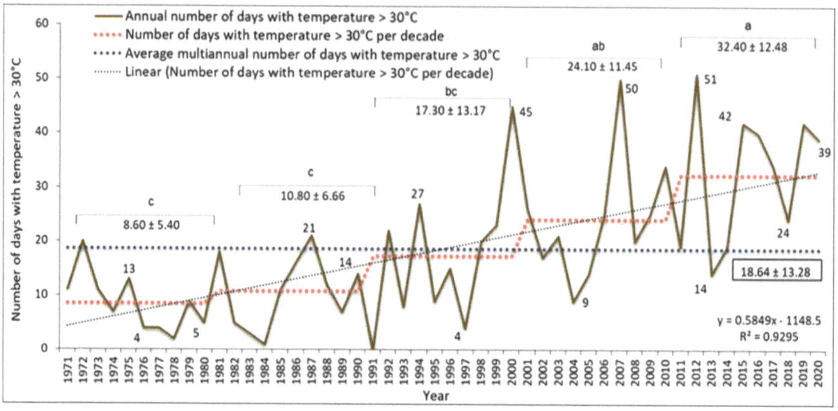

(a)

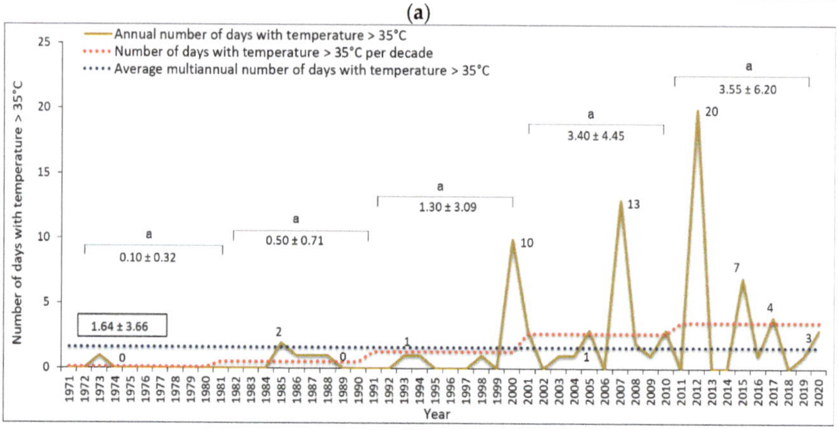

(b)

Figure 4. The average number of days with temperatures above 30 °C (**a**) and above 35 °C (**b**) in the Copou-Iaşi wine-growing center, NE of Romania (1971–2020). Note: The mean values of the decades are presented as the average of the annual data (n = 10) with standard deviation (±). Values with the same letter are not statistically significant ($p > 0.05$) using Tukey's test.

Low temperatures are among the abiotic factors with an important negative impact on the grapevine, depending on the time of occurrence. Thus, the number of days with frosts in winter (<−15 °C) and spring (<−2 °C) was evaluated. In the case of winter frosts (temperatures below −15 °C), a large fluctuation of years with very low temperatures was highlighted. Thus, decades with a higher number of days with frost (>4) alternated with decades with fewer days with frosts; however, the differences between decades were non-significant (Figure 5a). On average, in the period 1971–2020, in the Copou-Iaşi wine-growing center, there were 3.3 days of frost during winter (December, January, and February), a particularly important aspect when deciding on the cultivars to be planted in the area and wine type production.

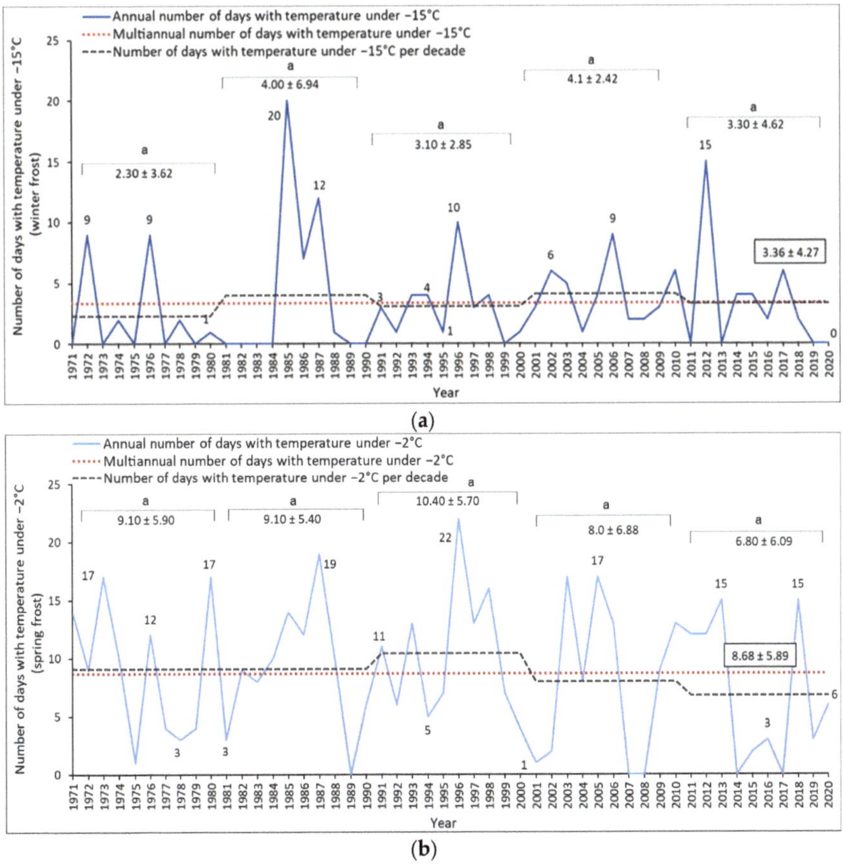

Figure 5. The average number of days with temperatures below −15 °C (December, January, and February; winter frost) (**a**) and below −2 °C (March, April, May; spring frost) (**b**) in the Copou-Iaşi wine-growing center, NE of Romania (1971–2020). Note: The mean values of the decades are presented as the average of the annual data (n = 10), with standard deviation (±). Values with the same letter are not statistically significant ($p > 0.05$) using Tukey's test.

Temperature is the main factor influencing the budburst in spring. Mild winters cause early budbursts, which can lead to devastating damage from late spring frost affecting the green shoots and young leaves, with significant harvest decreases [61]. Knowing the frequency of winter and spring frosts is essential in the selection process of cultivars that can better withstand the action of low temperatures. On average (1971–2020), in the Copou-Iaşi area, spring frosts occurred more than 8 days/year. If, in the period 1971–1990, the number of days with spring frost (<−2 °C) was on average 9–10, after the year 2000, the average annual number of days with freezing temperatures decreased non-significantly to 6–7, as well as the annual frequency of the phenomenon (in some years, no spring frosts occurred; two out of ten) (Figure 5b).

3.2. Bioclimatic Indices

To assess the climate characteristics of a vineyard, a series of bioclimatic indices must be calculated, based on the summation of active, useful, or global temperatures throughout the year or during the growing season (Winkler index, Huglin index, cool night index) or combinations of thermal and hydrological indices, such as the oenoclimatic aptitude index (IAOe) or grapevine bioclimatic index (Ibcv). These bioclimatic indicators are very effective in representing the climate suitability for wine production in various climates by integrating factors such as temperature, precipitation, or sunshine duration [62,63]. The Huglin Index (HI) is widely applied as an effective tool for viticultural zoning [34,64], while IAOe is largely used to analyze vineyard climate suitability for specific cultivars [62].

Hygroscopicity (relative air humidity), which influences the intensity of physiological processes, showed values between 84 and 88%, without significant variations between decades, ensuring the optimal level of air humidity throughout the annual biological cycle of plants. Also, the multiannual average of the actual sunshine duration (as the sum of the hours of effective sunshine during the growing season) was 2062 ± 120 h, with small non-significant variations between decades, the climate being favorable for grapevine growing from this point of view. In the Romanian wine regions, a sunshine duration of over 1520 h is considered sufficient for the production of red wines [43]. The values of the main bioclimatic indices in the Copou-Iaşi wine-growing center, for the period 1971–2020 (on decades), are shown in Table 1.

Table 1. Changes in the main bioclimatic indices of the Copou-Iaşi wine-growing center, NE of Romania, in the period 1971–2020.

Bioclimatic Indices		Decades					Average (1971–2020)
		1971–1980	1981–1990	1991–2000	2001–2010	2011–2020	
Σt°g (°C)		3020 ± 172 ^{c}	3083 ± 99 ^{c}	3159 ± 148 ^{bc}	3262 ± 140 ^{ab}	3398 ± 139 ^{a}	3185 ± 191
Σt°a (°C)		2864 ± 191 ^{c}	2958 ± 139 ^{c}	3021 ± 191 ^{bc}	3151 ± 130 ^{ab}	3291 ± 168 ^{a}	3057 ± 219
Σt°u (°C)		1251 ± 160 ^{c}	1308 ± 87 ^{c}	1379 ± 126 ^{bc}	1469 ± 136 ^{ab}	1609 ± 132 ^{a}	1403 ± 178
HC	Average	1.5 ± 0.4	1.3 ± 0.4	1.4 ± 0.5	1.3 ± 0.4	1.1 ± 0.3	1.3 ± 0.4
	Class	Moderate humidity	Moderate humidity	Moderate humidity	Moderate humidity	Insufficient humidity	Moderate humidity
IDM	Average	30 ± 8	27 ± 7	31 ± 8	30 ± 5	27 ± 6	29 ± 7
	Class	Humid	Semi-humid	Humid	Humid	Semi-humid	Humid
IHr	Average	1.8 ± 0.3 ^{bc}	1.8 ± 0.3 ^{cd}	2.1 ± 0.3 ^{c}	2.1 ± 0.2 ^{ab}	2.4 ± 0.3 ^{a}	2.1 ± 0.3
Ibcv	Average	7.0 ± 3.6	7.2 ± 2.7	7.2 ± 2.6	7.7 ± 2.9	9.2 ± 3.3	7.7 ± 3.0
IAOe	Average	4152 ± 280 ^{c}	4223 ± 301 ^{c}	4325 ± 399 ^{bc}	4482 ± 266 ^{cb}	4690 ± 252 ^{ab}	4375 ± 352
	Class	Unsuitable for red wine production	Unsuitable for red wine production	Medium favorability for red wine	Medium favorability for red wine	Very favorable for red wine	Medium favorability for red wine
HI	Average	1822 ± 181 ^{c}	1928 ± 130 ^{c}	1976 ± 190 ^{bc}	2098 ± 150 ^{ab}	2268 ± 161 ^{a}	2018 ± 222
	Class	HI_3—temperate climate	HI_3—temperate climate	HI_3—temperate climate	HI_4—temperate climate	HI_4—warm temperate climate	HI_3—temperate climate
WI	Average	1303 ± 159 ^{c}	1374 ± 78 ^{c}	1452 ± 133 ^{bc}	1535 ± 120 ^{ab}	1682 ± 139 ^{a}	1469 ± 181
	Class	Ib	Ib	II	II	III	II
CNI	Average	10.9 ± 1.2	10.9 ± 1.3	10.8 ± 1.8	11.1 ± 0.6	12.2 ± 1.2	11.2 ± 1.3
	Class	Very cold nights	Very cold nights	Very cold nights	Very cold nights	Cold nights	Very cold nights

Note: Σt°g—the sum of positive temperatures; Σt°a—the sum of active temperatures; Σt°u—the sum of effective temperatures; HC—hydrothermal coefficient; IDM—De Martonne aridity index; IHr—actual heliothermal index; Ibcv—grapevine bioclimatic index; IAOe—oenoclimate aptitude index; HI—Huglin index; WI—Winkler index; CNI—cool night index. Values with the same letter are not statistically significant ($p > 0.05$) using Tukey's test.

The sum of temperatures varies from one year to another, determining the different favorability of the years for grape production. The increase in Σt°g, Σt°a, and Σt°u was significant in the last 20 years. Regarding the Σt°u, in the decade 2011–2020, the values increased by 358 °C compared to the period 1971–1980, resulting in a multiannual average of 1403 ± 178 °C, which allows the efficient cultivation of grapevine. According to the studies conducted by Oşlobeanu et al. [65], in Romanian wine-growing regions, the Σt°g values are within the range 2700–3600 °C, Σt°a between 2600 and 3500 °C, while the Σt°u values vary between 1000 and 1700 °C. High values favor the ripening of grapes, and the accumulation of sugars and phenolic compounds, while the thermal deficit delays the ripening of grapes, determines a high level of total acidity, and limits the accumulation of useful organic substances.

Lower values (<0.6) of the hydrothermal coefficient (HC) indicate the need to irrigate the grapevine plantations [59,66]. In the Copou-Iaşi area, HC showed optimal values, varying between 1.1 and 1.5, decreasing non-significantly in recent decades. However, in the decade 2011–2020, a transition from a climate with moderate humidity to one with insufficient humidity was observed, indicating a stagnation of the precipitation regime in parallel with the increase in thermal resources.

The actual heliothermal index (IHr) reveals the availability of thermal resources in the analyzed viticultural area. In the reference interval, IHr values increased significantly from 1.8 (1971–1990) to 2.4 (2011–2020), with a multiannual average of 2.1 ± 0.3, offering after the year 2000 the possibility of obtaining quality red wines in the Copou-Iași center.

The values of the viticultural bioclimatic index (Ibcv), which integrates the influence of temperature, precipitation, and insolation (being recommended for the temperate climate), registered an upward trend, from 7.0 (1971–1980) to 9.2 (2011–2020), indicating, in this case, an increase in the humidity deficit, and at the same time, an abundance of heliothermic resources in the vineyard.

The oenoclimatic aptitude index (IAOe) expresses the combined action of insolation, temperature, and precipitation during the growing season and reveals the possibility of producing red wines in a specific vineyard [33]. Due to climate change, the IAOe values in the last 50 years showed a significant increase, evolving from 4152 (value unsuitable for the production of red wines) in the period 1971–1980, to 4325 (medium favorability for red wine production) in the period 1991–2000, and to 4690, in the period 2011–2020, placing the climate of the Copou- Iași wine-growing center in the class of very favorable areas for the ripening of grapevine cultivars for red wines. However, the IAOe showed a multiannual value of 4375 ± 352, corresponding to an area with medium favorability for the production of red wines, in which suitable conditions are met only in some years.

The Huglin index and the Winkler index are extensively used to determine the relationship between climate and the sugar content of grapes [67]. Grapevine cultivars need a certain amount of heat and precipitation to ripen the grapes. The Huglin heliothermal index (HI) provides information regarding the thermal potential of the vineyard and the possibility of growing certain cultivars with various periods of grape maturation. In the case of the Copou-Iaşi wine-growing center, the average value of HI in the period 1971–2020 was 2018 ± 222, corresponding to the theoretical class HI_3 (values between 1800 and 2100), respectively, a classic temperate climate. The significant increase in HI values in the last 40 years made it so that in the last decade (2011–2020), the Copou-Iaşi area became part of an upper climatic class, respectively, warm temperate climate—HI_4 (values between 2100 and 2400), in which there are no heliothermic constraints for grapevine, being recommended for planting cultivars regardless of the length of the vegetation period.

The Winkler index (WI) is used for classifying the climate of wine regions based on heat summation, the areas being divided into five climate regions based on temperature converted to growing degree-days [35]. According to WI climate classification, the Copou-Iaşi wine-growing center was initially (1971–1990) included in Region Ib (1111–1389 °C). In the interval 1991–2010, the recorded WI values framed the Copou-Iaşi wine-growing area in Region II (1389–1667), the continuous heating from the last decade leading to the integration of the monitored area in Region III (1667–1994 °C), with the recent climate change being significant. Thus, according to WI average values (1469 ± 181 °C), the

Copou-Iaşi wine-growing center climate is favorable for high production of standard to good-quality wines.

The cool night index showed a relatively constant value in the last 50 years, the slight increase from the decade 2011–2020 being sufficient for the transition of the Copou-Iaşi wine-growing center from the class with "very cold nights" to "cold nights" classification, daily temperatures increasing more than night temperatures. The obtained data are consistent with those reported by Bucur and Dejeu [57], who showed that in many Romanian vineyards, the values of the cool night index were not significantly affected by climate change. Also, the research carried out in the last 10 years in different areas of Romania revealed the increasing trend of extreme weather conditions or shifts in climate suitability of traditional vineyards [23,26,62].

3.3. Changes in Harvest Date

Among the most evident biological effects of global warming are the phenological shifts [54]. As previously shown, phenological stages are highly influenced by temperature, thermal amplitude, and solar radiation, which together determine the initiation and length of the phenophases for a certain cultivar [68–70]. Previous studies conducted in Copou-Iaşi wine-growing center showed that high temperatures and soil water deficit may determine a shift in phenological phases and a forced ripening of grapes, with negative impact on the yield of various cultivars [71]. Overall, the warming process makes the grapevine go through some phenological phases faster, starting with budburst and finishing with an earlier grape ripening and harvest. According to Ruml et al. [72], an increase in the average temperature by 1 °C is enough to advance average harvest times by up to 7.4 days. To highlight this aspect, four *V. vinifera* L. grape cultivars for white wines were monitored in the interval 1981–2020 (40 years), determining the date of grape harvesting, respectively, the time of reaching grape technological maturity, considering that the production direction and the obtained wine style were not modified during this time interval (Figure 6).

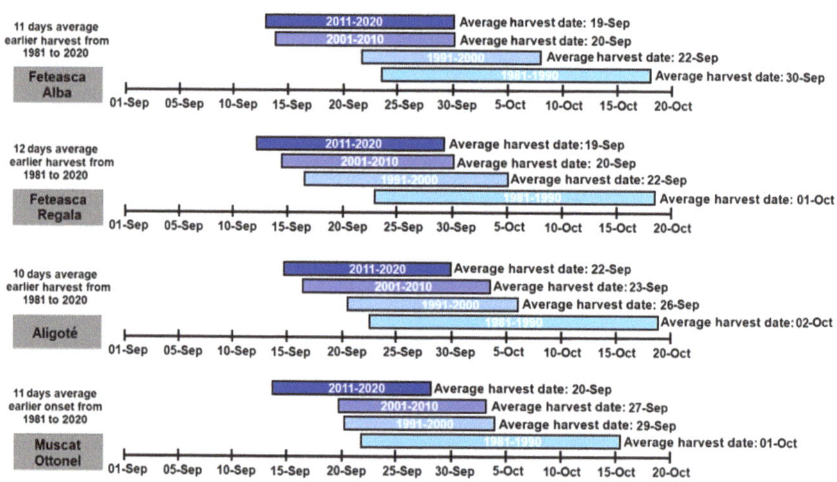

Figure 6. The comparative presentation of the grape harvest intervals (by decade), in the Copou-Iasi wine-growing center, NE of Romania (1981–2020). Note: The decade average was calculated as the mean value of the annual data (n = 10), for each cultivar.

Although the faster occurrence of grape maturity is obvious, each cultivar responded differently to climate changes, as proof of their different plasticity and adaptability. If, in the decade 1981–1990, the date of grape harvest varied in a wider range (21 September–20 October), the increase in temperatures in recent decades led to a shortening of this interval (12 September–

30 September), with a delay being determined in the average harvest date by about 11 days in the case of autochthonous cultivars (Fetească Albă and Fetească Regală). A lower advance in the date of grape harvest was recorded for the Aligoté cultivar, respectively, 10 days earlier than 40 years ago, a difference registered mainly in recent decades (2001–2020).

A similar trend was reported by Venios et al. [54] and Koch and Oehl [73], which showed a clear change in the date of harvest for five grapevine cultivars over approximately 40 years, in S-W Germany, significantly changing in the last two decades (about four weeks earlier comparing to the 1980–1990 decade). Trends related to earlier harvest dates have been frequently reported for various regions in Europe, confirming this process of continuous climate warming [26,74,75].

3.4. Changes in Grape Chemical Composition

Grape quality depends on the relationships between the plant and the environment. Long-term temperature increases all over the world had a direct impact on grape composition, the sugar content increasing and total acidity decreasing [67,76,77]. Climate influences grape quality by modifying the concentration of sugars, organic acids, and secondary compounds [9,78]. Overall, climate change accelerates the ripening process due to warming temperatures, and grapes reach maturity in a shorter time and accumulate large amounts of sugars [78,79]. However, grape sugar content can be indirectly affected by changes in water content and berry size [80].

Changes in sugars and total acidity content in matured grapes of Fetească Albă, Fetească Regală, Aligoté, and Muscat Ottonel cultivars, in the period 1981–2020, are shown in Figure 7 (as average values per decades). In the last 40 years, a significant increase in the average sugar content of grapes was observed, these increases varying depending on the cultivar. In the case of the autochthonous cultivars, the average sugar amount in grapes increased between +15.30 g/L (~8%) (Fetească Albă) and +20.56 g/L (~10%) (Fetească Regală) (Figure 7a,b).

Although the advance of the harvest date was almost similar for all four analyzed cultivars, the increase in sugar concentrations was much higher in the cultivars from the international assortment. Thus, the highest differences compared to the 1981–1990 decade in terms of sugar accumulation were recorded in grapes of Muscat Ottonel (+21.95 g/L; ~11%) and Aligoté (+25.4 g/L; ~12%) cultivars (Figure 7c,d). Regarding the multi-annual dynamics, grape sugar concentration has increased compared to the values of 40 years ago, but a slowing down of the process (Fetească Albă, Muscat Ottonel) or even a stagnation (Aligoté, Fetească Regală) was observed after the year 2000, even if the average air temperatures increased more during this period. Venios et al. [54] concluded that, in time, the grapevine might develop strategies to maintain homeostasis and cope with high-temperature stress, mechanisms that may include physiological adaptations and activation of signaling pathways and gene regulatory networks governing heat stress response and thermotolerance. However, the increase in sugar in grapes led to wines with higher alcohol concentrations or, in certain cases, wines with residual sugar (semisweet) [81].

Regarding the total acidity of mature grapes, a significant decrease was observed over the last 40 years. Unlike sugar accumulations, the decrease in total acidity was continuous. In grapes of the Fetească Albă cultivar, in the last decade (2011–2020), total acidity was lower by about 3.14 g/L tartaric acid (−33%) compared to the 1981–1990 interval. Along with the Fetească Albă cultivar, Aligoté (−3.50 g/L tartaric acid; −34%) suffered the most important losses of total acidity in the analyzed interval (1981–2020). Also, in the case of Fetească Regală (−3.24 g/L tartaric acid) and Muscat Ottonel cultivars (−2.00 g/L tartaric acid), the average acidity of matured grapes was lower by a percentage between 28 and 33% compared with the values of the 1981–1990 decade.

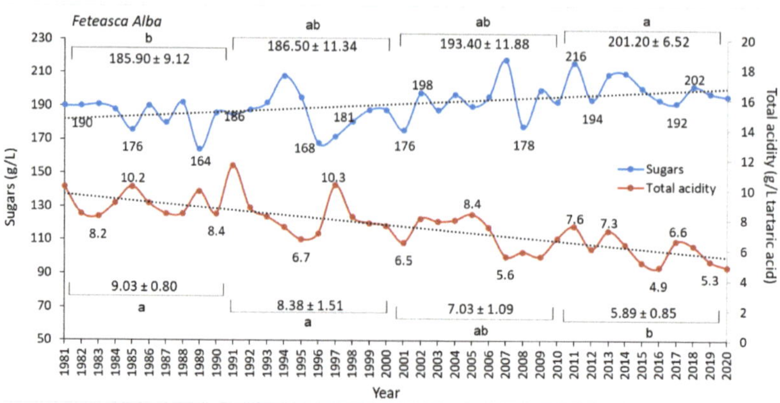

(**a**)

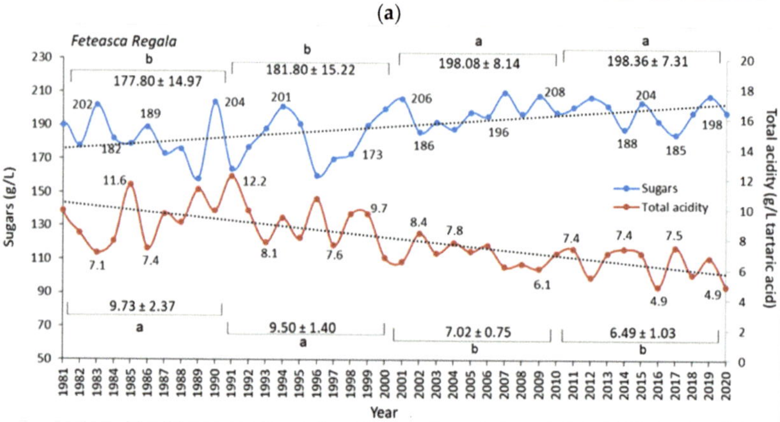

(**b**)

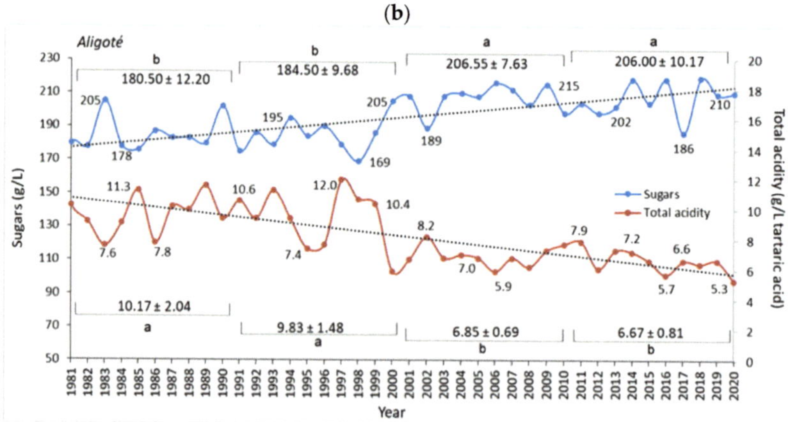

(**c**)

Figure 7. *Cont.*

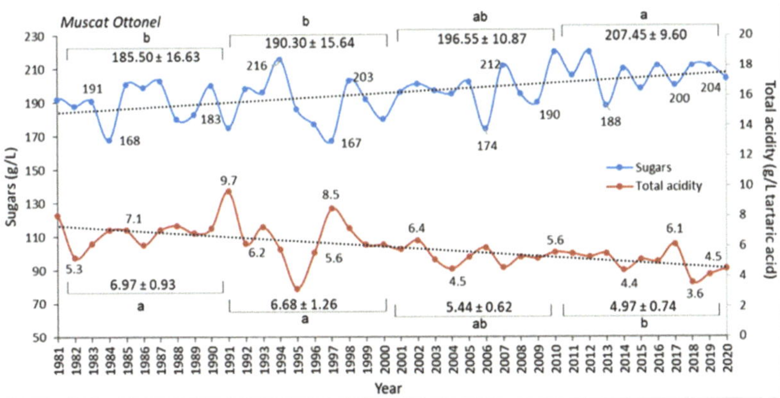

(d)

Figure 7. Changes in sugar amount (g/L) and total acidity (g/L as tartaric acid) in mature grapes of Fetească Albă (**a**), Fetească Regală (**b**), Aligoté (**c**), and Muscat Ottonel (**d**) cultivars in the period 1981–2020, in the Copou-Iaşi wine-growing center (NE of Romania). Note: The mean values of the decades are presented as the average of the annual data (n = 10), with standard deviation (±). Values with the same letter are not statistically significant ($p > 0.05$) using Tukey's test.

Climatic and biochemical data correlation for the interval 1981–2020 revealed the negative impact of the increasing annual and growing season temperatures on the date of the grape harvest, especially for the cultivars Aligoté (r > −0.9017) and Muscat Ottonel (r > −0.9832) (Table 2). For the same cultivars, a negative correlation between the high number of days with temperatures > 30 °C and the advance of the grape harvest date was highlighted (r = −0.9045–−0.9416). The annual average temperature (r = 0.8884–0.9851), growing season average temperature (r= 0.8891–0.9856), and the number of days with temperatures >30 °C (r = 0.9562–0.9815) were positively correlated ($p < 0.05$) with grape sugar content of all *V. vinifera* L. analyzed cultivars.

Also, the decrease in grapes' total acidity was inversely correlated with the increase in annual and growing season temperatures (r = −0.8966–−0.9854), as well as with the high number of days with temperatures >30 °C (r = −0.9094–−0.9940). Growing season precipitation showed a strong negative correlation with grape sugar concentrations (r = −0.9252–−0.9660) and a direct positive relationship with decreasing acidity values (r = 0.9128–0.9777). In the same context, the sugar content of grapes was negatively correlated with total acidity (r > −0.9416), meaning that a high concentration of sugar in grapes corresponded to a lower acidity. In the analyzed interval, increases in Σt°g, Σt°a, and Σt°u showed a direct influence on sugar accumulation (r > 0.90) and total acidity reduction in grapes of all analyzed cultivars (r > −0.90). Taking into account the calculation method, the main bioclimatic indices (IHr, Ibcv, IAOe, HI, WI) showed a positive correlation with carbohydrate accumulation (r = 0.8930–0.9924) and a negative relation with grape acidity (r > −0.9285), while hydrothermal coefficient (HC) and De Martonne aridity index (IDM), which involve the annual and growing season precipitation, were positively correlated with the acidity of the grapes (r > 0.85) (Table 2). Previously, Navrátilová et al. [67] revealed a higher correlation rate of the HI with the sugar content, while the WI proved to be less suitable for all viticultural areas. In our study, the Winkler index (WI) showed a high correlation with grape sugar content (r = 0.89–0.99), close to that of the Huglin index (r = 0.93–0.99).

Table 2. Pearson correlation coefficients of the relationships between climatic factors (1981–2020), phenological and chemical data of the analyzed *V. vinifera* L. cultivars (Copou-Iaşi wine-growing center; NE of Romania).

Parameters	No. Days T > 35 °C	No. Days T > 30 °C	GST (°C)	Annual T (°C)	GS PP (mm)	Annual PP (mm)	FA Sugars	FR Sugars	Al Sugars	MO Sugars	FA Acidity	FR Acidity	Al Acidity	MO Acidity
FA H-date	−0.6186	−0.6731	−0.6825	−0.6845	0.4956	−0.2119								
FR H-date	−0.7260	−0.8280	−0.8534	−0.8550	0.6604	−0.1025								
Al H-date	−0.8395	**−0.9045**	**−0.9017**	**−0.9026**	0.7800	0.0811								
MO H-date	−0.7806	**−0.9416**	**−0.9833**	**−0.9832**	0.8642	0.2408								
FA sugars	0.9040	**0.9815**	**0.9705**	**0.9696**	**−0.9660**	−0.4700	1							
FR sugars	0.9802	**0.9595**	**0.8983**	**0.8977**	**−0.9352**	−0.4413	0.8977	1						
Al sugars	0.9864	**0.9562**	**0.8891**	**0.8884**	**−0.9420**	−0.4728	0.8969	0.9993	1					
MO sugars	0.8522	**0.9710**	**0.9856**	**0.9851**	**−0.9252**	−0.3622	0.9917	0.8645	0.8595	1				
FA acidity	**−0.9094**	**−0.9895**	**−0.9854**	**−0.9854**	**0.9128**	0.2782	**−0.9520**	−0.9554	−0.9473	−0.9541	1			
FR acidity	**−0.9902**	**−0.9793**	**−0.9199**	**−0.9189**	**0.9777**	0.5237	−0.9456	**−0.9884**	−0.9904	−0.9107	0.9576	1		
Al acidity	**−0.9926**	**−0.9649**	**−0.8975**	**−0.8966**	**0.9609**	0.5101	−0.9168	−0.9962	**−0.9981**	−0.8782	0.9484	0.9967	1	
MO acidity	**−0.9720**	**−0.9940**	**−0.9534**	**−0.9527**	**0.9715**	0.4608	−0.9646	−0.9821	−0.9812	**−0.9416**	0.9796	0.9953	0.9877	1
Σt°g	0.8937	0.9925	0.9967	0.9964	−0.9308	−0.3255	**0.9828**	**0.9233**	**0.9168**	**0.9875**	**−0.9894**	**−0.9471**	**−0.9265**	**−0.9733**
Σt°a	0.9135	0.9967	0.9912	0.9908	−0.9482	−0.3718	**0.9880**	**0.9346**	**0.9298**	**0.9861**	**−0.9877**	**−0.9603**	**−0.9405**	**−0.9820**
Σt°u	0.8789	0.9863	0.9947	0.9943	−0.9319	−0.3444	**0.9900**	**0.9007**	**0.8949**	**0.9961**	**−0.9768**	**−0.9345**	**−0.9086**	**−0.9627**
HC	−0.9698	−0.9851	−0.9384	−0.9372	0.9942	0.5578	−0.9807	−0.9540	−0.9569	−0.9512	**0.9512**	**0.9878**	**0.9719**	**0.9895**
IDM	−0.9649	−0.9202	−0.8388	−0.8368	0.9925	0.7320	−0.9326	−0.9061	−0.9171	−0.8783	**0.8574**	**0.9555**	**0.9393**	**0.9393**
IHr	0.8890	0.9879	0.9944	0.9944	−0.9068	−0.2609	**0.9602**	**0.9354**	**0.9265**	**0.9682**	**−0.9978**	**−0.9441**	**−0.9300**	**−0.9712**
Ibcv	0.9269	0.9975	0.9861	0.9858	−0.9395	−0.3435	**0.9705**	**0.9585**	**0.9527**	**0.9666**	**−0.9972**	**−0.9702**	**−0.9579**	**−0.9890**
IAOe	0.9176	0.9936	0.9863	0.9862	−0.9249	−0.3064	**0.9605**	**0.9574**	**0.9504**	**0.9601**	**−0.9995**	**−0.9637**	**−0.9532**	**−0.9843**
HI	0.9290	0.9972	0.9820	0.9813	−0.9653	−0.4270	**0.9930**	**0.9386**	**0.9360**	**0.9842**	**−0.9800**	**−0.9692**	**−0.9489**	**−0.9863**
WI	0.8689	0.9844	0.9986	0.9984	−0.9169	−0.3005	**0.9818**	**0.9004**	**0.8930**	**0.9924**	**−0.9822**	**−0.9285**	**−0.9042**	**−0.9593**
CNI	0.7127	0.8700	0.9130	0.8820	−0.8842	−0.2936	0.9443	0.7063	0.7050	0.9278	−0.8802	−0.7946	−0.7305	−0.8251

Note: FA—Fetească Albă cv.; FR—Fetească Regală cv.; Al—Aligoté; MO—Muscat Ottonel cv.; Σt°g—the sum of positive temperatures; Σt°a—the sum of active temperatures; Σt°u—the sum of effective temperatures; HC—hydrothermal coefficient; IDM—De Martonne aridity index; IHr—actual heliotermal index; Ibcv—grapevine bioclimatic index; IAOe—oenoclimate aptitude index; HI—Huglin index; WI—Winkler index; CNI—cool night index; PP (mm)—precipitation; T (°C)—temperature; GS—growing season; H-date—harvest date. Correlation coefficients marked in bold are shown in the text ($p < 0.05$; Microsoft® Excel; data analysis).

Principal component analysis (PCA), as a multivariate technique, allows visual representation of the correlations between multiple variables, increasing data interpretability and explaining data variation [82]. PCA biplot showed the grouping of grape sugar of analyzed cultivars along with the annual and growing season temperatures during the last decade, while the number of days with freezing temperatures in winter (T < −15 °C) and spring (T < −2 °C) was paired with grape acidity in the period 1981–2000 (Figure 8a). WI, HI, IAOe, and IHr, together with the increasing number of days with high temperatures (T > 30 °C; T > 35 °C), were related to grape sugar concentrations, grouped especially in the quadrants of recent decades.

The cluster method admits the existence of the polythetic groups, whose elements are equivalent or similar for several criteria, but not for all characteristics. At the same time, the similitude of the elements from the group and the difference among groups are measured [83]. However, cluster analysis is a convenient method for identifying homogenous groups considering a multitude of factors. Agglomerative Hierarchical Clustering (AHC)—Ward's method—was used to group the analyzed decades (1981–2020), considering all the features analyzed in the study (climatic, phenological, and chemical factors). AHC analysis indicated the formation of two main clusters, grouping the decades 1981–1990 and 1991–2000 in a separate homogeneous node, these periods being more similar concerning the studied characteristics (Figure 8b). Also, the last two decades (2001–2020) were grouped in a separate branch, less homogenous, being aggregated at a higher dissimilarity index (77,378).

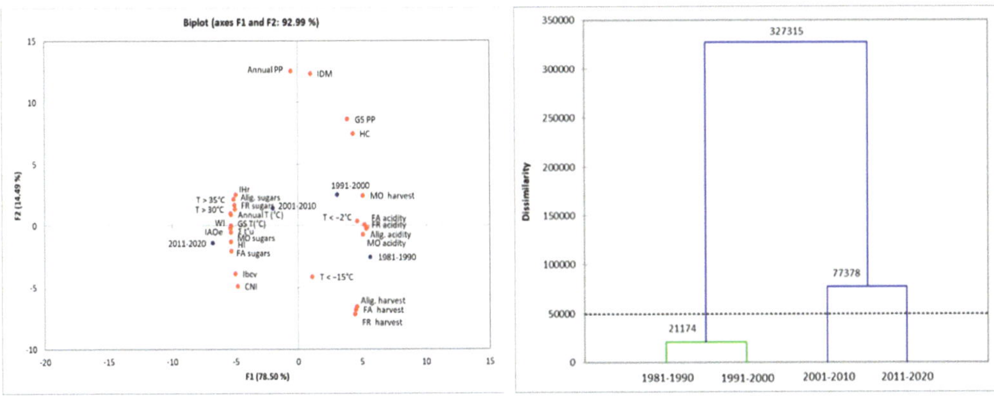

(**a**) (**b**)

Figure 8. Principal Component Analysis (PCA) biplot combining the output variables (**a**) and the Agglomerative Hierarchical Clustering (AHC) of the decades (**b**) for the interval 1981–2020, in the Copou-Iaşi wine-growing center. Note: FA—Fetească Albă cv.; FR—Fetească Regală cv.; Alig.—Aligoté; MO—Muscat Ottonel cv.; GS—growing season; T—temperature; Harvest—harvest date; Σt°u—the sum of active temperatures; HC—hydrothermal coefficient; IDM—De Martonne aridity index; IHr—actual heliotermal index; Ibcv—grapevine bioclimatic index; IAOe—oenoclimate aptitude index; HI—Huglin index; Wi—Winkler index; CNI—cool night index.

Similar studies concluded that increasing temperatures accelerate the rate of sugar accumulation in grapes, forcing growers to harvest earlier [78]. Thus, it is necessary for producers to sync grapevine cultivars with climates that are most compliant to their specific growing conditions. Harvest dates have shifted earlier historically, and climate models predict the further acceleration of grape ripening [10,78,84]. Extreme temperatures during grape maturation reduce grapevine metabolism, resulting in higher sugar levels and lower total acidity, affecting taste balance and the overall quality of the grapes [77]. According to Iland et al. [85], grapes from a hot environment are likely to possess lower acidity compared to fruit from a cool environment. For precision viticulture and the economic success of the grapevine plantation, it is necessary to know and select cultivars with a growing season length suitable for the type of climate and with biological resistance to extreme temperatures and erratic precipitation. Research on climate suitability must continue permanently, for a better understanding of what we are facing, testing the ability of plants to adapt to the new conditions and finding the best measures to ensure the sustainable cultivation of the grapevine and support its resilience to climate change.

4. Conclusions

Climate change poses a major challenge for viticulture. Global warming affects both grapevine physiology and biochemistry, changing grapevine phenology and berry composition. For the Copou-Iaşi wine-growing area, in northeastern Romania, the study revealed a significant gradual increase in the average air temperature in the last 50 years (+1.71 °C), more pronounced in the last 10 years (+0.61 °C). In the last decade, the number of days with extremely hot temperatures (>30 °C) was over 3.5 times higher compared with the first decades (1971–1990), in parallel with a fluctuating precipitation regime. The increase in the average air temperature in the last 40 years (1981–2020) was highly correlated with the advancement of grape maturity and harvesting date of *V. vinifera* L. white grape cultivars (up to 12 days), a significant increase in grape sugar accumulation (+15–25 g/L), and a drastic decrease in total acidity (−2.0–3.5 g/L as tartaric acid). Increasing positive, active, and effective sum of temperatures directly influences sugar accumulation and total acidity decrease in grapes of all analyzed cultivars (r > 0.90). However, climate changes exerted a distinctive impact on grapevine physiology and biochemistry depending on the cultivar. In the viticultural area of northeastern Romania, the last decade (2011–2020) made the transition from a humid temperate climate, unsuitable for red wine production (Winkler Region Ib-II), to a semi-humid warm temperate climate (Winkler Region III), more suitable for cultivars with a longer growing season or intended for red wine production. However, the transition from a climate with very cold nights to a climate with cold nights was noticed. Long-term climate change analysis, as part of the precision viticulture strategy and efficient management of vineyards, is of particular importance for grape producers and winemakers, to be prepared and to take action to counter the effects of global warming, to choose the most effective measures to maintain economic and sustainable grape growing, and to increase the resilience of viticulture to climate change.

Author Contributions: Conceptualization, R.M.F., R.V.F. and C.I.B.; methodology, R.M.F., R.V.F. and F.D.B.; software, R.V.F.; validation, D.D. and C.I.B.; formal analysis, R.M.F. and F.D.B.; investigation, R.M.F., R.V.F. and D.D.; resources, C.I.B., D.D. and R.V.F.; data curation, R.V.F. and F.D.B.; writing—original draft preparation, R.V.F.; writing—review and editing, R.M.F., D.D. and C.I.B.; visualization, D.D. and C.I.B.; supervision, R.M.F. and R.V.F.; funding acquisition, C.I.B. and F.D.B. All authors have read and agreed to the published version of the manuscript.

Funding: This research was funded by Ministry of Agriculture and Rural Development of Romania, project number ADER 6.3.10/21.07.2023.

Data Availability Statement: All data are contained within the article.

Acknowledgments: The authors gratefully acknowledge the University of Agricultural Sciences and Veterinary Medicine Cluj-Napoca for the financial and scientific support given in carrying out the research.

Conflicts of Interest: The authors declare no conflicts of interest.

References

1. OIV (International Organization of Vine and Wine). State of the World Vine and Wine Sector in 2023. 2024. Available online: https://www.oiv.int/sites/default/files/2024-04/OIV_STATE_OF_THE_WORLD_VINE_AND_WINE_SECTOR_IN_2023.pdf (accessed on 5 April 2024).
2. IPCC (Intergovernmental Panel on Climate Change). Climate change 2014: Synthesis Report. In *Contribution of Working Groups I, II and III to the fifth Assessment Report of the Intergovernmental Panel on Climate Change*; Pachauri, R.K., Meyer, L.A., Eds.; IPCC: Geneva, Switzerland, 2014; Available online: https://www.ipcc.ch/site/assets/uploads/2018/02/AR5_SYR_FINAL_Front_matters.pdf (accessed on 11 February 2024).
3. Santos, J.A.; Fraga, H.; Malheiro, A.C.; Moutinho-Pereira, J.; Dinis, L.-T.; Correia, C.; Moriondo, M.; Leolini, L.; Dibari, C.; Costafreda-Aumedes, S.; et al. A review of the potential climate change impacts and adaptation options for European viticulture. *Appl. Sci.* **2020**, *10*, 3092. [CrossRef]
4. Irimia, L.M.; Patriche, C.V.; Roșca, B. Climate change impact on climate suitability for wine production in Romania. *Theor. Appl. Climatol.* **2018**, *133*, 1–14. [CrossRef]
5. Ilina-Dumitru, A.M.; Manolescu, A.E.; Sumedrea, D.I.; Florea, A.C.; Cosmulescu, S.N. Climate influence on the phenology of table grapes in Stefanesti, Romania. *Acta H ortic.* **2024**, *1385*, 181–188. [CrossRef]
6. Lorenzo, M.N.; Taboada, J.J.; Lorenzo, J.F.; Ramos, A.M. Influence of climate on grape production and wine quality in the Rías Baixas, north-western Spain. *Reg. Environ. Chang.* **2013**, *13*, 887–896. [CrossRef]
7. Palliotti, A.; Tombesi, S.; Silvestroni, O.; Lanari, V.; Gatti, M.; Poni, S. Changes in vineyard establishment and canopy management urged by earlier climate-related grape ripening: A review. *Sci. Hortic.* **2014**, *178*, 43–54. [CrossRef]
8. Filimon, V.R.; Rotaru, L.; Filimon, R. Quantitative investigation of leaf photosynthetic pigments during annual biological cycle of *Vitis vinifera* L. table grape cultivars. *S. Afr. J. Enol. Vitic.* **2016**, *37*, 1–14. [CrossRef]
9. Jones, G.V.; White, M.A.; Cooper, O.R.; Storchmann, K. Climate change and global wine quality. *Clim. Chang.* **2005**, *73*, 319–343. [CrossRef]
10. Mira de Orduña, R. Climate change associated effects on grape and wine quality and production. *Food Res. Int.* **2010**, *43*, 1844–1855. [CrossRef]
11. Droulia, F.; Charalampopoulos, I. Future climate change impacts on European viticulture: A review on recent scientific advances. *Atmosphere* **2021**, *12*, 495. [CrossRef]
12. Van Leeuwen, C.; Darriet, P. The impact of climate change on viticulture and wine quality. *J. Wine Econ.* **2016**, *11*, 150–167. [CrossRef]

13. Dequin, S.; Escudier, J.L.; Bely, M.; Noble, J.; Albertin, W.; Masneuf-Pomarède, I.; Marullo, P.; Salmon, M.J.; Sablayrolles, J.M.; Ollat, N. How to adapt winemaking practices to modified grape composition under climate change conditions. *Oeno ONE* **2017**, *51*, 205–214. [CrossRef]
14. Jones, G.V.; Edwards, E.J.; Bonada, M.; Sadras, V.O.; Krstic, M.P.; Herderich, M.J. Climate change and its consequences for viticulture—Chapter 17. In *Managing Wine Quality, Vol. 1: Viticulture and Wine Quality*, 2nd ed.; Reynolds, A.G., Ed.; Woodhead Publishing: London, UK, 2022; pp. 727–778. [CrossRef]
15. OIV (International Organization of Vine and Wine). Resolution OIV/VITI 333/2010. *Definition of Vitivinicultural "Terroir"*. 2010. Available online: https://www.oiv.int/public/medias/379/viti-2010-1-en.pdf (accessed on 12 February 2024).
16. Brillante, L.; Bonfante, A.; Bramley, R.G.V.; Tardaguila, J.; Priori, S. Unbiased scientific approaches to the study of terroir are needed! *Front. Earth Sci.* **2020**, *8*, 539377. [CrossRef]
17. Laget, F.; Tondut, J.L.; Deloire, A.; Kelly, T.M. Climate trend in a specific Mediterranean viticultural area between 1950 and 2006. *J. Int. Sci. Vigne Vin.* **2008**, *42*, 113–123. [CrossRef]
18. Filimon, V.R.; Sandu-Ville, Ș.; Bora, F.D.; Tudor, G.; Filimon, R.; Nechita, A.; Damian, D. Methods for producing low-alcohol wine I. Viticultural and pre-fermentation strategies. *Lucr. Şt. USAMV Iaşi-Serie Hortic.* **2020**, *63*, 95–100. Available online: https://repository.uaiasi.ro/xmlui/handle/20.500.12811/1372 (accessed on 16 February 2024).
19. Filimon, V.R.; Sandu-Ville, Ș.; Bora, F.D.; Tudor, G.; Filimon, R.; Nechita, A.; Zaiț, M. Methods for producing low-alcohol wine II. Fermentation and post-fermentation strategies. *Lucr. Şt. USAMV Iaşi-Serie Hortic.* **2020**, *63*, 101–106. Available online: https://repository.uaiasi.ro/xmlui/handle/20.500.12811/1373 (accessed on 18 February 2024).
20. Pickering, G.J.; Heatherbell, D.A.; Barnes, M.F. The production of reduced alcohol wine using glucose oxidase treated juice. Part I. Composition. *Am. J. Enol. Vitic.* **1999**, *50*, 291–298. [CrossRef]
21. Longo, R.; Blackman, J.W.; Antalick, G.; Torley, P.J.; Rogiers, S.Y.; Schmidtke, L.M. A comparative study of partial dealcoholisation versus early harvest: Effects on wine volatile and sensory profiles. *Food Chem.* **2018**, *261*, 21–29. [CrossRef] [PubMed]
22. Ganichot, B. Évolution de la date des vendanges dans les Côtes-du-Rhône méridionales. In Proceedings of the 6th Rencontres rhodaniennes; Institut Rhodanien Editeur: Orange Avignon, France, 2002; pp. 38–41.
23. Bucur, G.M.; Cojocaru, G.A.; Antoce, A.O. The climate change influences and trends on the grapevine growing in Southern Romania: A long-term study. *BIO Web Conf.* **2019**, *15*, 01008. [CrossRef]
24. IPCC (Intergovernmental Panel on Climate Change). Summary for Policymakers. In *Climate Change 2023: Synthesis Report. In Contribution of Working Groups I, II and III to the Sixth Assessment Report of the Intergovernmental Panel on Climate Change*; Lee, H., Romero, J., Eds.; IPCC: Geneva, Switzerland, 2023; pp. 1–34. [CrossRef]
25. WMO (World Meteorological Organization). Temperatures in Europe Increase more Than Twice Global Average. 2022. Available online: https://wmo.int/news/media-centre/temperatures-europe-increase-more-twice-global-average (accessed on 20 March 2024).
26. Colibaba, L.C.; Bosoi, I.; Pușcalău, M.; Bodale, I.; Luchian, C.; Rotaru, L.; Cotea, V.V. Climatic projections vs. grapevine phenology: A regional case study. *Not. Bot. Horti Agrobot.* **2024**, *52*, 13381. [CrossRef]
27. Beck, H.E.; McVicar, T.R.; Vergopolan, N.; Berg, A.; Lutsko, N.J.; Dufour, A.; Zeng, Z.; Jiang, X.; van Dijk, A.I.J.M.; Miralles, D.G. High-resolution (1 km) Köppen-Geiger maps for 1901-2099 based on constrained CMIP6 projections. *Sci. Data* **2023**, *10*, 724. [CrossRef] [PubMed]
28. Google Earth. Map Showing Location of Iasi County, Romania. 2024. Available online: https://earth.google.com/web/search/Rom%C3%A2nia (accessed on 12 April 2024).
29. Selyaninov, G.L. O selskogo khozyaistvennoi otsenka klimata [About the agricultural evaluation of the climate]. *Trudy GGO.* **1928**, *20*, 177–185.
30. De Martonne, E. Aerisme et índices d'aridite. *Comptesrendus de L'Academie Des. Sci.* **1926**, *182*, 1395–1398.
31. Branas, J.; Bernon, G.; Levadoux, L. *Éléments de Viticulture Générale*; Impr. Delmas: Montpellier, France, 1946.
32. Constantinescu, G. Méthodes et principes de détermination des aptitudes viticoles d'une région et du choix des cépages appropriés. *Bull. OIV* **1967**, *40*, 1179–1205.
33. Teodorescu, S.; Popa, A.I.; Sandu, G.H.N. *Oenoclimatul României [The Oenoclimate of Romania]*; Ştiinţifică şi Enciclopedică: Bucharest, Romania, 1987.
34. Huglin, P. *Biologie et Ecologie de la Vigne*; Edition Payot: Lausanne, Switzerland, 1986.
35. Winkler, A.J.; Cook, A.; Kliewer, W.M.; Lider, I.A. *General Viticulture*; University of California Press: Berkeley, CA, USA, 1974.
36. Tonietto, J. Les Macroclimats Viticoles Mondiaux et L'influence du Mésoclimat sur la Typicité de la Syrah et du Muscat de Hambourg dans le sud de la France: Méthodologie de caractérisation. Ph.D. Dissertation, Ecole Nationale Supérieure Agronomique, Montpellier, France, 1999.
37. Eichhorn, K.W.; Lorenz, D.H. Phänologische entwicklungsstadien der rebe [Phenological stages of the vine]. *Nach Deut. Pflanzenschutzd* **1977**, *29*, 119–120. Available online: https://www.openagrar.de/servlets/MCRFileNodeServlet/openagrar_derivate_00037935/1977-032.pdf (accessed on 10 February 2024).
38. OIV (International Organization of Vine and Wine). *Compendium of International Methods of Wine and Must Analysis*; OIV: Paris, France, 2019; Volume 1, Available online: https://www.oiv.int/public/medias/7907/oiv-vol1-compendium-of-international-methods-of-analysis.pdf (accessed on 10 February 2024).

39. Antón, J.; Cattaneo, A. Chapter 15—Agricultural Risk Management and Climate Change: What role for policy? In *Handbook of International Food and Agricultural Policies. Vol. IV: Global Challenges for Future Food and Agricultural Policies*; Blandford, D., Hassapoyannes, K., Eds.; World Scientific Publishing: Hackensack, NJ, USA, 2019; pp. 281–306. [CrossRef]
40. Antón, J.; Cattaneo, A.; Kimura, S.; Lankoski, J. Agricultural risk management policies under climate uncertainty. *Glob. Environ. Chang.* **2013**, *23*, 1726–1736. [CrossRef]
41. Chan, E.Y. Climate change is the world's greatest threat—In Celsius or Fahrenheit? *J. Environ. Psychol.* **2018**, *60*, 21–26. [CrossRef]
42. Dimri, A.P.; Kumar, D.; Choudhary, A.; Maharana, P. Future changes over the Himalayas: Mean temperature. *Glob. Planet. Chang.* **2018**, *162*, 235–251. [CrossRef]
43. Irimia, L.M. *Biologia, Ecologia şi Fiziologia Viţei de vie [Biology, Ecology and Physiology of the Grapevine]*; Ion Ionescu de la Brad: Iaşi, Romania, 2012.
44. Jones, G.V. Climate, grapes, and wine: Structure and suitability in a changing climate. *Acta Hortic.* **2012**, *931*, 19–28. [CrossRef]
45. US EPA (U.S. Environmental Protection Agency). Climate Change Indicators: U.S. and Global Temperature. 2022. Available online: https://www.epa.gov/climate-indicators/climate-change-indicators-us-and-global-temperature (accessed on 10 April 2024).
46. US EPA (U.S. Environmental Protection Agency). Climate Change Indicators: U.S. and Global Precipitation. 2022. Available online: https://www.epa.gov/climate-indicators/climate-change-indicators-us-and-global-precipitation (accessed on 10 April 2024).
47. Galet, P. *Précis de Viticulture Pratique*, 6th ed.; Pierre Galet: Montpellier, France, 1993.
48. Williams, L.E. Grapevine water relations. In *Raisin Production Manual;* Christensen, L.P., Ed.; DANR Publications, University of California: Oakland, CA, USA, 2000; pp. 121–126.
49. Burzo, I.; Dejeu, L.; Şerdinescu, A.; Bădulescu, L. *Fiziologia plantelor de Cultură. Fiziologia viţei de vie [The Physiology of crop Plants. Grapevine Physiology]*; Elisavaros: Bucharest, Romania, 2005; Volume 3.
50. Asseng, S.; Zhu, Y.; Wang, E.; Zhang, W. Chapter 20—Crop modeling for climate change impact and adaptation. In *Crop Physiology*, 2nd ed.; Sadras, V.O., Calderini, D.F., Eds.; Academic Press: Amsterdam, The Netherlands, 2015; pp. 505–546. [CrossRef]
51. Chen, G.; Liu, H.; Zhang, J.; Liu, P.; Dong, S. Factors affecting summer maize yield under climate change in Shandong Province in the Huanghuaihai Region of China. *Int. J. Biometeorol.* **2012**, *56*, 621–629. [CrossRef] [PubMed]
52. Dore, M.H.I. Climate change and changes in global precipitation patterns: What do we know? *Environ. Int.* **2005**, *31*, 1167–1181. [CrossRef]
53. Caloiero, T.; Caloiero, P.; Frustaci, F. Long-term precipitation trend analysis in Europe and in the Mediterranean basin. *Water Environ. J.* **2018**, *32*, 433–445. [CrossRef]
54. Venios, X.; Korkas, E.; Nisiotou, A.; Banilas, G. Grapevine responses to heat stress and global warming. *Plants* **2020**, *9*, 1754. [CrossRef] [PubMed]
55. Zhang, K.; Chen, B.-H.; Hao, Y.; Yang, R.; Wang, Y.-A. Effects of short-term heat stress on PSII and subsequent recovery for senescent leaves of *Vitis vinifera* L. cv. Red Globe. *J. Integr. Agric.* **2018**, *17*, 2683–2693. [CrossRef]
56. Berry, J.; Bjorkman, O. Photosynthetic response and adaptation to temperature in higher plants. *Annu. Rev. Plant Physiol.* **2003**, *31*, 491–543. [CrossRef]
57. Bucur, G.M.; Dejeu, L. Research on phenotyping and eno-carpological traits of twenty-three new Romanian table grape varieties (*Vitis vinifera* L.). *Conf. Proceeding Agric. Life. Life Agric.* **2018**, *1*, 268–275. [CrossRef]
58. Ren, C.; Fan, P.; Li, S.; Liang, Z. Advances in understanding cold tolerance in grapevine. *Plant Physiol.* **2023**, *192*, 1733–1746. [CrossRef] [PubMed]
59. Rotaru, L.; Colibaba, L.C. *Elemente de Geografie Viticolă [Elements of Viticultural Geography]*; Ion Ionescu de la Brad: Iaşi, Romania, 2022.
60. Ding, Y.; Shi, Y.; Yang, S. Advances and challenges in uncovering cold tolerance regulatory mechanisms in plants. *New Phytol.* **2019**, *222*, 1690–1704. [CrossRef] [PubMed]
61. Marcon Filho, J.L.; Allebrandt, R.; de Bem, B.; Mudrei, P.I.; Macedo, T.A.; Schlemper, C.; Lerin, S.; Outemane, M.; Kretzschmar, A.A.; Rufato, L. Damage to 'Cabernet Sauvignon' after late frost in the southern Brazilian highlands. *Acta Hortic.* **2016**, *1115*, 211–216. [CrossRef]
62. Irimia, L.M.; Patriche, C.V.; Roşca, B. Changes in oenoclimate aptitude index characterizing climate suitability of Romanian wine growing regions. *Appl. Ecol. Environ. Res.* **2017**, *15*, 755–767. [CrossRef]
63. Rotaru, L.; Filimon, V.R.; Filimon, R.M.; Mustea, M.; Bernardis, R.R.; Colibaba, L.C. Preliminary studies on some morpho-structural and biochemical characterization of some genotypes of *Vitis vinifera* L. cultivated in northeast Romania. *J. Appl. Life Sci. Environ.* **2024**, *57*, 69–90. [CrossRef]
64. Jones, G.V.; Duff, A.A.; Hall, A.; Myers, J.W. Spatial analysis of climate in wine grape growing regions in the Western United States. *Am. J. Enol. Vitic.* **2010**, *61*, 313–326. [CrossRef]
65. Oşlobeanu, M.; Macici, M.; Georgescu, M.; Stoian, V. *Zonarea Soiurilor de viţă de vie în România [The Zoning of Grapevine Varieties in Romania]*; Ceres: Bucharest, Romania, 1991.
66. Voropay, N.N.; Ryazanova, A.A. A comparative assessment of the aridity indices for analysis of the hydrothermal conditions. *IGCCMR IOP Publ. IOP Conf. Ser. Earth Environ. Sci.* **2018**, *190*, 012041. [CrossRef]
67. Navrátilová, M.; Beranová, M.; Severová, L.; Šrédl, K.; Svoboda, R.; Abrhám, J. The impact of climate change on the sugar content of grapes and the sustainability of their production in the Czech Republic. *Sustainability* **2021**, *13*, 222. [CrossRef]

68. Zapata, D.; Salazar-Gutierrez, M.; Chaves, B.; Keller, M.; Hoogenboom, G. Predicting key phenological stages for 17 grapevine cultivars (*Vitis vinifera* L.). *Am. J. Enol. Vitic.* **2016**, *68*, 60–72. [CrossRef]
69. Nechita, A.; Zaldea, G.; Damian, D.; Filimon, R.M.; Pisticiuc, I.; Filimon, V.R. The evolution over time of the development of vegetation phenophases of the vine varieties cultivated in the Copou Iaşi wine center in the context of climate change. *Lucr. Şt. USAMV Iaşi Ser. Hortic.* **2020**, *63*, 89–94. Available online: https://repository.uaiasi.ro/xmlui/handle/20.500.12811/1369 (accessed on 10 March 2024).
70. Ausseil, A.E.; Law, R.M.; Parker, A.K.; Teixeira, E.I.; Sood, A. Projected wine grape cultivar shifts due to climate change in New Zealand. *Front Plant Sci.* **2021**, *12*, 618039. [CrossRef]
71. Zaldea, G.; Nechita, A.; Damian, D.; Ghiur, A.D.; Cotea, V.V. Climate changes in recent decades, the evolution of the drought phenomenon and the impact on vineyards in north-eastern Romania. *Not. Bot. Horti Agrobot.* **2021**, *49*, 12448. [CrossRef]
72. Ruml, M.; Korać, N.; Vujadinović, M.; Vuković, A.; Ivanišević, D. Response of grapevine phenology to recent temperature change and variability in the wine-producing area of Sremski Karlovci, Serbia. *J. Agric. Sci.* **2016**, *154*, 186–206. [CrossRef]
73. Koch, B.; Oehl, F. Climate change favors grapevine production in temperate zones. *Agric. Sci.* **2018**, *9*, 247–263. [CrossRef]
74. Duchêne, E.; Schneider, C. Grapevine and climatic changes: A glance at the situation in Alsace. *Agron. Sustain. Dev.* **2005**, *25*, 93–99. [CrossRef]
75. Schultz, H.R.; Jones, J.V. Climatic induced historic and future changes in viticulture. *J. Wine Res.* **2010**, *21*, 137–145. [CrossRef]
76. Schultz, H.R. Global climate change, sustainability, and some challenges for grape and wine production. *J. Wine Econ.* **2016**, *11*, 181–200. [CrossRef]
77. Hussain Shah, M.; Rafique, R.; Rafique, T.; Naseer, M.; Khalil, U.; Rafique, R. *Effect of Climate Change on Polyphenols Accumulation in Grapevine*; IntechOpen: London, UK, 2022. [CrossRef]
78. Stanfield, R.C.; Forrestel, E.J.; Elmendorf, K.E.; Bagshaw, S.B.; Bartlett, M.K. Phloem anatomy predicts berry sugar accumulation across 13 wine-grape cultivars. *Front. Plant Sci.* **2024**, *15*, 1360381. [CrossRef] [PubMed]
79. Rogiers, S.Y.; Greer, D.H.; Liu, Y.; Baby, T.; Xiao, Z. Impact of climate change on grape berry ripening: An assessment of adaptation strategies for the Australian vineyard. *Front. Plant Sci.* **2022**, *13*, 1094633. [CrossRef] [PubMed]
80. Arias, L.A.; Berli, F.; Fontana, A.; Bottini, R.; Piccoli, P. Climate change effects on grapevine physiology and biochemistry: Benefits and challenges of high altitude as an adaptation strategy. *Front. Plant Sci.* **2022**, *13*, 835425. [CrossRef]
81. Filimon, V.R.; Filimon, R.M.; Nechita, A.; Bora, F.D.; Cotea, V.V. Compositional characteristics of low-alcohol wines obtained by staggered grape harvesting technology. *Lucr. Şt. USAMV Iaşi Ser. Agron.* **2021**, *64*, 69–74. Available online: https://repository.uaiasi.ro/xmlui/handle/20.500.12811/3000 (accessed on 12 March 2024).
82. Filimon, R.M.; Rotaru, L.; Filimon, V.R. Effects of exogenous growth regulators on agrobiological, technological and physiological characteristics of an interspecific grapevine cultivar. *Biol. Agric. Hortic.* **2023**, *39*, 91–114. [CrossRef]
83. Rotaru, L. Identifying the phenotypic resemblances of the vine breeds by means of cluster analysis. *Not. Bot. Hort. Agrobot.* **2009**, *37*, 249–252.
84. Delrot, S.; Grimplet, J.; Carbonell-Bejerano, P.; Schwandner, A.; Bert, P.F.; Bavaresco, L.; Costa, L.D.; Di Gaspero, G.; Duchêne, E.; Hausmann, L.; et al. Genetic and genomic approaches for adaptation of grapevine to climate change. In *Genomic Designing of Climate-Smart Fruit Crops*; Kole, C., Ed.; Springer: Cham, Switzerland, 2020. [CrossRef]
85. Iland, P.; Dry, P.; Proffit, T.; Tyerman, S.; Collins, C.; Pagay, V.; Steel, C. *The Grapevine: From the Science to the Practice of Growing Vines for Wine*; Patrick Iland Wine Promotions: Adelaide, Australia, 2024.

horticulturae

MDPI

Article

The Impact of Climatic Warming on Earlier Wine-Grape Ripening in Northeastern Slovenia

Stanko Vršič [1,*], Borut Pulko [1], Tadeja Vodovnik-Plevnik [2] and Andrej Perko [1]

[1] Faculty of Agriculture and Life Sciences, University of Maribor, Pivola 10, 2331 Hoče, Slovenia; a.perko@um.si (A.P.)

[2] The Maribor Agriculture and Forestry Institute, Vinarska 14, 2000 Maribor, Slovenia; tadeja.vodovnik-plevnik@kmetijski-zavod.si

* Correspondence: stanko.vrsic@um.si

Abstract: In this study, the development trends of bioclimatic parameters recorded at the Maribor and Murska Sobota climate stations from 1952 to 2022 and the dynamics of grape ripening in early-, medium-, and late-ripening grape varieties in the Podravje wine-growing region in Slovenia (north-eastern Slovenia) from 1980 to 2022 were investigated. Based on the data on soluble solids content, total acidity, and the recommended harvest date per year (until the technological ripeness of the grapes; 76°Oe), trends for shortening the growing period of the vines were calculated. Temperature changes have been more pronounced since 1980. The number of so-called hot days (with a maximum of T > 30 °C) has increased the most, which has the greatest impact on other bioclimatic parameters, e.g., the average temperature and growing degree days (GDDs) and the Huglin index (HI). For the period of 1980 to 2022, the trends were 0.44 °C (Murska Sobota) and 0.51 °C (Maribor) per decade, respectively. The trends were more pronounced for the average temperature in the period of May–June (TMJ). After 1980, the HI increased by about 10 units per year. As a result of the climate warming, grapes in north-eastern Slovenia ripened 26 ('Sauvignon Blanc') to 35 ('Welschriesling') days earlier. The trends showed a decrease in total acidity, which can be attributed to the higher temperatures during the growing season period, especially during the ripening period of the grapes (véraison). After 2010, the average temperatures during the growing season (1 April to 31 October) in Podravje were 1.6 °C higher than in the 1980s. In line with the earlier ripening of the grapes, the actual average temperature from 1 April to the harvest date was a further 1.0 °C higher. The higher temperatures in the late-ripening varieties 'Riesling' and 'Furmint' had a positive effect on the lower total acidity. Total annual precipitation and precipitation in the growing season for the period 1980 to 2022 in the Maribor area show decreasing trends of 6 mm/m^2 (p = 0.001) and 4 mm/m^2 (p = 0.012), respectively. In the eastern sub-wine-growing region of Podravje (Murska Sobota), the trends in precipitation were not significant.

Keywords: grapevine; climate change; bioclimatic parameters; grape ripeness

Citation: Vršič, S.; Pulko, B.; Vodovnik-Plevnik, T.; Perko, A. The Impact of Climatic Warming on Earlier Wine-Grape Ripening in Northeastern Slovenia. *Horticulturae* **2024**, *10*, 611. https://doi.org/10.3390/horticulturae10060611

Academic Editors: Sergio Ruffo Roberto, Sarita Leonel and Simone Rodrigues da Silva

Received: 13 May 2024
Revised: 4 June 2024
Accepted: 4 June 2024
Published: 8 June 2024

1. Introduction

Many researchers have studied climate change in different regions [1–5] and its impact on viticulture [6–8], often with a particular focus on assessing and predicting the impact on grapevine development and the wine industry [9,10]. Warming has the potential to bring numerous risks and challenges that affect both the quality and quantity of grape production [11]. While changes in average temperatures are obvious and important, increasing attention is being paid to the analysis of extreme events due to their potential impact on viticulture [12–14]. The results show that the risk of unfavorable conditions during vine flowering decreases, while the risk of late frosts in spring increases due to the shift in the timing of bud break [15–17].

Throughout history, one of the main objectives of wine-growers in the different wine-growing regions has been to achieve maximum soluble solids contents without causing the berries to shrivel. In recent times, more consumers prefer lighter wines with moderate alcohol contents. In addition to the scenario described above, there is also global warming [18]. Worldwide, the phenology of grapevines has developed in step with the temperature trends caused by climate change in recent decades [19–21]. Analysis of the meteorological data shows a clear rise in temperatures. However, it is more likely that the most significant impacts will result from an increase in temperature, even if other climate variables, such as precipitation, are also considered [22]. A comparison of climatic and phenological data shows that the period between bud break and harvest has become earlier and shorter [23,24], and grapes ripen earlier under increasingly warmer conditions, which often has undesirable effects [25]. Post-flowering water requirements tend to increase due to climate, and as there is no clear evidence of a change in precipitation, the risks associated with dry summers are likely to increase in the future [26]. However, to fully understand how climate change contributes to changes in harvest dates, grapevine phenology and its relationship with climate must be analyzed over a longer time period, including data that predate anthropogenic interventions in the climate system [21,27,28]. During the growing season, grapevines require sustained daily average temperatures above 10 °C to initiate growth, followed by sufficient heat accumulation for fruit ripening [29]. However, temperature extremes during berry growth lead to stress, premature ripening, berry shedding, enzyme activation, and reduced flavor development [30]. Frost occurrence and timing are also important for grapevines, which benefit from a low risk of frost in spring and fall and a long frost-free season of 160 to 200 days or more. In terms of moisture requirements, grapevines should ideally have sufficient soil moisture for initial growth at the beginning of the growing season and then receive nominal amounts throughout the growing season (either naturally or through irrigation) [10].

The impact of climate change on the viticulture sector varies from region to region. The vines need both sufficient cold periods for hardening and fruit formation and sufficient warm periods to ripen quality fruit at an economically viable level without being overly stressed. The grapevine is therefore a model system for monitoring the effects of climate change because it has a long history, because it is grown in narrow climatic zones for which the individual varieties are best suited, and because the wines are tasted and rated for quality [27]. The predicted rise in global temperatures over the next half century could ultimately prove problematic for the wine industry. Minor changes in growing season temperatures could lead to shifts in varietal suitability in many regions [19] or require costly adaptation measures, particularly in soil management [31]. In addition, in regions such as Europe, where vines are not irrigated, either due to legal restrictions or for supply reasons, changes in the total amount of precipitation or its distribution over the year can have a significant impact on water availability for plants, especially in the warmer seasons [10].

Wine-growers traditionally select different wine-grape varieties according to the phenotypic characteristics that best suit their microclimates [32] and soils. They retain those varieties that produce consistent yields under local climatic conditions and have an appropriate balance of sugar, acidity, and other compounds [33]. Climate change, with its extreme weather conditions, will make it necessary to change varieties in many wine-growing regions, especially in regions with disease pressure and difficult growing conditions. New varieties (PIWI) are therefore better suited to climate-adapted and sustainable viticulture than traditional varieties are [34].

Climatic conditions during grape ripening have already changed, leading to a change in the composition of the grapes at harvest [17]. Grapes are being harvested with ever higher sugar levels, resulting in wines with higher alcohol levels [26,35]. In wine-growing regions around the world, the rise in temperature associated with climate change is responsible for earlier harvests. Determining the suitability of grape varieties in existing or new wine-growing regions is often based on temperature, without considering other factors. Sugar accumulation characteristics are also influenced by antecedent and concurrent climatic factors, such as the photosynthetically active radiation, temperature, and water status of the vine, regardless of whether this occurs before or after the mid-veraison [36]. Sugar is one of the most important metabolites in grape berries used for winemaking. Sugar concentration is a determining factor in the alcohol content of the finished wine, and its content during berry ripening is also involved in regulating the development of phenolic compounds that give the wine its color, flavor, and tannin structure [37,38].

Slovenia is a very small wine-growing region with different climatic zones (Mediterranean, continental, and Pannonian climates), where most of the world's important grape varieties for quality wines are grown. Quantified data on climate development and bioclimatic indices during the growing season of grapevines and their possible effects on the earlier ripening of grapes are presented and discussed.

2. Data and Methods

2.1. Study Area

The long-term data available from the two meteorological stations (Maribor and Murska Sobota) in the Podravje wine-growing region in Slovenia were used for this study (Figure 1). The Podravje wine-growing region lies between the river Sava (SW) and the Hungarian border (NE). Geologically, the area is part of the former Pannonian Sea Basin and has a Pannonian continental transitional climate. The continental climate characteristics increase with increasing distance from the Alps. The vineyards (6000 ha) are predominantly planted with white grape varieties, on steep slopes with inclinations of 30–50%, and at altitudes of 250 to 350 m. The long-term average (1952–2022) for precipitation during the growing season (1 April to31 October) varies between 574 mm/m^2 (Murska Sobota) and 701 mm/m^2 (Maribor), and precipitation is very unevenly distributed throughout the year [39].

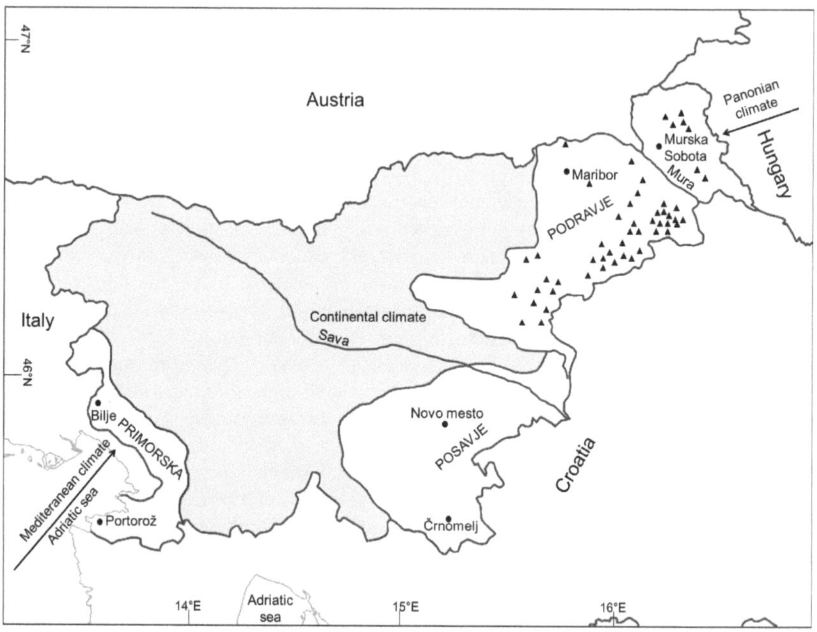

Figure 1. Map with study regions and climate stations (Maribor and Murska Sobota) and vineyard locations for the weekly monitoring of grapevine ripening (black triangles) in the Podravje wine-growing region in north-eastern Slovenia.

2.2. Climate Parameters and Grapevine Growing

Daily precipitation and temperature data (mean, maximum, and minimum) recorded at two meteorological stations (1952–2022) were used for the analysis. The data were taken from the archives of the Slovenian Environment Agency (SEA).

An analysis of the observed climate was carried out for the periods 1952–2022 and 1980–2022. The daily data from each station were divided into annual and vegetation periods and used to derive bioclimatic indices (Table 1). For the growing season (1 April to 31 October), precipitation and temperature data (average, minimum, and maximum) of each station were summarized, as growing season averages trend to correlate significantly with wine production and quality [27]. To assess the signs of heat stress, the number of days with temperatures above 30 °C was determined. This temperature leads to the premature ripening of the grapes (a shorter vegetation period), lower total acidity, and fewer aroma compounds [24].

Table 1. Analyzed bioclimatic parameters.

Parameter	Parameter Description
Tavg	Average annual temperature, °C
Tmax	Average annual maximum temperature, °C
Tmin	Average annual minimum temperature, °C
GSTavg	Average growing season temperature (1 April to 31 October), °C
GSTmax	Average growing season maximum temperature (1 April to31 October), °C
GSTmin	Average growing season minimum temperature (1 April to 31 October), °C
TMJ	Average temperature May–October, °C
HI	Huglin index (1 April to 30 September) °C units
GDD	Growing degree days °C units
NDT30	Number of days with maximum temperature > 30 °C
NDF	Number of days with a minimum temperature <0 °C (frost occurrence)
NDFF	Number of days between the last frost and the first frost (length of frost-free period)
AP	Total annual precipitation, mm/m^2
GSP	Total growing season precipitations (April to October), mm/m^2

To simplify the global description of the weather during the growing season, there are several climatic indices. These indices enable qualified estimates of the effects of climate change on the development of the grapevine and the characteristics of the wine.

To obtain more information about the wine region and to determine general guidelines for the potential quality and style of the wine, the GDDs (growing degree days) [40] and the Huglin index (HI) [41] were calculated by summing the daily average temperatures above a base value of 10 °C (the sum of the effective temperatures), with values below 10 °C set to zero.

GDDs were calculated for the period of 1 April to 31 October (Winkler index—WI). Navratilova et al. (2021) [42] used the "shortened Winkler index", which was recalculated for the period corresponding to the Huglin index. They believe that the original WI index is less indicative of the soluble solids content, according to their calculations. For the purposes of this article, GDDs were also calculated from 1 April to the harvest date (76°Oe; see Section 2.4), which better reflects the impact of warming on the earlier harvest date.

The Huglin index (HI) for the Northern Hemisphere is calculated using the following formula:

$$HI = \sum_{01.04}^{30.09} d \cdot \left[\frac{(Tavg - 10) + (Tmax - 10)}{2} \right],$$

where Tavg is the daily average air temperature (°C), Tmax is the daily maximum air temperature (°C), and d is the day length coefficient, which lies between 1.02 and 1.06 and between 40° and 50° north latitude. Base temperature = 10 °C. This index makes it possible to classify the wine-growing regions according to the sum of the temperatures required for the development of the vines and the ripening of the grapes.

Temperature extremes were calculated by the number of days with maximum temperatures >30 °C (NDT30) and average temperatures for the May–June period (TMJ). This parameter is important for predicting an increase in disease pressure (e.g., downy mildew), as more severe epidemics can be a direct result of more favorable temperature conditions in May and June [43].

2.3. Evaluated Varieties of *Vitis vinifera* L.

A long-term dataset (1980–2022) for several *Vitis vinifera* L. varieties has enabled a comprehensive assessment of varietal differences in terms of ripening time and relationships with climate and climate change in the Slovenian wine-growing region of Podravje. Data from the weekly monitoring of grape ripening in the period from 1980 to 2022 were collected and statistically analyzed for early-, medium-, and late-ripening grape varieties. Only in the Podravje wine-growing region were ripening data available for 15 varieties over such a long period. The data were collected at permanent locations in this region (Figure 1) and recorded by the Institute of Agriculture and Forestry in Maribor. Six varieties were selected as model grape varieties to assess the impact of climate change on vine ripening: 'Bouvier', the local early-ripening variety, 'Chardonnay' and 'Sauvignon Blanc', the two globally widespread varieties with medium–late ripening, 'Blaufränkisch', the local red variety, 'Welschriesling', the most widespread variety in Slovenia (especially in the Podravje wine-growing region), and 'Furmint', the late-ripening variety.

2.4. Monitoring of Grapevine Ripening

In the Podravje wine-growing region (Slovenia), the ripening of grapes has been monitored since 1980. For each variety, 100 berries were randomly sampled from about 25 plants (distance between plants was 1 m) from mid-véraison until technological maturity. The berries were always taken from the middle part of the bunch, with one from the inside and one from the outside of the bunch, so that there was a total of 50 berries on each side of the row. The degree of ripeness was determined according to the Slovenian wine law at the point when the total soluble solids had reached about 76°Oe (i.e., 76° on the Oechsle scale, a hydrometer scale that measures the density of grape must, which around 18°Brix; the limit for quality wine).

In poor vintages, the harvest date was determined according to the soluble solids content for quality wine or, in very poor vintages, according to the health of the grapes (vintages in the early 1980s). The focus was on the relationship between the bioclimatic and ripening parameters (total acidity) at the recommended harvest dates (76°Oe). Based on the data of soluble solids and total acidity in the grape juice and the recommended harvest date, the tendency towards an earlier ripening of the grapes and the correlations between the ripening parameters and the bioclimatic parameters were calculated for six varieties.

2.5. Data Evaluation and Statistical Analyses

The variables were evaluated using descriptive statistics and trend analysis. The average values of the parameters, the range of minimum and maximum values, and the processing of the linear trends of temperature, bioclimatic indices, and harvest date, as well as the linear correlations of temperatures, bioclimatic indices, and harvest date, were calculated. Since some of the parameters examined in the study were not normally distributed, a more stringent non-parametric Mann–Kendall trend test (MK test) with a significance level of 95% was applied to all series [44] implemented in R/Kendall [45]. The Mann–Kendall test, like other distribution-free or parametric tests, is very sensitive to the autocorrelation effect (persistence).

3. Result and Discussion

3.1. Climatic Structure and Temperatures Trends

The general climate for the period of 1952–2022 for the inland wine-growing region of Podravje in Slovenia was moderately continental, characterized by considerable seasonal temperature variations, cold winters, and moderately hot summers, with an average annual temperature of 10.3 °C (5.7 to 15.5 °C) for Maribor and 9.9 °C (4.8 to 15.3 °C) for Murska Sobota, with annual precipitation amounts of 998 mm/m^2 and 801 mm/m^2, respectively (Table 2). As far as the ripening potential of the grapes is concerned, the location was in the middle range (15.5 to 15.8 °C) (Table 2) [46]. The variability in temperature in the growing season (GSTavg, GSTmax, and GSTmin), the average temperature in the period of May to June (TMJ), the number of days with temperatures of <0 °C (NDF) and >30 °C (NDT30), and the frost-free period (NDFF) were similar at both stations (Table 2).

Table 2. Bioclimatic parameters for the two meteorological stations (Maribor and Murska Sobota) in the Slovenian wine-growing region of Podravje for the long-term period 1952–2022 and for the period 1980–2022, for which data on the ripening of grape are available. Figures in bold indicate significant trends ($p \leq 0.05$).

Station/Period	Maribor 1952–2022					Maribor 1980–2022				
	Variables					Variables				
Parameters	Mean	SD	Trend yr^{-1}	Tau	p	Mean	SD	Trend yr^{-1}	Tau	p
T_{avg}	10.3	0.99	**0.037**	0.628	**0.001**	10.8	0.82	**0.048**	0.524	**0.001**
T_{max}	15.5	1.14	**0.038**	0.542	**0.001**	16.0	1.09	**0.064**	0.561	**0.001**
T_{min}	5.7	1.06	**0.042**	0.66	**0.001**	6.4	0.71	**0.034**	0.420	**0.001**
GST_{avg}	15.8	0.99	**0.037**	0.589	**0.001**	16.4	0.80	**0.044**	0.482	**0.001**
GST_{max}	21.5	1.15	**0.038**	0.491	**0.001**	22.1	1.09	**0.063**	0.526	**0.001**
GST_{min}	10.6	1.02	**0.041**	0.615	**0.001**	11.3	0.66	**0.024**	0.300	**0.005**
TMJ	17.0	1.28	0.038	0.48	0.001	17.34	1.25	**0.054**	0.471	**0.001**
HI	1839	206	**7.03**	0.559	**0.001**	1947	187	**10.08**	0.491	**0.001**
GDD	1325	186	**6.88**	0.599	**0.001**	1432	154	**8.10**	0.480	**0.001**
NDT30	13.2	11.8	**0.57**	0.502	**0.001**	18	12.5	**0.68**	0.540	**0.001**
NDF	95	19	**−0.56**	−0.411	**0.001**	87	15.1	**−0.39**	−0.221	**0.040**
NDFF	206	22	0.53	0.340	0.310	214	20.8	0.30	0.118	0.272
AP	998	150	**−2.88**	−0.252	**0.002**	973	148	**−5.8**	−0.344	**0.001**
GSP	700	124	**−1.68**	−0.214	**0.008**	685	127	**−3.9**	−0.268	**0.012**
	Murska Sobota, 1952–2022					Murska Sobota, 1980–2022				
T_{avg}	9.9	0.99	**0.034**	0.565	**0.001**	10.4	0.9	**0.059**	0.601	**0.001**
T_{max}	15.3	1.16	**0.036**	0.520	**0.001**	15.8	1.1	**0.066**	0.528	**0.001**
T_{min}	4.8	1.05	**0.038**	0.599	**0.001**	5.4	0.9	**0.057**	0.566	**0.001**
GST_{avg}	15.5	1.0	**0.033**	0.511	**0.001**	16.0	0.9	**0.051**	0.530	**0.001**
GST_{max}	21.6	1.2	**0.038**	0.446	**0.001**	22.2	1.1	**0.058**	0.464	**0.001**
GST_{min}	9.7	1.0	**0.037**	0.576	**0.001**	10.3	0.8	**0.048**	0.585	**0.001**
TMJ	16.8	1.25	0.037	0.475	0.001	17.48	1.28	**0.058**	0.375	**0.001**
HI	1831	213	**6.43**	0.478	**0.001**	1933	199.2	**10.5**	0.455	**0.001**
GDD	1278	186	**6.21**	0.532	**0.001**	1373	166.5	**9.6**	0.530	**0.001**
NDT30	14.1	12.1	**0.37**	0.447	**0.001**	19.1	12.5	**0.55**	0.428	**0.001**
NDF	110	18	**−0.43**	−0.374	**0.050**	105	17.4	**−0.84**	−0.431	**0.001**
NDFF	188	18	0.41	0.359	0.111	194	15.9	**0.61**	0.362	**0.001**
AP	801	112	−0.17	0.021	0.800	801	111.0	−0.07	0.013	0.908
GSP	574	94	0.28	0.037	0.655	576	92.5	0.52	0.069	0.523

The values of growing degree days (GDDs) from 1 April to 31 October (Winkler index—WI) total 1278 to 1325 units, which places Podravje in Winkler region I (very cool—WI ≤ 1390), indicating a generally favorable climate for early-ripening grape varieties suitable for producing high-quality wines [40]. The average Huglin index (HI) values, which may be more appropriate than the WIs for European regions [47], date from 1831 to 1839 and place Podravje in Huglin's at the beginning of the temperate climate type (HI-1; 1800 < HI ≤ 2100), which is suitable for 'Pinot Noir', 'Traminer', 'Chardonnay', 'Riesling', 'Sauvignon Blanc', and 'Cabernet Franc', for example [41].

The values for growing season precipitation (GSP) generally indicate that precipitation amounts decreased slightly for the entire period (1952–2022), but the trends for the Murska Sobota site were not significant. The variability in the GSP over this long-term period shows a variation of ~16% between years at both locations. The Murska Sobota site, with a total GSP amount of 574 mm, was drier (influenced by the Pannonian climate) than Maribor with 700 mm (Table 3). However, drier conditions with more frequent and longer dry periods were more likely at both locations, as higher temperatures probably led to a higher evapotranspiration rates, as noted by Ramos et al. (2008) [10]. Precipitation becomes more intense with more intense dry spells. During the growing season, precipitation reaches ~70% of its annual amount.

Table 3. Mean values and trends of total acidity in g/L and day in the year when the soluble solids content in grapes of 15 varieties was 76°Oe in the period 1980–2022 in the Podravje wine-growing region in Slovenia.

Parameters	Total Acidity g/L					Day in Year				
Variety/Variable	Mean	± SD	Trend yr^{-1}	Tau	p	Mean	± SD	Trend yr^{-1}	Tau	p
'Bouvier'	7.5	±1.10	−0.040	−0.273	**0.010**	247	±16.3	−0.99	−0.498	**0.001**
'Müller Thurgau'	7.1	±1.16	−0.056	−0.421	**0.001**	258	±13.0	−0.65	−0.415	**0.001**
'Muscat Ottonel' *	6.0	±0.88	−0.035	−0.304	**0.014**	256	±13.5	−0.79	−0.364	**0.003**
'Pinot Blanc' **	8.9	±1.23	−0.036	−0.192	0.109	255	±14.5	−0.83	−0.414	**0.001**
'Chardonnay'	10.2	±1.47	−0.072	−0.377	**0.001**	256	±15.5	−0.85	−0.482	**0.001**
'Pinot Gris'	9.2	±1.57	−0.084	−0.458	**0.001**	253	±14.8	−0.76	−0.451	**0.001**
'Sylvaner' ***	8.6	±1.31	−0.029	−0.162	0.145	263	±16.7	−0.98	−0.490	**0.001**
'Sauvignon Blanc'	10.3	±1.42	−0.053	−0.198	0.062	258	±16.2	−0.82	−0.455	**0.001**
'Traminer'	8.2	±1.09	−0.035	−0.245	**0.021**	254	±15.7	−0.79	−0.469	**0.001**
'Yellow Muscat'	8.3	±1.69	−0.095	−0.464	**0.001**	264	±17.5	−0.96	−0.502	**0.001**
'Kerner' ****	9.0	±1.32	−0.046	−0.180	0.138	255	±12.9	−0.76	−0.394	**0.001**
'Blaufränkisch'	9.4	±1.08	−0.019	−0.057	0.601	266	±14.9	−0.83	−0.488	**0.001**
'Welschriesling'	8.7	±1.39	−0.077	−0.508	**0.001**	271	±18.2	−1.05	−0.527	**0.001**
'Riesling'	11.0	±1.76	−0.068	−0.270	**0.011**	271	±15.1	−0.85	−0.496	**0.001**
'Furmint'	10.5	±1.88	−0.088	−0.412	**0.001**	279	±17.9	−1.06	−0.514	**0.001**

Data available after * 1990, ** 1988, *** 1982, and **** 1989; bold numbers indicate significant trends.

The trends in the increase in the average temperature of the growing season (GSTavg) for the period 1952–2022 were 0.37 °C (Maribor) and 0.33 (Murska Sobota), and for the period 1980–2022, 0.44 to 0.51 °C per decade. The average temperature in the period of May to June (TMJ) shows the same significant trends for both locations ($p = 0.001$), namely, the TMJ increased more than 0.37 °C per decade in the whole study period and more than 0.54 °C per decade for the period of 1980–2022 (Table 2). These more favorable temperature conditions in May and June may lead to higher disease pressure (earlier powdery mildew infections). There were even more pronounced trends for NDT30, namely, 5.7 and 3.7 days and 6.8 and 5.5 days per decade, respectively (Table 2). An increased number of days with daily maximum temperatures of 30 °C is critical for optimal vine development and can lead to a reduction in photosynthesis, greater water deficiency, the premature ripening of grapes, and the drying of berries in early varieties, especially in the early 'Bouvier' variety in Slovenia [24]. Further, some days with temperatures above 30 °C during the ripening period could have been beneficial [9,11], especially for late-ripening varieties [24], such as 'Riesling' and 'Furmint', in this region.

After 2010, the average values of NDT30 increased three- to fourfold compared to the first decade after the 1980s (Figure 2). This was reflected in the trends of heat accumulation indices (HI and GDDs), which increased by about 6 (1952–2022) and even by about 10 units per year after 1980 (Table 2). Their values increased on average by around 100 units per decade (Figure 2).

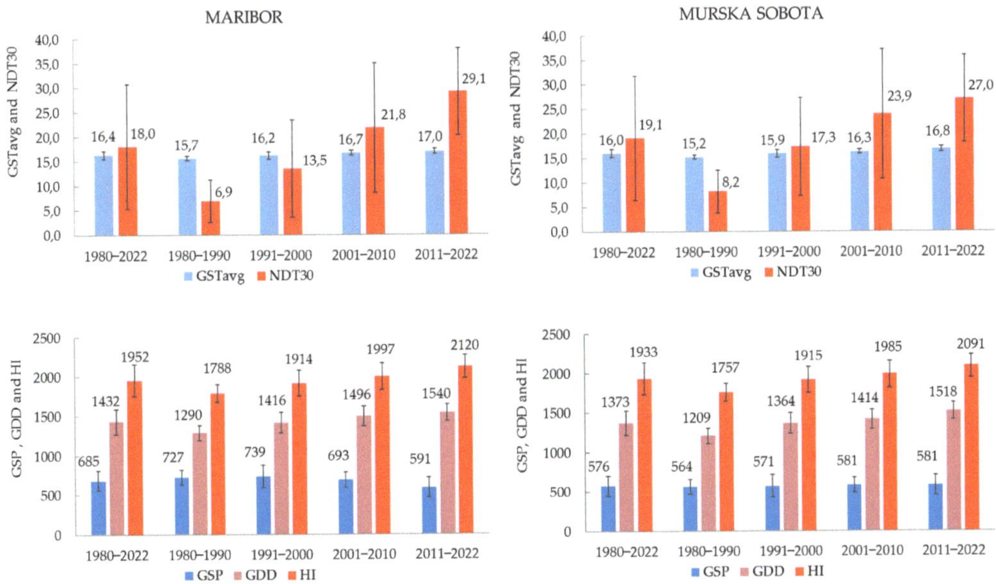

Figure 2. Average values of bioclimatic parameters for the meteorological stations Maribor and Murska Sobota (wine-growing Podravje region) for the period 1980–2022 and for individual decades in the period 1980–2022.

The number of days with temperatures <0 °C (NDF) showed a decreasing trend, and the number of days between the last spring frost and the first fall frost (NDFF) increased (Table 2). In the period of 1980–2022, the trends in the average temperatures of the growing season (GSTavg) show a warming of 1.9 °C (Maribor) to 2.2 °C (Murska Sobota). Similar results were found in other European wine-growing regions, where growing seasons warmed by 1.7 °C and heat accumulation increased by 250–300 units in the 30–50 years. In Spain, heat accumulation (WI and HI) increased in the inland wine-growing regions, but not in the coastal regions [27].

Based on the categorization of wine-growing regions into climatic groups [48] and the increase in GSTavg in the last decade of the observation period, we can conclude that this wine-growing region is becoming suitable for the cultivation of some wine varieties from the warm-climate-ripening group, such as 'Cabernet Franc' and 'Merlot'. If the warming trend continues in the next 30 years in a similar way to how it has since the 1990s (Figure 2), we can assume that the Podravje wine-growing region will completely transition to the warm-climate grape variety group [49]. Similar trends in bioclimatic parameters can also be observed for the period of 1980–2022 (Table 2), although some of them are more pronounced. In the Podravje wine-growing region, data on grape ripening are available for this period.

Precipitation trends were significant only for the Maribor site, i.e., for the period of 1952–2022, total annual precipitation (AP), and growing season precipitation (GSP) decreased by −2.9 and −1.7 mm/m^2 per year, respectively. After 1980, this downward trend was even more pronounced, with −5.8 L per year for AP and −3.9 L per year for GSP. After 2010, the total amount of GSP at the Maribor site decreased by 136 L and matched the amount at the Murska Sobota site (Figure 2), where the amount was stable throughout the study period. This indicates that the precipitation pattern of the Pannonian climate extends from east to west into the interior of the region. The seasonal precipitation amounts changed significantly. This could not be confirmed before 2010 [24], which was also pointed out by Tomasi et al. (2011) [49].

3.2. Grapevine Reactions to Climate Changes in the Podravje Wine-Growing Region

In the period of 1980–2022, all varieties show a trend towards earlier grape ripening (total soluble solids contents reached about 76°Oe) by slightly less than one day per year, with the exception of the varieties 'Welschriesling' and 'Furmint', where the trend is slightly more than one day per year (Table 3). The more pronounced trend for 'Welschriesling' in this region is not only due to climate change. In the last 15 to 20 years, this variety has been planted in sun-exposed sites, where it has mainly displaced the aromatic varieties. 'Sauvignon Blanc', for example, began to retreat from the sunniest locations to less sunny ones, mainly in order to preserve its primary aromas.

In the period of 1980–2022, trends towards an earlier harvest time of 43 days ('Bouvier'), 37 days ('Chardonnay'), 35 days ('Sauvignon Blanc'), 36 days ('Blaufränkisch'), 45 days ('Welschriesling'), and 46 days ('Furmint') were observed, i.e., by 8–10 days per decade. Earlier grape ripening has been observed in many wine-growing regions [10,23,26]. In South Australia, ripening has advanced by 8 days per decade [25,26], while other values are estimated at 0.5 to 3 days per year [50], but most studies were conducted before 2010 and for a shorter period. After 2010, the harvest date for six varieties is on average one month earlier than in the period of 1980–1990. Taking into account the increase in GSTavg (Maribor station; 1.9 °C) in the studied period of 1980–2022, the combined trends of harvest date and climate result in an average shift of 18 ('Sauvignon Blanc') to 24 days ('Furmint') per 1.0 °C warming. Ramos et al. (2008) [10], found an earlier harvest for the 'Chardonnay' grape variety by about 5 days per 1 °C of warming in the growing season for the shorter period studied (1997–2006).

In the period of 1980–2022, the total acidity per decade decreased from 0.19 g/L for 'Blaufränkisch' to 0.95 g/L for 'Gelber Muskateller', while no trends were discernible for 'Pinot Blanc', 'Sylvaner', 'Sauvignon Blanc', 'Kerner', or 'Blaufränkisch' (Table 3). For the six varieties analyzed in more detail ('Bouvier', 'Chardonnay', 'Sauvignon 'Blanc', 'Blaufränkisch', 'Welschriesling', and 'Furmint'), the decreasing trends in total acidity were significant for 'Chardonnay' (R^2 = 0.374), 'Welschriesling' (R^2 = 0.477), and 'Furmint' (R^2 = 0.336), while the trends for earlier grape ripening were significant for all six varieties, from R^2 = 0.395 in 'Sauvignon Blanc' to R^2 = 0.566 for 'Bouvier' (Figure 3). The total acidity in grape juice decreased on average from 0.43 g/L ('Blaufränkisch') to 1.99 g/L ('Furmint') per 1 °C of warming. Figure 4 shows various bioclimatic parameters and the total acidity for six varieties (early-, medium-, and late-ripening varieties) for the first and last decades in the period of 1980–2022.

Even though the grapes are harvested at least one month earlier than in the period of 1980–1990, the total acidity content has fallen sharply as a result of the higher air temperatures. This has so far proved to be positive for late-ripening varieties ('Furmint') in particular, while it is negative for early varieties ('Bouvier'). Early-ripening varieties are often subjected to greater dehydration and so-called forced ripening during the ripening period. As a result, undesirable astringency notes can occur later in the wine tasting. In cold climate regions such as Podravje, total acidity reduction can lead to a better balance between sugar and acidity, while in early-ripening varieties, acid reduction without acid correction in the cellar (as in warmer regions) can lead to less fruitiness in the wine. A warmer growing season usually results in an earlier harvest and a lower yield (possibly also due to spoilage, as was the case in 2022) as well as better wine quality if there was no excessive heat stress [5,10,25].

From the GSTavg values for each variety, it can be concluded that the grapes developed and ripened at a higher average temperature in the growing season after 2010, as shown by the GSTavg values for the meteorological growing season (1 April to 31 October) (Table 4). At the beginning of the observation period (1980–1990), the GSTavg was lower than in the last decade (2011–2022), as the grapes were ripe at the end of September or even in mid-

October. After 2011, the average temperature in the vegetation period from 1 April to the ripening of the grapes is about 1 °C higher than the official GSTavg for the meteorological period (1 April to 31 October). For the period of 2011–2022, for example, the GSTavg is 17.0 °C (Table 4), but all varieties ripened at an average temperature of over 18 °C. This was mainly influenced by the so-called hot days. In the period of 1980–1990, there were 6 to 7 such days per year, and after 2010 there were 25 to 29 per year (Figure 4).

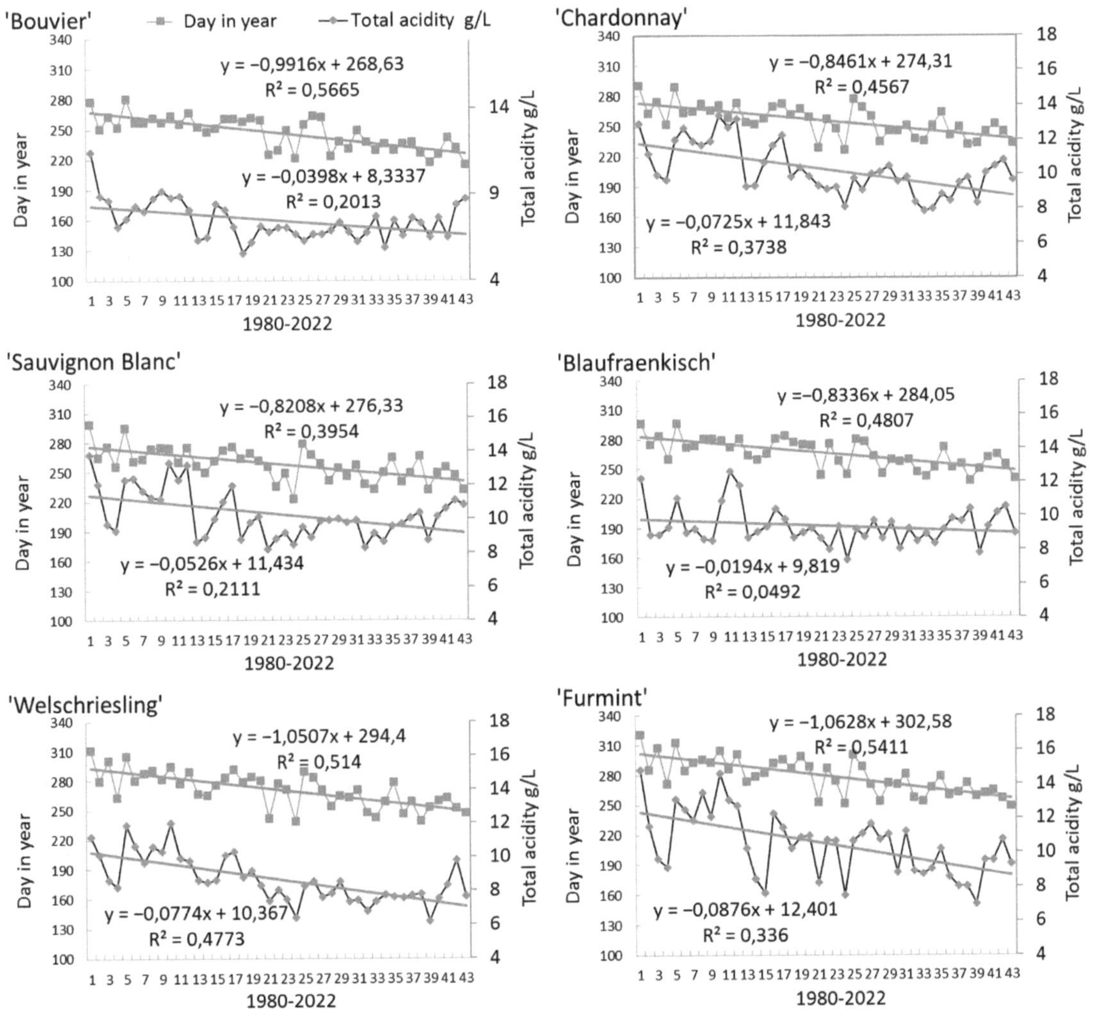

Figure 3. Trends of total acidity (g/L) and the day of the year when the soluble solids content was 76°Oe for six varieties ('Bouvier', 'Chardonnay', 'Sauvignon Blanc', 'Blaufränkisch', 'Welschriesling', and 'Furmint') in the period of 1980–2022 in the Podravje wine-growing region in Slovenia.

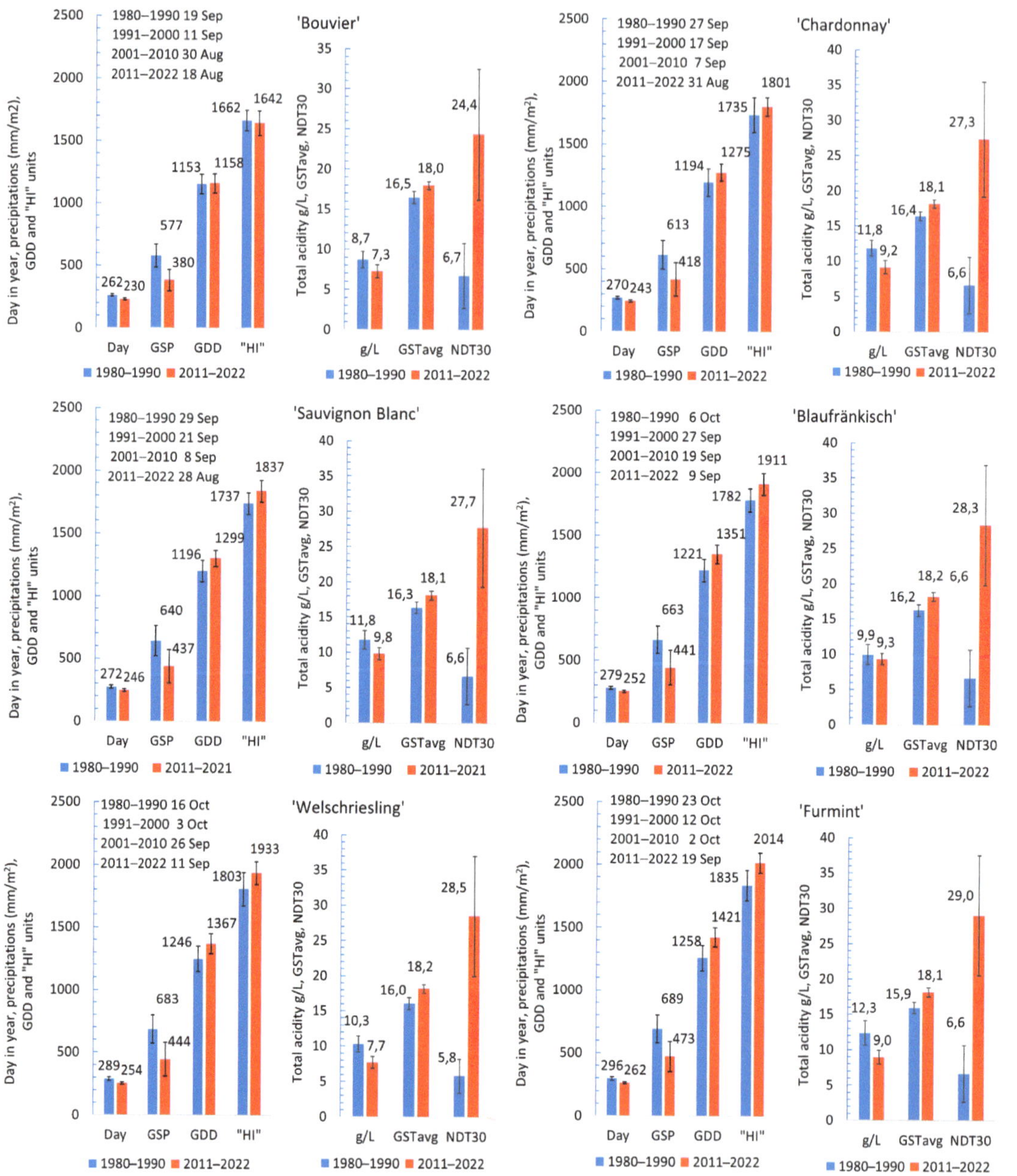

Figure 4. Mean values of various bioclimatic parameters (growing season temperature—GSTavg, total growing season precipitation—GSP, number of days with maximum temperature > 30°—NDT30, growing degree days—GDDs, and Huglin index—HI) and total acidity contents in g/L and day of the year on which these values were reached (soluble solids 76°Oe) in the first and last decades, with the harvest date in the individual decade in the period of 1980–2022.

Table 4. Mean temperature (GSTavg) and total precipitation (GSP) in the growing season (1 April to 31 October) for the Maribor meteorological station and from 1 April to the day in the year with technological grape ripeness (76°Oe) and total acidity g/L content for six varieties in the Podravje wine-growing region in Slovenia in the period from 1980 to 2022, and by decades in this period ($p \leq 0.05$).

Location-Variety	Variables/Period	GSTavg ±SD		GSP ±SD		Total Acid. g/L ±SD		Day in yr ±SD	
Maribor	1980–2022	16.4	±0.80	685	±127	1 April to 31 October			
	1980–1990	15.7	±0.56	727	±90				
	1991–2000	16.2	±0.75	739	±142				
	2001–2010	16.7	±0.48	693	±97				
	2011–2022	17.0	±0.57	591	±129				
'Bouvier' 1 April to 76°Oe	1980–2022	17.4	±0.89	499	±150	7.5	±1.1	247	±16.3
	1980–1990	16.5	±0.73	577	±91	8.7	±1.1	262	±9.1
	1991–2000	17.3	±0.66	567	±170	7.0	±0.9	255	±11.2
	2001–2010	18.0	±0.65	495	±148	6.8	±0.3	243	±15.0
	2011–2022	18.0	±0.48	375	±83	7.3	±0.8	230	±8.0
'Chardonnay' 1 April to 76°Oe	1980–2022	17.5	±0.98	534	±163	10.2	±1.5	256	±15.5
	1980–1990	16.4	±0.60	613	±116	11.8	±1.1	270	±11.2
	1991–2000	17.3	±0.75	591	±180	10.6	±1.3	260	±12.4
	2001–2010	18.0	±0.78	529	±148	9.5	±0.6	251	±14.9
	2011–2022	18.1	±0.60	418	±133	9.2	±1.0	243	±9.3
'Sauvignon Blanc' 1 April to 76°Oe	1980–2022	17.4	±1.04	547	±164	10.3	±1.4	258	±16.2
	1980–1990	16.3	±0.83	640	±118	11.8	±1.3	272	±13.3
	1991–2000	17.3	±0.79	603	±175	10.1	±1.5	264	±8.4
	2001–2010	18.0	±0.76	520	±145	9.5	±0.6	251	±14.5
	2011–2022	18.1	±0.61	437	±132	9.8	±0.9	246	±11.4
'Blaufrankisch' 1 April to 76°Oe	1980–2022	17.4	±1.12	570	±171	9.4	±1.1	266	±14.9
	1980–1990	16.2	±0.79	663	±108	9.9	±1.4	279	±10.3
	1991–2000	17.2	±0.90	614	±193	9.5	±0.9	270	±11.7
	2001–2010	17.9	±0.92	580	±156	8.8	±0.7	262	±12.4
	2011–2022	18.2	±0.63	441	±138	9.3	±0.8	252	±9.8
'Welschriesling' 1 April to 76°Oe	1980–2022	17.3	±1.17	587	±179	8.7	±1.4	271	±18.2
	1980–1990	16.0	±0.84	683	±112	10.3	±1.1	289	±13.2
	1991–2000	17.1	±0.92	630	±206	9.0	±0.8	276	±14.0
	2001–2010	17.8	±0.96	610	±166	7.8	±0.6	269	±15.1
	2011–2022	18.2	±0.62	444	±134	7.7	±0.8	254	±10.2
'Furmint' 1 April to 76°Oe	1980–2022	17.2	±1.20	608	±169	10.5	±1.9	279	±17.9
	1980–1990	15.9	±0.80	689	±110	12.3	±1.7	296	±13.6
	1991–2000	16.9	±0.98	665	±200	10.3	±1.6	285	±13.6
	2001–2010	17.6	±0.98	623	±151	10.4	±1.3	275	±15.0
	2011–2022	18.1	±0.64	473	±119	9.0	±1.0	262	±7.8

Precipitation during the growing season (1 April–31 October) at the Maribor site decreased from 727 mm/m^2 in the period of 1980–1990 to 591 mm/m^2 in the period of 2011–2022 (Figure 4), but the actual amount of precipitation during the growing season was lower for each variety (Table 4). From 1 April to the day of ripening of the individual varieties (76°Oe), the amount of precipitation was even lower, namely by an additional 202, 195, 203, 222, 239, and 216 mm/m^2 for the six varieties 'Bouvier', 'Chardonnay', 'Sauvignon Blanc', 'Blaufränkisch', 'Welschriesling', and 'Furmint', which corresponds to 135, 115, 113, 111, 109, and 98 mm/m^2 per 1 °C increase in temperature during the growing season, respectively (Table 4). For example, after 2010, the available precipitation in the growing season (1 April to harvest) was only 375 mm/m^2 for the early variety 'Bouvier' and 473 mm/m^2 for the late-ripening variety 'Furmint' (Table 4). Authors who studied climate change two decades ago found that precipitation would not decrease significantly in most cases [49]. Today, these estimates are too general, as they refer to the amount of precipitation in the growing season (1 April to 31 October) and do not reflect the actual amount of precipitation available to the individual varieties. In the Podravje wine-growing region, the amount of precipitation from 1 April to the technological ripeness of each variety was one-third less (Table 4) than in the period from 1 April to 31 October (which is mainly used in climate change research).

In the Podravje wine-growing region, the established method of soil care is the permanent maintenance of green cover on the soil in the vineyards, which has so far proven to be the most suitable method of soil care, both in terms of soil life [51] and erosion, as most of the vineyards are located on steep slopes [31]. Given climate change and trends in precipitation patterns, this method of soil management may be less suitable in the future.

When describing the positive effects of climate change on wine quality, particularly in the case of late-ripening grape varieties, it should not be overlooked that the development of viticulture and winemaking practices and the reduction in yields over the last 43 years have also had a significant influence on the improvement in wine quality. A one-sided assessment of the effects of climate change on wine quality is therefore inadmissible. However, should the trend of regional warming continue, as predicted by climate models [46], and should it continue in this region with the same dynamics as in the last 43 years, the Podravje wine-growing region will most likely experience poorer vintages, mainly due to lower total acidity, very high alcohol content, and other less desirable characteristics of the wine, possibly resulting in less balanced wines. A serious question arises as to whether it will be possible to maintain the same varieties in the future by adapting viticultural and oenological practices, as reported by Seguin and Garcia de Cortazar (2005) [22]. Due to the climatic changes observed in the wine-growing region that is the subject of this research, it will be necessary to exceed certain limits for the grape varieties that are currently defined for this region ('terroir').

For the available data (1980–2022), the relationships between the GDDs, the average temperature in the growing season (GSTavg), and the HI with the total acidity were calculated. The linear relationships between GDDs, GSTavg, and HI and total acidity are shown as individual variables in Figure 5. Total acidity showed high negative correlations with all climatic parameters for all six varieties except 'Bouvier'. The most pronounced correlation was found between GDDs and total acidity (Figure 5).

The recalculated heat sum values of GDDs* (1 April to harvest date) and HI* (1 April to harvest date) for the individual varieties remain in the same range according to the Winkler (1974) [40] and Huglin (1978) [41] ripeness group classifications (Table 5), but the trends after 2010 show that these values are reached 26 to 34 days earlier than in the 1980s.

Table 5. Mean values of GDDs* and HI* from 1 April to the day in the year with technological grape maturity (76°Oe) for six varieties ('Bouvier', 'Chardonnay', 'Sauvignon Blanc', 'Blaufränkisch', 'Welschriesling', and 'Furmint') in the Podravje wine-growing region in Slovenia in the period from 1980 to 2022 ($p \leq 0.05$).

Variety/Parameters	GDD*	±SD	HI*	±SD
'Bouvier'	1191	±87.4	1683	±106.0
'Chardonnay'	1259	±84.8	1782	±96.9
'Sauvignon Blanc'	1275	±85.7	1804	±94.3
'Blaufränkisch'	1320	±99.4	1873	±103.0
'Welschriesling'	1341	±99.3	1902	±114.8
'Furmint'	1371	±110.3	1954	±122.0

GDDs* and HI* recalculated for the period from 1 April to technological maturity (76°Oe).

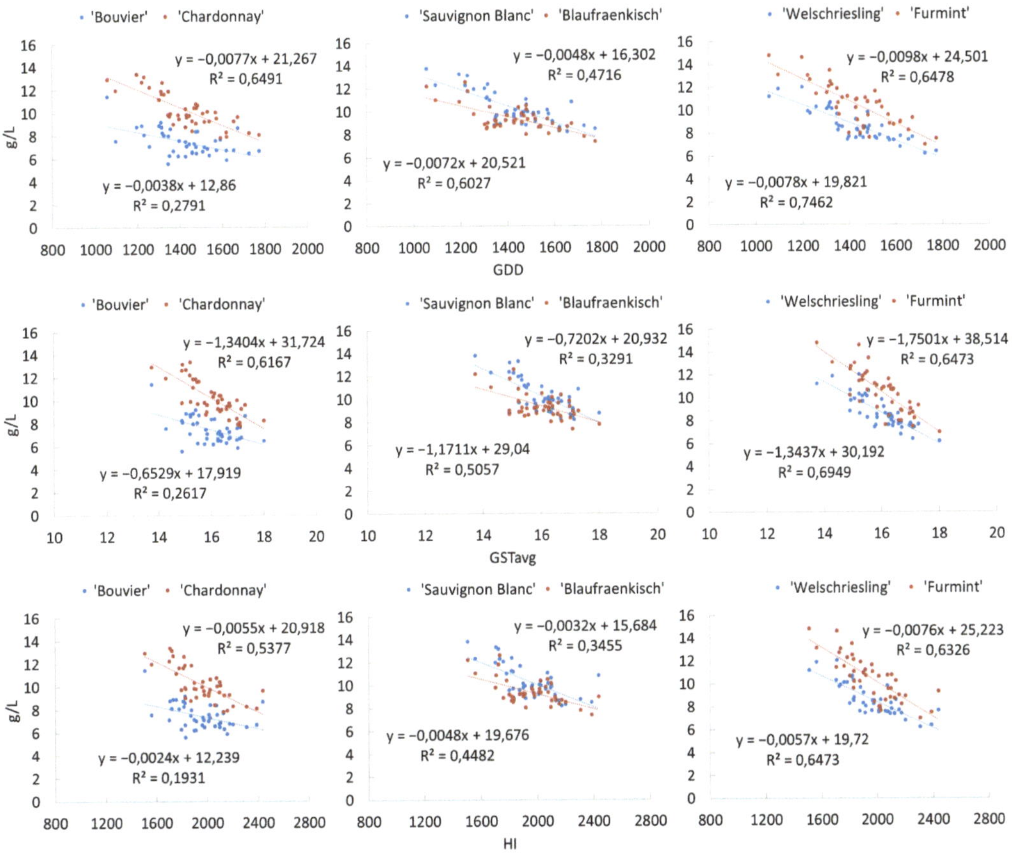

Figure 5. Correlation between GSTavg and GDDs (1 April to 31 October) and HI (1 April to 30 September) with total acidity (g/L) for six varieties ('Bouvier', 'Chardonnay', 'Sauvignon Blanc', 'Blaufränkisch', 'Welschriesling', and 'Furmint') in the Podravje wine region in Slovenia in the period of 1980–2022.

4. Conclusions

Global warming affects the growth of the vines and the composition of the berries, i.e., the onset of the phenological phase of ripening occurs earlier. When analyzing the trends of the time series of up to 43 years observed in the Podravje wine-growing region, we found that the average temperature in the growing season has increased by 1.6 °C since the 1980s. The increase in average temperature was most pronounced in the May–June period (TMJ), which can lead to an earlier and more intense occurrence of diseases, especially downy mildew. Climate warming has caused grapes to ripen about a month earlier than in the 1980s. The earlier ripening of the grapes is generally accelerated by the increased accumulation of sugar in the berries, which leads to a higher alcohol content in the wine. Due to the higher temperature during grape ripening, a downward trend in total acids was also observed. This was confirmed by correlations between the climate indices (GSTavg, GDDs, and HI) and total acidity. The current climate change in this wine-growing region has had a positive effect on the late-ripening variety 'Furmint', leading to a lower (more balanced) total acidity.

Author Contributions: Conceptualization, S.V.; methodology, S.V. and B.P.; software, A.P.; validation S.V. and A.P.; formal analysis, S.V. and A.P.; investigation, S.V.; resources, S.V.; data curation, S.V., B.P., A.P. and T.V.-P.; writing—original draft preparation, S.V.; writing—review and editing, S.V.; visualization, S.V.; supervision, S.V. and B.P.; project administration, S.V.; funding acquisition, S.V. All authors have read and agreed to the published version of the manuscript.

Funding: This research was funded by the Slovenian national research agency and the Ministry of Agriculture Forestry and Food of the Republic of Slovenia (grant no. CRP V4-2202) via the project "Guidelines for adapting grape and wine production to climate change and market demands".

Data Availability Statement: The raw data supporting the conclusions of this article will be made available by the authors on request.

Acknowledgments: We thank to Slovenian Environment Agency for the availability of climate data.

Conflicts of Interest: The authors declare no conflicts of interest. The funders had no role in the design of the study; in the collection, analyses, or interpretation of data; in the writing of the manuscript; or in the decision to publish the results.

References

1. Jones, P.D.; Conway, D. Precipitation in the British Isles: An analysis of area-average data updated to 1995. *Int. J. Climatol.* **1997**, *17*, 427–438. [CrossRef]
2. Salinger, M.J. Climate variability and change: Past, present and future—An overview. *Clim. Chang.* **2005**, *70*, 9–29. [CrossRef]
3. Kenny, G.J.; Harrison, P.A.; Parry, M.L.; Porter, J.R. The effects of climate change on agriculture and horticulture in Europe. *Eur. J. Agron.* **1993**, *2*, 243–246. [CrossRef]
4. Chloupek, O.; Hrstkova, P.; Schweigert, P. Yield and its stability, crop diversity, adaptability and response to climate change, weather and fertilisation over 75 years in the Czech Republic in comparison to some European countries. *Field Crops Res.* **2004**, *85*, 167–190. [CrossRef]
5. Maracchi, G.; Sirotenko, O.; Bindi, M. Impacts of present and future climate variability on agriculture and forestry in the temperate regions: Europe. *Clim. Chang.* **2005**, *70*, 117–135. [CrossRef]
6. Jones, G.V. Climate change in the western United States grape growing regions. *Acta Hortic.* **2005**, *689*, 41–60. [CrossRef]
7. Lobell, D.B.; Field, C.B.; Cahill, K.N.; Bonfils, C. Impacts of future climate change on Californian perennial crop yields: Model projections with climate and crop uncertainties. *Agric. For. Meteorol.* **2006**, *141*, 208–218. [CrossRef]
8. Kenny, G.J.; Harrison, P.A. The effects of climate variability and change on grape suitability in Europe. *J. Wine Res.* **1992**, *3*, 163–183. [CrossRef]
9. Jones, G.V.; Davis, R.E. Climate Influences on Grapevine Phenology, Grape Composition, and Wine Production and Quality for Bordeaux, France. *Am. J. Enol. Vitic.* **2000**, *51*, 249–261. [CrossRef]
10. Ramos, M.C.; Jones, G.V.; Martínez-Casasnovas, J.A. Structure and trends in climate parameters affecting winegrape production in northeast Spain. *Clim. Res.* **2008**, *38*, 1–15. [CrossRef]
11. White, M.A.; Diffenbaugh, N.S.; Jones, G.V.; Pal, J.S.; Giorgi, F. Extreme heat reduces and shifts United States premium wine production in the 21st century. *Proc. Natl. Acad. Sci. USA* **2006**, *103*, 11217–11222. [CrossRef] [PubMed]
12. Martínez-Casasnovas, J.A.; Ramos, M.C.; Ribes-Dasi, M. Soil erosion caused by extreme rainfall events: Mapping and quantification in agricultural plots from very detailed digital elevation models. *Geoderma* **2002**, *105*, 125–140. [CrossRef]
13. Michael, A.; Schmidt, J.; Enke, W.; Deutschländer, T.H.; Malitz, G. Impact of expected increase in precipitation intensities on soil loss—Results of comparative model simulations. *Catena* **2005**, *61*, 155–164. [CrossRef]
14. Vršič, S.; Ivančič, A.; Pulko, B.; Valdhuber, J. Effect of soil management systems on erosion and nutrition loss in vineyards on steep slopes. *J. Environ. Biol.* **2011**, *32*, 289–294. [PubMed]
15. Mosedale, J.R.; Wilson, R.J.; Maclean, I.M.D. Climate change and crop exposure to adverse weather: Changes to frost risk and grapevine flowering conditions. *PLoS ONE* **2015**, *10*, e0141218. [CrossRef] [PubMed]
16. Venios, X.; Korkas, E.; Nisiotou, A.; Banilas, G. Grapevine Responses to Heat Stress and Global Warming. *Plants* **2020**, *9*, 1754. [CrossRef] [PubMed]
17. Fraga, H.; Malheiro, A.C.; Moutinho-Pereira, J.; Santos, J.A. An overview of climate change impacts on European viticulture. *Food Energy Secur.* **2013**, *1*, 94–110. [CrossRef]
18. Palliotti, A.; Tombesi, S.; Silvestroni, O.; Lanari, V.; Gatti, M.; Poni, S. Changes in vineyard establishment and canopy management urged by earlier climate-related grape ripening: A review. *Sci. Hortic.* **2014**, *178*, 43–54. [CrossRef]
19. Jones, G.V.; Duchêne, E.; Tomasi, D.; Yuste, J.; Braslavska, O.; Schultz, H.; Martinez, C.; Boso, S.; Langellier, F.; Perucho, C.; et al. Changes in European winegrape phenology and relationships with climate. In Proceedings of the GESCO 2005, Geisenheim, Germany, 23–27 August 2005; pp. 55–61.

20. Kast, W.K.; Rupp, D. Effects of climate change on phenology and ripening conditions of grapevine. *Mitt. Klosterneubg.* **2009**, *59*, 3–7.
21. Cook, B.; Wolkovich, E. Climate change decouples drought from early wine grape harvests in France. *Nat. Clim. Chang.* **2016**, *6*, 715–719. [CrossRef]
22. Seguin, B.; de Cortazar, I.G. Climate warming: Consequences for viticulture and the notion of 'terroirs' in Europe. *Acta Hortic.* **2005**, *689*, 61–69. [CrossRef]
23. Webb, L.B.; Whetton, P.H.; Barlow, E.W.R. Observed trends in winegrape maturity in Australia. *Glob. Chang. Biol.* **2011**, *17*, 2707–2719. [CrossRef]
24. Vršič, S.; Vodovnik-Plevnik, T. Reactions of vines varieties to climate changes in NE Slovenia. *Plant Soil Environ.* **2012**, *58*, 34–41.
25. Webb, L.B.; Whetton, P.H.; Bhend, J.; Darbyshire, R.; Briggs, P.R.; Barlow, E.W.R. Earlier wine-grape ripening driven by climatic warming and drying and management practices. *Nat. Clim. Chang.* **2012**, *2*, 259–264. [CrossRef]
26. Duchêne, E.; Schneider, C. Grapevine and climatic changes: A glance at the situation in Alsace. *Agron. Sustain. Dev.* **2005**, *25*, 93–99. [CrossRef]
27. Jones, G.V.; White, M.A.; Cooper, O.R.; Storchmann, K. Climate change and global wine quality. *Clim. Chang.* **2005**, *73*, 319–343. [CrossRef]
28. Bernáth, S.; Paulen, O.; Šiška, B.; Kusá, Z.; Tóth, F. Influence of Climate Warming on Grapevine (*Vitis vinifera* L.) Phenology in Conditions of Central Europe (Slovakia). *Plants* **2021**, *10*, 1020. [CrossRef]
29. Huglin, P. *Biologie et Écologie de La Vigne*; Lavoisier: Paris, France, 1986; p. 372.
30. Coombe, B.G. Influence of temperature on composition and quality of grapes. *Acta Hortic.* **1987**, *206*, 23–35. [CrossRef]
31. Stanchi, S.; Zecca, O.; Hudek, C.; Pintaldi, E.; Viglietti, D.; D'Amico, M.E.; Colombo, N.; Goslino, D.; Letey, M.; Freppaz, M. Effect of Soil Management on Erosion in Mountain Vineyards (N-W Italy). *Sustainability* **2021**, *13*, 1991. [CrossRef]
32. Wolkovich, E.M.; García de Cortázar-Atauri, I.; Morales-Castilla, I.; Nicholas, K.A.; Lacombe, T. From Pinot to Xinomavro in the world's future wine-growing regions. *Nat. Clim. Chang.* **2018**, *8*, 29–37. [CrossRef]
33. van Leeuwen, C.; Seguin, G. The concept of terroir in viticulture. *J. Wine Res.* **2006**, *17*, 1–10. [CrossRef]
34. Töpfer, R.; Trapp, O. A cool climate perspective on grapevine breeding: Climate change and sustainability are driving forces for changing varieties in a traditional market. *Theor. Appl. Gen.* **2022**, *135*, 3947–3960. [CrossRef] [PubMed]
35. Godden, P.; Wilkes, E.; Johnson, D. Trends in the composition of Australian wine 1984–2014: Composition of Australian wine 1984–2014. *Aust. J. Grape Wine Res.* **2015**, *21*, 741–753. [CrossRef]
36. Suter, B.; Destrac Irvine, A.; Gowdy, M.; Dai, Z.; van Leeuwen, C. Adapting Wine Grape Ripening to Global Change Requires a Multi-Trait Approach. *Front. Plant Sci.* **2021**, *12*, 624867. [CrossRef] [PubMed]
37. Conde, C.; Silva, P.; Fontes, N.; Dias, A.C.P.; Tavares, R.M.; Souza, M.J.; Agasse, A.; Delrot, S.; Geros, H. Biochemical changes throughout grape berry development and fruit and wine quality. *Food* **2007**, *1*, 1–22.
38. Brandt, M.; Scheidweiler, M.; Rauhut, D.; Patz, C.D.; Will, F.; Zorn, H.; Stoll, M. The influence of temperature and solar radiation on phenols in berry skin and maturity parameters of *Vitis vinifera* L. cv. Riesling. *Oeno One* **2019**, *53*. [CrossRef]
39. Vršič, S.; Pulko, B.; Perko, A. Climate change trends in the wine-growing regions of Slovenia. In *Proceedings of 6th Slovenian Viticulture and Wine Congress, Ptuj, Slovenia, 21–22 April 2023*; Vršič, S., Ed.; Agricultural Forestry Institute Ptuj: Ptuj, Slovenia, 2023; pp. 97–111.
40. Winkler, A.J.; Cook, J.A.; Kliewer, W.M.; Lider, L.A. *General Viticulture*; University of California Press: Berkeley, CA, USA, 1974.
41. Huglin, M.P. Nouveau mode d'évaluation des possibilités héliothermiques d'un milieu viticole. *Comptes Rendus Séances L'acad. D'agric. De. Fr.* **1978**, *64*, 1117–1126.
42. Navrátilová, M.; Beranová, M.; Severová, L.; Šrédl, K.; Svoboda, R.; Abrhám, J. The Impact of Climate Change on the Sugar Content of Grapes and the Sustainability of their Production in the Czech Republic. *Sustainability* **2021**, *13*, 222. [CrossRef]
43. Salinari, F.; Giosuè, S.; Tubiello, F.N.; Rettori, A.; Rossi, V.; Spanna, F.; Rosenzweig, C.; Gullino, M.L. Downy mildew (*Plasmopara viticola*) epidemics on grapevine under climate change. *Glob. Chang. Biol.* **2006**, *12*, 1299–1307.
44. Hirsch, R.M.; Alexander, R.B.; Smith, R.A. Selection of methods for the detection and estimation of trends in water quality. *Water Resour. Res.* **1991**, *27*, 803–813. [CrossRef]
45. McLeod, A.I.; McLeod, M.A. Package 'Kendall'. R Software: London, UK, 2015.
46. Jones, G. Climate change and wine: Observations, impacts and future implications. *Wine Ind. J.* **2006**, *21*, 21–26.
47. Blanco-Ward, D.; Garcia-Queijeiro, J.M.; Jones, G.V. Spatial climate variability and viticulture in the Miño River Valley of Spain. *Vitis* **2007**, *46*, 63–70.
48. Jones, G.V.; Duff, A.A.; Hall, A.; Myers, J. Spatial analysis of climate in winegrape growing regions in the western United States. *Am. J. Enol. Vitic.* **2010**, *61*, 313–326. [CrossRef]
49. Tomasi, D.; Jones, G.V.; Giust, M.; Lovat, L.; Gaiotti, F. Grapevine Phenology and Climate Change: Relationships and Trends in the Veneto Region of Italy for 1964–2009. *Am. J. Enol. Vitic.* **2011**, *62*, 329–339. [CrossRef]

50. Petrie, P.R.; Sadras, V.O. Advancement of grapevine maturity in Australia between 1993 and 2006: Putative causes, magnitude of trends and viticultural consequences. *Aust. J. Grape Wine Res.* **2008**, *14*, 33–45. [CrossRef]
51. Vršič, S.; Breznik, M.; Pulko, B.; Rodrigo-Comino, J. Earthworm Abundance Changes Depending on Soil Management Practices in Slovenian Vineyards. *Agronomy* **2021**, *11*, 1241. [CrossRef]

horticulturae

MDPI

Article

Hop Tropicalization: Chemical Compositions of Varieties Grown under Organic and Conventional Systems in Subtropical Conditions

Gabriel Cássia Fortuna [1], Caio Scardini Neves [1], Olivia Pak Campos [1], Jordany Aparecida Oliveira Gomes [1], Júlio César Rodrigues Lopes Silva [1], Amauri Alves Souza [1], Cristiano Soleo de Funari [1], Márcia Ortiz Mayo Marques [2] and Filipe Pereira Giardini Bonfim [1,*]

1 Department of Plant Production, School of Agronomic Sciences, Botucatu Campus, Júlio de Mesquita Filho São Paulo State University, Botucatu 18610-034, SP, Brazil; gabriel.cassia.fortuna@gmail.com (G.C.F.); caioscardini@gmail.com (C.S.N.); olivia.pakc@gmail.com (O.P.C.); jordanyoliveira327@gmail.com (J.A.O.G.); julio.lopes@unesp.br (J.C.R.L.S.); amaurialvejunior@gmail.com (A.A.S.); cristiano.funari@unesp.br (C.S.d.F.)

2 Center for Research on Plant Genetic Resources, Agronomic Institute, Avenida Barão de Itapura, 1481, Botafogo, Campinas 13020-902, SP, Brazil; marcia.marques@sp.gov.br

* Correspondence: filipe.giardini@unesp.br

Abstract: The interest in the production of hops in Brazil, motivated by the third position in the world ranking of beer producers and the growth of the craft brewery business, justifies the intensification of studies into its adaptation to local growing conditions. Due to the high internal demand, the aim of this study was to evaluate the phytochemical profiles of hop varieties grown in subtropical conditions under different cropping systems. Studies that promote the expansion of cultivation areas in distinct climate conditions and ensure quality are very important. A randomized block design was adopted with a 2×5 subdivided plot. The main factor was the cropping system (organic and conventional), and the secondary factor was the hop variety (Columbus, Chinook, Nugget, Cascade and Hallertau Mittelfrüeh), with four blocks and four plants per plot. The quality parameters monitored in this work were the contents of alpha and beta acids, and xanthohumol in the inflorescences of hops, as well as the relative composition of their essential oils. The variations in the chemical profiles of essential oils showed differences between some varieties, and the different compositions and levels resulting from the two cropping systems show that management and cultural practices can influence the aromatic characteristics of hops; in total, 23 compounds were found. The terpene fraction represented 79.67% of the oil in Hallertau and 93.63% in Cascade, with myrcene being the main compound. The levels of bitter acids and xanthohumol did not differ statistically as a function of the treatments. This study contributes the first records of the chemical profiles of hops grown in subtropical conditions in Brazil, in general, the Nugget variety had the highest qualitative potential

Keywords: alpha and beta acids; Brazilian hops; essential oil; management; phytochemical profile; xanthohumol

Citation: Fortuna, G.C.; Neves, C.S.; Campos, O.P.; Gomes, J.A.O.; Silva, J.C.R.L.; Souza, A.A.; Funari, C.S.d.; Marques, M.O.M.; Bonfim, F.P.G. Hop Tropicalization: Chemical Compositions of Varieties Grown under Organic and Conventional Systems in Subtropical Conditions. *Horticulturae* **2023**, *9*, 855. https://doi.org/10.3390/horticulturae9080855

Academic Editor: Rita Maggini

Received: 12 April 2023
Revised: 10 July 2023
Accepted: 14 July 2023
Published: 27 July 2023

1. Introduction

Hop (*Humulus lupulus* L.) is a perennial, herbaceous and dioecious vine that is considered a horticultural plant due to its agricultural suitability and multiple uses. The hop plant produces inflorescences annually; lupulin glands develop in these structures, also called cones, and biosynthesize specialized metabolites such as terpenoids, alpha and beta acids, and phenolic compounds, among other substances whose properties characterize hop quality [1,2].

In brewing, hop essential oils contribute to beer aroma and flavor, which can confer a range of notes (woody, citrus, spicy, floral, fruity, sulfurous, tangy, herbal, resinous, and earthy), according to the chemical profile of a cone [3]. Alpha acid is related to bitterness, with values ranging from 2–8% for aromatic varieties to 12–18% for bitter varieties [4]. Beta acids are less important to the brewing process; however, they present high levels of antimicrobial activity due to the presence of three isoprenyl groups that act as antioxidant and preservative agents [5].

Xanthohumol is a relevant hop polyphenol that has been widely studied as a prenylflavonoid that can act as an inhibitor in the initiation, promotion and progression of carcinogenesis [6].

Brazil is the third-largest beer producer in the world; however, it stands out as the largest importer of hop in South America. In 2019, Brazil imported 3.24 thousand tons of hop at a cost of approximately US $57 million [7] because hop is predominantly cultivated in temperate climates. There is potential to cultivate this crop in other regions due to the many varieties adapted to different climatic conditions [8]. Different performances, compositions and levels of metabolites are found within and among varieties since plant expression depends on interactions with external edaphoclimatic and biotic factors [9].

Climatic and geographical characteristics, plant genetics, age and health, environment, plant interactions, cropping conditions, cultural practices, management, and postharvest practices are sources of variability in hop cone chemical composition [10–14]. In this sense, the cropping systems can influence productivity and quality, the basic differences between organic and conventional systems are the fertilization sources and plant protection protocols, which according to Grzyb et al. [15] affect the composition of plants. Solarska and Sosnowska [16] report in their studies that some hop varieties perform better under organic cropping systems than conventional systems.

Clinical studies with small animals support that consumption of organically produced food is better for human health than conventionally produced [16]. There is a growing concern with nutrition linked to health, in addition to the alarming necessity maintaining environmental sustainability, and also the economic interest. The conventional agricultural practices use levels of inputs that can result in a disruption of the natural production of specialized metabolites in the plants, so, this management affect the nutrients levels in plants [15].

Therefore, in view of the high internal demand and expansion of the national brewing market with interest in hops with peculiar phytochemical profiles [17], studies that promote the expansion of new cultivation zones and that ensure quality are very important. Thus, the aim of this study was to evaluate the chemical profiles of essential oils and alpha and beta acids and xanthohumol contents of five hop varieties cultivated in subtropical conditions (Botucatu-SP, Brazil) in organic and conventional cropping systems.

2. Materials and Methods

2.1. Experimental Area

The experiment was conducted at the Department of Horticulture of the School of Agronomic Sciences of UNESP in the municipality of Botucatu-SP, Brazil (latitude, 22°50′ S; longitude, 48°26′ W; elevation, 791 m). According to Köppen [18], the climate is classified as subtropical with hot summers (Cfa), and the soil in the study area is clayey dystroferric Red Latosol. The hops were planted in November 2018, and the research data were collected in the second year of production (November 2019 to March 2020). In this period, the minimum average temperature was 17.94 °C, the maximum average temperature was 28.45 °C, and the cumulative rainfall was 1257.61 mm. The annual average minimum and maximum temperatures were 15.83 °C and 25.91 °C, respectively, and the annual average precipitation was 100.23 mm.

2.2. Treatments and Experimental Design

A randomized block design was adopted with a 2 × 5 split plot; the main factor was the cropping systems (organic and conventional) and the secondary factor was the hop varieties (Columbus, Chinook, Nugget, Cascade and Hallertau Mittelfrüeh), with four blocks and four plants per plot. The organic and conventional management systems were differentiated mainly by fertilization and phytosanitary control, following Brazilian legislation and the recommendations established for hop in international literature [19]. The elements used in each cropping system are described in the next section.

2.2.1. Description of Varieties

As described above, the varieties analyzed in this research were Columbus, Chinook, Nugget, Cascade and Hallertau Mittelfrüeh. They were used in this research because they have characteristics that distinguish them from each other, such as their profile and contents of essential oils, or that generate distinct aromatic characteristics and percentages of alpha and beta acids. These factors determine the potential styles of beers that can be produced with each variety. Finally, the varieties have characteristics that determine whether they are more or less bitter and have a more intense or lighter aromatic profile. Therefore, each variety is characterized by its unique characteristics, a fact that justifies the development of new varieties and their commercial planting.

Columbus is a variety that has dual aptitude, and its alpha acid level and percentage of essential oils characterizes it for use as a bittering or aroma hop. Its aroma is pungent with citrus notes. The alpha and beta acid levels of this variety vary between 14 and 18% and 4.5 and 6%, respectively, and the total essential oil contents vary between 1.5 and 4.5 mL/100 g [20].

Chinook also has dual functions, as it has a high load of alpha acids (12–14%) and is widely used to provide bitterness to beers; however, due to the composition of its oils (1.5–2.7 mL/100 g) and its aroma, which is characterized by pine and spices, it is also used to provide aroma in certain styles of beers [20,21].

Nugget is a variety that has an intense herbal aroma, light flavor and marked bitterness and is used both to provide aroma and bitterness to beers, so it also has dual functions; it has approximately 9.5–14% alpha acids and 1.5–3 mL/100 g total essential oils [20,22].

Cascade is one of the most popular and widely cultivated varieties in the world due to its excellent development and vigor. It has dual functions but is most commonly used to provide aroma to beers, and its aromatic profile has floral notes with citrus and grapefruit elements. The alpha acid content is lower, from 4.5–8.9%, and therefore, it is not commonly used to provide bitterness to beers, and its total oils are between 0.8 and 1.5 mL/100 g [20,21,23].

Hallertau is one of the varieties that is considered noble; it has been cultivated for more than 100 years in Germany. It is used only for providing aroma to beers and has a slightly floral and spicy aromatic characteristic. The contents of the total essential oils is between 0.6 and 2 mL/100 g, and the alpha acid content is 3.5% [20,24].

Thus, each variety has unique characteristics that allow the use of its cones in different styles of beers. For example, Columbus is commonly used in more distinctive styles, such as imperial stout, stout and American ales; Chinook in pale ale, India pale ale, porter, stout, lager, American lager and others; Cascade mainly in American pale ale, ale and lager; and finally, Hallertau in German pilsner, pale ale, wheat and American lager [20].

Several styles of beer are produced with each variety, and the number of uses for each variety can increase further since there is no fixed rule about the use of a certain hop variety for a particular style of beer, leaving it open to the creativity of the brewmaster.

2.2.2. Conventional Cultivation

Fertilizer was applied according to the needs determined by soil analyses (Table A1). In the first year, topdressing containing calcium nitrate (375 kg ha^{-1}), urea (94 kg ha^{-1}), potassium chloride (186 kg ha^{-1}) and micronutrients with MIB® (20 kg ha^{-1}) was added. Phytosanitary control included applications of abamectin (Abamex®) for streaked mites (*Tetranychus urticae*), fipronil (Regent®) for leaf-cutting ants, and tebuconazole (Folicur®) for powdery mildew (*Podosphaera macularis*), which was identified in the first year. In the second year, the same fertilizers were used, and conventional poultry litter (3.12 t ha^{-1}) was added. Borate fertilization was done with boric acid (4 kg ha^{-1}), and leaf fertilization was done with zinc sulfate (5 kg ha^{-1}). The phytosanitary control of mites and ants was identical to that described above, and *Bacillus thuringiensis* (Dipel®) was applied following the appearance of caterpillars.

2.2.3. Organic Farming

Fertilizer was applied according to the needs determined by soil analyses (Table A1). In the first year of cultivation, cattle manure, castor bean cakes and potassium sulfate were used. For phytosanitary control, sulfur–calcium spray solutions were applied for streaked mites (*T. urticae*), organic formicides (Bioisca®) for ants, and raw milk and Bordeaux mixture for powdery mildew (*P. macularis*). In the second year, fertilization was done with bokashi (1.5 t ha^{-1}), castor bean cakes (1.4 t ha^{-1}), and organic poultry litter (2 t ha^{-1}). Potassium sulfate (94 kg ha^{-1}), potassium silicate (312 kg ha^{-1}), thermophosphate (203 kg ha^{-1}), boric acid (4 kg ha^{-1}) and bone meal (1 t ha^{-1}) were also used. Spraying was performed with SuperMagro biofertilizer, and biological activation of the soil was performed with effective microorganisms (EM). *Metarhizium anisopliae* + *Beuaveria bassiana* (B Ex-change®) were applied for preventative pest control.

2.3. *Evaluations*

2.3.1. Chemical Composition of the Essential Oil

The hop samples analyzed were collected when the cones reached the mature stage (February–March 2020). The plants were harvested in full and taken to the laboratory to remove the cones; these were dried in a forced air ventilation oven at 40 °C for a variable time from 24 to 48 h until they reached approximately 10% moisture.

The extraction of essential oils was performed by hydrodistillation in a Clevenger apparatus from 50 g cones for 1 h and 30 min.

The determination of the essential oil chemical profiles was performed at the Instituto Agronômico de Campinas (IAC) in a gas chromatograph coupled to a mass spectrometer (CG-EM, QP 5000–Shimadzu) equipped with an OV-5 MS capillary column and helium as the capillary gas.

The system was operated in full scan mode with electron impact (70 eV), and ranged from 40 to 450 m/z. The injector was kept at 220 °C, with a carrier gas flow rate of 1:20 and temperature programming of 60 °C–240 °C (3 °C min^{-1}). The interface temperature was maintained at 240° C. Oil samples were diluted in ethyl acetate, and 1 µL of solution was injected.

For quantitative analysis, a gas chromatograph with a flame ionization detector (CG-DIC, Shimadzu, CG-2010/AOC-20i) was used. The system was equipped with an OV-5 capillary column, helium as the carrier gas, injector at 280 °C, detector at 300 °C, 1:20 split and the same temperature program as the GC-MS system.

The identification of chemical constituents was performed by comparing the mass spectra of the substances with the National Institute of Standards and Technology library (Nist 62.lib) and the substance retention indices [25]; these indices were obtained from the injection of a mixture of n-alkanes (C9–C24, Sigma, St. Louis, MO, USA) under the same chromatographic conditions as the samples, applying the equation by Van Den Dool and Kratz [26].

2.3.2. Quantification of Alpha Acids, Beta Acids and Xanthohumol

Quantifications were based on methods reported by Prencipe [27]. Exactly 50 mg of ground cones were extracted by dynamic maceration at 1000 rpm for 30 min (Heidolph MR Hei-Tec, Germany) with 2.0 mL of MeOH-HCOOH (99:1 *v:v*). After filtration (22 µm PTFE syringe filter), 1.5 µL was injected into an ultrahigh-performance liquid chromatograph coupled to a UV/Vis spectrophotometer (Shimadzu Nexera UC, Kyoto, Japan). Separations were achieved in a C18 column of 150 × 2.1 mm and 1.7 µm (Kinetex, Phenomenex, Torrance, CA, USA). The mobile phase was composed of H_2O and acetonitrile (ACN), both with 0.25% HCOOH, in the following gradient: 35–75% ACN (0–12 min), 75–100% ACN

(12–25 min), and 100% ACN (25–28 min). The analysis temperature and flow rate were 30 °C and 0.39 mL min^{-1}, respectively. Online UV spectra were recorded from 200 to 400 nm. Data were processed using LabSolutions LCMS software version 5.96 (Shimadzu Corporation, Japan).

2.4. Statistical Analysis

The chemical composition data of the essential oils were assessed with multivariate principal component analysis (PCA) using Minitab® Statistical Software [28]. Alpha and beta acid and xanthohumol data were subjected to analysis of variance, and means were compared using the Scott–Knott test at 5% significance using Sisvar software [29].

3. Results

In this study, the chemical compositions of hop varieties grown in a subtropical environment in Brazil under organic and conventional systems were elucidated for the first time.

3.1. Chemical Composition of the Essential Oils

The chemical composition of the essential oils is described in Table 1; the volatile fraction showed variations mainly in terms of relative percentages, and in total, 23 substances were identified. The main classes were monoterpenes, oxygenated monoterpenes, sesquiterpene hydrocarbons, oxygenated sesquiterpenes, esters, aldehydes and ketones; the main volatile components were myrcene, linalool, caryophyllene, farnesene and ledene.

The terpene fraction was relatively large, accounting for 79.67% of Con Hal essential oil and 96.63% of Org Cas essential oil, with myrcene being the major compound in all varieties in both cropping systems, except for Con Hal, whose most abundant terpene compound was beta-farnesene (38.43%). Org Hal had a 62% higher myrcene content (39.26%) than the same variety grown under the conventional system (14.83%); furthermore, Con Hal showed a myrcene content that was 71% lower than Con Cas (51.63%).

Linalool is the most abundant oxygenated terpene alcohol; the highest content of this compound was found in Con Nug (0.91% ± 0.26) and the lowest content was found in Org Chi (0.52% ± 0.19).

Beta-caryophyllene had a higher percentage in the organic cropping system than in the conventional system, with averages of 6.04% and 3.79%, respectively; among the varieties, Columbus had the highest content of this compound, being 77%, 38%, 29% and 20% higher than the contents in Hallertau Mittelfrüeh, Cascade, Chinook, and Nugget, respectively.

Beta-farnesene was the sesquiterpene present at the highest percentage in the five varieties; in Hallertau Mittelfrüeh, this content of this compound was higher than in other varieties, with an average content of 38.43% in the conventional system. This variety presented 12.27% more beta-farnesene than Con Nug, which presented the second highest percentage (26.16% ± 8.78) and 12.59% more than Org Hal (38.43% ± 2.33).

The sesquiterpenes ledene and beta-selinene stood out among the chemical compositions of the varieties, with percentages ranging from 6.70% (Org Hal) to 11.72% (Org Col) and 6.77% (Org Hal) to 10.76% (Org Col), respectively.

Through PCA (Figure 1), it was possible to verify the similarities between the varieties in the cropping systems. Figure 1 shows the PCA, with 76.80% of the total variance explained by the first two principal components. The main difference was observed for Hallertau Mittelfrüeh, the chemical characteristics of which were distinguished from the other varieties in both cropping systems; in particular, the presence of substances from the ketone and aldehyde groups differentiated this variety. The substances methyl heptanone, 2-nonanal, n-nonanal, methyl octanoate, 2-undecanone, undecanal, undec-9E-em-1-al, 2-methyl-lavandula butanoate, germacrene B and eudesmol were observed only in the Hallertau Mittelfrüeh variety.

Table 1. Essential oil chemical constituents of *Humulus lupulus* L. varieties grown in conventional and organic cropping systems under subtropical conditions.

	Substance (%)	LRI Cal	LRI Lit	Org Cas	Con Cas	Org Nug	Con Nug	Org Chi	Con Chi	Org Hal	Con Hal	Org Col	Con Col
1	β-Pinene	976	974	0.90 ± 0.06	0.68 ± 0.32	0.48 ± 0.30	0.52 ± 0.37	0.57 ± 0.39	0.82 ± 0.02	0.68 ± 0.06	0.23 ± 0.00	0.50 ± 0.18	0.50 ± 0.19
2	Myrcene	991	988	55.92 ± 7.21	51.63 ± 19.08	41.40 ± 13.49	43.20 ± 16.43	43.97 ± 19.06	57.65 ± 2.68	39.26 ± 3.15	14.83 ± 5.65	38.27 ± 11.82	44.84 ± 3.83
3	Methylheptanone	1021	1021	–	–	–	–	–	–	0.38 ± 0.11	0.27 ± 0.07	–	–
4	2-Nonanal	1089	1087	–	–	–	–	–	–	0.43 ± 0.14	0.28 ± 0.10	–	–
5	Linalool	1098	1095	0.65 ± 0.08	0.79 ± 0.27	0.62 ± 0.13	0.91 ± 0.26	0.52 ± 0.19	0.68 ± 0.04	0.62 ± 0.12	0.66 ± 0.00	0.59 ± 0.25	0.72 ± 0.11
6	n-Nonanal	1102	1100	–	–	–	–	–	–	0.63 ± 0.34	0.54 ± 0.13	–	–
7	Methyl octanoate	1123	1123	–	–	–	–	–	–	0.24 ± 0.04	0.23 ± 0.02	–	–
8	2-Decanone	1191	1192	0.24 ± 0.01	0.32 ± 0.18	–	–	–	–	1.09 ± 0.92	0.49 ± 0.05	–	–
9	Methyl nonanoate	1222	1223	0.60 ± 0.26	1.03 ± 0.60	0.40 ± 0.19	1.39 ± 0.60	0.45 ± 0.15	0.69 ± 0.32	0.56 ± 0.22	1.16 ± 0.18	0.23 ± 0.00	0.23 ± 0.00
10	2-Undecanone	1191	1293	–	–	–	–	–	–	0.64 ± 0.55	2.42 ± 0.31	–	–
11	Undecanal	1307	1305	–	–	–	–	–	–	0.56 ± 0.22	1.16 ± 0.18	–	–
12	Undec-9E-em-1-al	1312	1311	–	–	–	–	–	–	0.60 ± 0.25	1.47 ± 0.30	–	–
13	(*E*)-Caryophyllene	1417	1417	4.99 ± 1.85	3.82 ± 1.24	7.49 ± 1.85	3.89 ± 0.88	7.20 ± 4.07	2.81 ± 0.15	2.13 ± 0.34	2.60 ± 0.18	8.37 ± 1.52	5.85 ± 1.05
14	α-(*E*)-Bergamotene	1434	1432	0.78 ± 0.10	0.93 ± 0.40	1.00 ± 0.23	1.11 ± 0.37	0.97 ± 0.30	0.80 ± 0.05	1.43 ± 0.27	1.89 ± 0.15	1.04 ± 0.22	0.98 ± 0.11
15	(*E*)-β-Farnesene	1456	1452	17.24 ± 1.16	22.85 ± 9.31	23.98 ± 5.52	26.16 ± 8.78	22.69 ± 6.75	19.16 ± 1.10	25.84 ± 2.12	38.43 ± 2.33	25.29 ± 5.66	24.01 ± 2.65
16	β-Charmigrene	1474	1476	1.23 ± 0.25	1.23 ± 0.49	1.67 ± 0.40	1.63 ± 0.45	1.63 ± 0.59	1.09 ± 0.09	1.23 ± 0.20	1.87 ± 0.19	1.76 ± 0.35	1.56 ± 0.11
17	β-Selinene	1485	1489	7.37 ± 1.45	7.40 ± 2.95	10.05 ± 2.43	9.73 ± 2.77	9.72 ± 3.42	6.63 ± 0.62	6.77 ± 1.15	10.56 ± 1.22	10.66 ± 2.07	9.43 ± 0.75
18	Ledene	1494	1496	8.08 ± 1.57	8.12 ± 3.24	11.09 ± 2.57	10.65 ± 2.99	10.72 ± 3.86	7.41 ± 0.67	6.70 ± 1.03	9.26 ± 0.57	11.72 ± 2.31	10.19 ± 0.65
19	2-Methyl-lavandula butanoate	1513	1511	–	–	–	–	–	–	0.74 ± 0.17	0.77 ± 0.13	–	–
20	α-Cadinene	1532	1537	0.30 ± 0.08	–	0.46 ± 0.16		0.33 ± 0.20	0.37 ± 0.11	0.81 ± 0.14	1.22 ± 0.13	0.27 ± 0.08	–
21	mi	1554	1559	–						0.81 ± 0.13	1.20 ± 0.09		
22	α-Eudesmol	1650	1652	–	–	–	–	–	–	0.88 ± 0.20	1.49 ± 0.20	–	–
23	6-Z-Pentadecen-2-o	1667	1667	0.63 ± 0.32	0.68 ± 0.23	1.04 ± 0.27	0.82 ± 0.25	1.00 ± 0.43	0.55 ± 0.04	0.90 ± 0.26	1.47 ± 0.23	0.94 ± 0.17	0.87 ± 0.11
	Monoterpene hydrocarbons			56.82	52.31	41.88	43.72	44.54	57.68	39.94	15.06	38.77	45.34
	Oxygenated monoterpenes			0.65	0.79	0.62	0.91	0.52	0.68	0.62	0.66	0.59	0.72
	Sesquiterpene hydrocarbons			39.81	44.35	55.28	53.17	53.27	38	44.10	64.61	59.94	52.02
	Oxygenated sesquiterpenes			–	–	–	–	–	–	0.88	1.49	–	–
	Esters			0.60	1.03	0.40	1.39	0.45	0.69	1.18	1.66	0.23	0.23
	Ketones			–	–	–	–	–	–	2.22	3.45	–	–
	Aldehydes			0.24	0.32	–	–	–	–	1.73	2.91	–	–

LRI Cal = Calculated linear retention index; LRI Lit = linear retention index stated in the literature. (Org Cas—Organic Cascade, Con Cas—conventional Cascade, Org Nug—organic Nugget, Con Nug—conventional Nugget, Org Chi—organic Chinook, Con Chi—conventional Chinook, Org Hal—organic Hallertau Mittelfrüeh, Con Hal—conventional Hallertau Mittelfrüeh, Org Col—organic Columbus, Con Col—conventional Columbus).

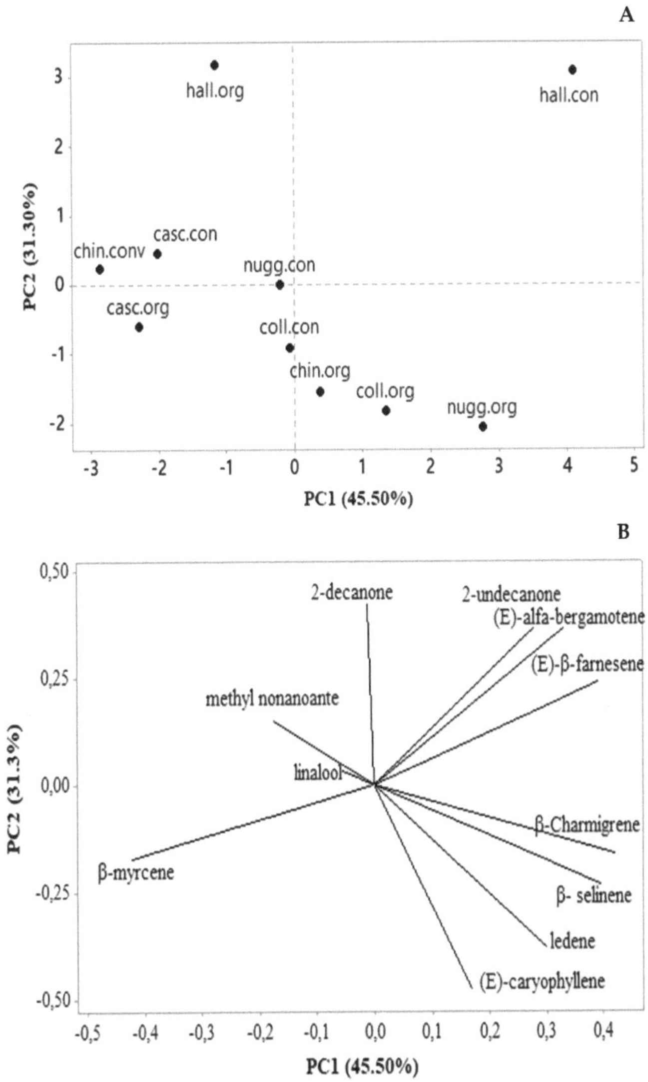

Figure 1. Plots of the principal component analysis of *Humulus lupulus* L. varieties cultivated in organic and conventional cultivation systems and their chemical constituents. Score plot (**A**) and loadings plot (**B**). See Table 1 for trait labels.

The Cascade, Chinook, Nugget and Columbus varieties showed greater similarity; however, small variations within varieties were observed based on the cropping system. The monoterpene myrcene showed the highest correlation with Cascade grown in both systems and Chinook grown in the conventional system.

3.2. Alpha acids and Beta Acids Contents

The fixed fraction of inflorescences of *H. lupulus* is mainly composed of xanthohumol and bitter acids (alpha and beta acids). The major alpha acids in these materials are cohumulone, humulone and adhumulone, while the major beta acids are colupulone, lupulone and adlupulone [30]. Therefore, these six bitter acids and xanthohumol were quantified here and summarized in Table 2. The total concentration of bitter acids ranged from 1.08 ± 0.21% (Org Col) to 6.16 ± 6.93% (Org Nug), and the total xanthohumol content ranged from 0.38 ± 0.03 mg g^{-1} (Org Cas) to 1.57 ± 1.40 mg g^{-1} (Org Nug).

Table 2. Alpha and beta acids and xanthohumol contents of *Humulus lupulus* L. varieties grown in conventional and organic cropping systems under subtropical conditions.

Substance	Org Cas	Con Cas	Org Nug	Con Nug	Org Chi	Con Chi	Org Hal	Con Hal	Org Col	Con Col
n-Humulone (mg g^{-1})	0.77 ± 1.42	0.98 ± 1.26	8.17 ± 9.53	2.98 ± 1.81	1.38 ± 1.51	0.87 ± 1.56	2.39± 0.60	3.19 ± 2.33	0.69 ± 1.49	2.22 ± 3.09
Cohumulone (mg g^{-1})	2.86 ± 0.92	3.46 ± 1.72	28.78 ± 32.47	10.13 ± 7.79	3.45 ± 0.13	3.16 ± 0.78	8.38± 3.88	10.43 ± 5.02	2.34 ± 0.32	6.62 ± 7.64
Adhumulone (mg g^{-1})	0.68 ± 0.21	0.91 ± 0.35	6.08 ± 2.78	2.04 ± 1.66	0.85 ± 0.03	0.76 ± 0.19	1.84 ± 0.90	2.38 ± 1.24	0.58 ± 0.10	2.06 ± 1.75
Colupulone (mg g^{-1})	4.79 ± 0.94	6.78 ± 2.58	8.41 ± 6.31	6.27 ± 3.50	4.38 ± 0.39	4.41 ± 1.70	2.48 ± 0.95	3.18 ± 1.41	4.10 ± 0.92	6.31 ± 3.57
n-Lupulone (mg g^{-1})	2.68 ± 1.09	3.64 ± 0.88	7.48 ± 5.50	3.70 ± 0.84	2.37 ± 1.00	2.41 ± 0.87	2.46± 0.47	3.03 ± 1.97	2.08 ± 1.31	3.31 ± 1.60
Adlupulone (mg g^{-1})	1.26 ± 0.39	1.84 ± 0.47	2.69 ± 1.80	1.67 ± 0.68	1.19 ± 0.37	1.20 ± 0.39	0.89± 0.18	1.12 ± 0.73	1.01 ± 0.58	1.63 ± 0.87
Alpha acids (%)	0.43 ± 0.68	0.54 ± 0.59	4.31 ± 4.85	1.51 ± 0.70	0.53 ± 0.73	0.48 ± 0.78	1.26± 0.24	1.60 ± 1.15	0.36 ± 0.72	1.06 ± 1.39
Beta acids (%)	0.87 ± 0.26	1.23 ± 0.33	1.86 ± 1.23	1.16 ± 0.45	0.79 ± 0.27	0.80 ± 0.26	0.58± 0.11	0.73 ± 0.49	0.72 ± 0.38	1.12 ± 0.56
Alpha + beta acids (%)	1.31 ± 0.33	1.76 ± 0.69	6.16 ± 6.93	2.68 ± 1.25	1.32 ± 0.03	1.28 ± 0.17	1.84 ± 0.82	2.33 ± 1.11	1.08 ± 0.21	2.28 ± 1.84
Xanthohumol (mg g^{-1})	0.38 ± 0.03	0.56 ± 0.20	1.57 ± 1.40	0.66 ± 0.08	0.79 ± 0.97	0.52 ± 0.23	0.72± 0.27	1.05 ± 0.31	0.44 ± 0.21	0.40 ± 0.05

(Org Cas—Organic Cascade, Con Cas—conventional Cascade, Org Nug—organic Nugget, Con Nug—conventional Nugget, Org Chi—organic Chinook, Con Chi—conventional Chinook, Org Hal—organic Hallertau Mittelfrüeh, Con Hal—conventional Hallertau Mittelfrüeh, Org Col—organic Columbus, Con Col—conventional Columbus).

4. Discussion

This work showed, for the first time, the characteristics of different hop varieties in subtropical conditions (Botucatu, SP-Brazil, latitude 22°50′ S) under conventional and organic cropping systems, focusing on specialized metabolites associated with the sweetening characteristics of hop cones for beer production.

4.1. Chemical Composition of the Essential Oils

Myrcene is a monoterpene and is considered the primary compound of hop essential oils [31]; because it is the major compound, myrcene levels may vary to a greater degree than those of other compounds. It is responsible for the aroma of green hops and is related to resinous, pine and pungent flavors, providing interesting flavors for the preparation of fruity beers and IPAs [32,33].

The values of myrcene in Cascade were within the standard range of 45 to 60% in both cropping systems (Org Cas 55.92, Con Cas 51.63), those in Chinook were in the range of 35 to 40% in both systems (Org Chi 43.97, Con Chi 57.65), those in Hallertau Mittelfrüeh ranged from 20 to 28% only in the organic system (Org Hal 39.26, Con Hal 14.83), those in Columbus reached the standard range of 25 to 40% in the organic system (38.27) and exceeded this range in the conventional system (44.84), and those in Nugget did not reach the range of 48 to 59% in either system (Org Nug 41.40, Con Nug 43.20) [34]. The low myrcene content observed in Con Hal was expected because this compound appears in smaller quantities in European hop varieties, such as Hallertau Mittelfrüeh [35].

The autooxidation of myrcene gives rise to several cyclic reaction products (e.g., alpha-pinene, beta-pinene, camphene, and r-cymene) and forms terpenoids such as linalool, nerol, geraniol, citral, alpha-terpineol and carvone [12]. Linalool and beta-pinene were observed in this study and may have been formed by the oxidation of myrcene; it is emphasized that a wide range of distinct aromas and flavors can be obtained with the compounds originating from myrcene.

Linalool is one of the most important indicators of beer aroma quality and hop freshness and is among the most interesting oxygenated aromatic compounds for the brewing market, with great sensory activity even though it is normally present in essential oil in proportions smaller than 1% [32]. In the present study, Con Nug stood out among the varieties for having a higher linalool content.

Alpha-humulene, along with beta-caryophyllene, is the most abundant sesquiterpene in hops and it has positive impacts on beer aroma, being desired in a 3:1 ratio (alpha-humulene: beta-caryophyllene) to provide a more refined aromatic character with a strong emphasis on herbal, floral and spicy notes [32]. Alpha-humulene was not detected in this study, which may be related to the tropical cultivation conditions.

Increases in the percentage of beta-caryophyllene are commonly related to attack by pests and pathogens [36], justifying the higher percentage of this compound in the organic system, in which preventive control was limited, with management instead focused on maintaining insect populations at acceptable levels that did not result in crop damage.

Beta-farnesene, the sesquiterpene that was present at the highest percentage in all the evaluated varieties, especially Con Hal, can reach up to 30% of the total essential oils in noble hops, which are widely used in Bohemian pilsner style beers [32].

The ledene and beta-selinene contents contributed to the high contents of terpene hydrocarbons observed in this study, which exceeded the values reported by studies performed in temperate climates, which ranged from 50 to 80% [32,37].

Despite the influence of major compounds, the entire set of essential oils determines the formation of aromas of interest to the brewing industry. In hops, the presence of terpene hydrocarbons and substances such as esters, alcohols and ketones provide different aromatic ranges. In this context, the PCA showed similarities between the varieties in both cropping systems.

The differences observed may have occurred due to the environment and management practices (e.g., fertilization and phytosanitary control). These factors differed from those established in traditional cultivation sites and exerted a great influence on the hop essential oil chemical composition [38,39], thus allowing the expression of Brazilian "terrior" in the chemical composition of the essential oils.

Organic cropping adds value to the product, in addition, brings environmental [40] and social benefits, and, as observed in this study and confirmed by others [15,41], can increase the levels of some chemical compounds of interest to the brewing and medicinal market, i.e., it is economically interesting.

4.2. Alpha Acid and Beta Acid Contents

Overall, all varieties cultivated in this work presented lower average levels of bitter acids than those described in the literature for *H. lupulus* cultivated in temperate zones (Table 2) (refer to "Description of varieties" subsection in the Materials and methods). It is also important to highlight that no statistical differences were observed among varieties or cropping systems in this work. This was expected due to the high relative standard deviations (RSD) observed among replicates for bitter acid contents (Table 2). As the HPLC-UV method used here was a validated method [27], being its reproducibility further confirmed in our laboratory from four commercial hop pellets (all with RSD $\leq$ 4.4%), it was concluded that the observed RSD evidence the high variability among specimens belonging to the same variety and cropping system. Both, the average low levels of bitter acids and the observed high RSD, can be at least partially explained by the fact that the evaluated plants were in their second year of cultivation. That is because the expected physiological maturation of plants should occur only in the third to fifth year of cultivation, when the biochemical machinery for the production of bitter acid tends to me more efficient and stable [42,43]. On the other hand, it is important to monitor the contents of such compounds throughout the development of the plants as carried out in this work, since it can give important feedbacks regarding the development of the plants and indicate tendencies.

Finally, the lower average levels of bitter acids found here compared to those found in temperate zones might be related to what was reported by Mozny et al. [44], who found that the increasing in temperatures observed in recent years in Czech Republic is inducing early flowering of the 'Saaz' hop variety, with reduced levels of alpha acids. Early flowering was observed for the plants investigated here, probably due to both warmer days and shorter photoperiods found in Botucatu city when compared to traditional temperate cultivation sites [44].

5. Conclusions

The cultural practices and management adopted in this work altered the composition of hop volatile compounds. The same could not be concluded for bitter acids due to the high relative standard deviation found between the analyzes of the same field replicates. However, this evidenced a high variability among specimens belonging to the same variety and cultivation system regarding the production of bitter acids. As a consequence, such a high variability could mask any possible effect from the different managements and varieties of hops adopt in this work on the production of bitter acids. As it can be related with the fact that the evaluated plants were only in their second year of cultivation, new studies aiming to know their phytochemical evolution throughout their physiological maturation are being carried. Such studies are necessary to eventually establish a scientific basis that would allow the expansion of new areas of cultivation of hops in Botucatu city, which hosts several craft breweries that are highly dependent on hop imports from other countries.

Author Contributions: Conceptualization, G.C.F.; Data curation, G.C.F., J.A.O.G., C.S.d.F., M.O.M.M. and F.P.G.B.; Formal analysis, J.A.O.G., J.C.R.L.S., A.A.S., C.S.d.F., M.O.M.M. and F.P.G.B.; Funding acquisition, G.C.F., C.S.N., O.P.C., J.A.O.G. and F.P.G.B.; Investigation, G.C.F., C.S.N., O.P.C., J.A.O.G., J.C.R.L.S., A.A.S., M.O.M.M. and F.P.G.B.; Methodology, G.C.F., C.S.N. and O.P.C.; Project administration, G.C.F. and F.P.G.B.; Software, J.A.O.G.; Supervision, C.S.d.F., M.O.M.M. and F.P.G.B.; Validation, F.P.G.B.; Visualization, G.C.F.; Writing—original draft, G.C.F., J.A.O.G. and F.P.G.B.; Writing—review & editing, O.P.C. and F.P.G.B. All authors have read and agreed to the published version of the manuscript.

Funding: The authors are grateful to the Fundação de Amparo à Pesquisa do Estado de São Paulo (FAPESP) for financial support to implement and maintain the experimental area and for the purchase of equipment and inputs for analyses (2018 01786-1 and 2019/27066-8) and to the Coordenação de Aperfeiçoamento de Pessoal de Nível Superior (CAPES) and Conselho Nacional de Desenvolvimento Científico e Tecnológico (CNPq) for financial support through postgraduate scholarships.

Conflicts of Interest: The authors declare no conflict of interest.

Appendix A

Table A1. Complete soil chemical analysis of organic and conventional cropping systems of *Humulus lupulus* L. between November 2018 and March 2020, Botucatu-SP.

		pH	OM	P	K	Ca	Mg	CEC	V%	S	B	Cu	Fe	Mn	Zn
		$CaCl_2$	$G{\cdot}dm^{-3}$	$mg{\cdot}dm^{-3}$		$mmol_c{\cdot}dm^{-3}$			%			$mg{\cdot}dm^{-3}$			
Date	**Sys./Depth (cm)**														
Nov. 2018	Org 0–20	5.4	29	49	13.5	32	15	84	72	124	0.44	4.7	34	6.9	3.0
	Org 20–40	5.0	22	35	7.5	25	12	84	53	77	0.63	4.9	34	5.3	1.6
	Conv 0–20	5.6	19	29	9.6	26	9	65	68	139	0.38	5.4	27	4.7	0.6
	Conv 20–40	5.3	18	12	2.6	14	5	51	42	52	0.31	5.5	39	4.6	0.4
Apr. 2019	Org 0–20	5.3	17	27	1.9	25	11	63	60	4	0.43	2.0	34	6.8	1.3
	Org 20–40	5.0	17	23	2.9	14	11	67	42	25	0.55	1.2	34	5.7	1.1
	Conv 0–20	5.4	22	34	1.4	29	10	70	58	27	0.48	1.9	35	4.5	0.5
	Conv 20–40	4.5	16	10	1.2	12	7	63	33	43	0.44	0.5	36	2.1	0.4
Aug. 2019	Org 0–20	5.7	21	14	1.6	33	12	64	72	17	0.34	3.6	18	2.1	1.2
	Org 20–40	4.4	15	3	0.5	11	5	64	26	67	0.33	4.5	18	0.8	0.1
	Conv 0–20	5.0	19	17	0.3	19	7	58	45	34	0.30	4.1	21	2.4	0.2
	Conv 20–40	4.3	15	4	0.4	11	4	70	21	48	0.35	4.8	17	1.7	0.2

Table A1. *Cont.*

		pH	OM	P	K	Ca	Mg	CEC	V%	S	B	Cu	Fe	Mn	Zn
		$CaCl_2$	$G{\cdot}dm^{-3}$	$mg{\cdot}dm^{-3}$		$mmol_c{\cdot}dm^{-3}$			%			$mg{\cdot}dm^{-3}$			
Date	Sys./Depth (cm)														
Nov. 2019	Org 0–20	6.0	24	39	4.3	47	13	84	76	101	0.39	4.2	24	3.7	3.9
	Org 20–40	5.6	22	40	2.8	40	17	83	71	80	0.63	3.8	24	3.4	3.4
	Conv 0–20	5.8	25	44	4.0	80	10	119	79	422	0.63	3.6	19	3.4	3.3
	Conv 20–40	5.6	25	49	5.5	66	8	107	74	305	0.58	4.0	26	3.9	2.8
Mar. 2020	Org 0–20	5.0	25	56	3.4	39	11	77	70	71	1.00	4.7	19	5.1	3.8
	Org 20–40	4.8	19	30	2.2	23	10	70	50	52	0.90	5.4	17	3.0	1.7
	Conv 0–20	4.9	19	16	2.6	20	5	64	42	42	1.09	6.1	19	3.2	0.9
	Conv 20–40	4.4	15	6	1.4	13	4	63	29	148	0.86	5.3	16	3.1	0.4

References

1. Denby, C.M.; Li, R.A.; Vu, V.T.; Costello, Z.; Lin, W.; Chan LJ, G.; Keasling, J.D. Industrial brewing yeast engineered for the production of primary flavor determinants in hopped beer. *Nat. Commun.* **2018**, *9*, 965. [CrossRef] [PubMed]
2. Bocquet, L.; Sahpaz, S.; Hilbert, J.L.; Rambaud, C.; Rivière, C. *Humulus lupulus* L., a very popular beer ingredient and medicinal plant: Overview of its phytochemistry, its bioactivity, and its biotechnology. *Phytochem. Rev.* **2018**, *17*, 1047–1090. [CrossRef]
3. Bamforth, W.C. *Beer is Proof God Loves Us*; FT Press: Upper Saddle River, NJ, USA, 2010.
4. Pavolovič, V.; Čerenak, A.; Pavolovič, M.; Košir, I.J.; Rozman, Č.; Bohanec, M.; Naglič, B. Modelling of Quality Prediction for Hops (*Humulus lupulus* L.) in Relation to Meteorological Variables. *Balwois* **2010**, 25–29. Available online: https://balwois.com/wp-content/uploads/old_proc/ffp-1920.pdf (accessed on 1 June 2023).
5. Boulton, C. *Encyclopedia of Brewing*, 1st ed.; John Wiley & Sons: Hoboken, NJ, USA, 2013.
6. Gerhauser, C.; Alt, A.; Heiss, E.; Gamal-Eldeen, A.; Klimo, K.; Knauft, J.; Becker, H. Cancer chemopreventive activity of Xanthohumol, a natural product derived from hop. *Mol. Cancer Ther.* **2002**, *1*, 959–969. Available online: https://www.researchgate.net/publication/10992279_Cancer_chemopreventive_activity_of_Xanthohumol_a_natural_product_derived_from_hop (accessed on 15 July 2021).
7. COMEXSTAT. Dados Estatísticos do Comércio Exterior Brasileiro. Available online: http://comexstat.mdic.gov.br/pt/home (accessed on 20 February 2021).
8. Kavalier, A.R.; Litt, A.; Ma, C.; Pitra, N.J.; Coles, M.C.; Kennelly, E.J.; Matthews, P.D. Phytochemical and Morphological Characterization of Hop (*Humulus lupulus* L.) Cones over Five Developmental Stages Using High Performance Liquid Chromatography Coupled to Time-of-Flight Mass Spectrometry, Ultrahigh Performance Liquid Chromatography Photodiode Array Detection, and Light Microscopy Techniques. *J. Agric. Food Chem.* **2011**, *59*, 4783–4793. [CrossRef]
9. Savithramma, N.; Linga Rao, M.; Suhrulatha, D. Screening of medicinal plants for secondary metabolites. *Middle East J. Sci. Res.* **2011**, *8*, 579–584.
10. Matsui, H.; Inui, T.; Ishimaru, M.; Hida, Y. The influence of the age of a hop plant on the quality of hop aromas in beer. *Acta Hortic.* **2013**, *1010*, 171–182. [CrossRef]
11. Van Holle, A.; Van Landschoot, A.; Roldán-Ruiz, I.; Naudts, D.; De Keukeleire, D. The brewing value of Amarillo hops (*Humulus lupulus* L.) grown in northwestern USA: A preliminary study of terroir significance. *J. Inst. Brew.* **2017**, *123*, 312–318. [CrossRef]
12. Rettberg, N.; Biendl, M.; Garbe, L. Hop Aroma and Hoppy Beer Flavor: Chemical Backgrounds and Analytical Tools—A. *Review, J. Am. Soc. Brew. Chem.* **2018**, *76*, 1402574. [CrossRef]
13. Rodolfi, M.; Chiancone, B.; Liberatore, C.M.; Fabbri, A.; Cirlini, M.; Ganino, T. Changes in chemical profile of Cascade hop cones accordingto the growing area. *J. Sci. Food Agric.* **2019**, *99*, 6011–6019. [CrossRef]
14. Campos, O.P.; Leme, F.M.; Fortuna, G.C.; Gomes, J.A.O.; Neves, C.S.; Arruda, R.C.O.; Bonfim, F.P.G. Morphological characteristics, trichomes, and phytochemistry of inflorescences of *Humulus lupulus* L: Comparison of cropping systems and varieties. *Aust. J. Crop Sci.* **2023**, *17*, 263–274. [CrossRef]
15. Grzyb, Z.S.; Piotrowski, W.; Bielicki, P.; Sas Paszt, L. Quality of Apple Maidens as Influenced by the Frequency of Application of Different Fertilizers in the Organic Nursery—Preliminary Results. *J. Fruit Ornam. Plant Res.* **2012**, *20*, 41–49. [CrossRef]
16. Solarska, E.; Sosnowska, B. The impact of plant protection and fertilization on content of bioactive substances in organic hops. *Acta Sci. Pol. Hortorum Cultus* **2015**, *14*, 93–101. Available online: https://www.cabdirect.org/cabdirect/abstract/20153213020 (accessed on 25 July 2021).
17. Jastrombek, J.M.; Faguerazzi, M.M.; de Cássio Pierezan, H.; Rufato, L.; Sato, A.J.; da Silva Ricce, W.; Marques, V.V.; Leles, N.R.; Roberto, S.R. Hop: An Emerging Crop in Subtropical Areas in Brazil. *Horticulturae* **2022**, *8*, 393. [CrossRef]
18. Köppen, W. *Climatologia: Con Un Studio De Los Climas De La Tierra*; Fundo de Cultura Econômica: Mexico City, México, 1948.
19. Gingrich, C.; Hart, J.; Christensen, N. *Hops Fertilizer Guide*; OSU Extension Catalog, Oregon State University: Corvallis, OR, USA, 2018.
20. Healey, J. *The Hops List: 265 Beer Hop Varieties from Around The World*; 2016; pp. 94–95, 109–110, 124–125, 205–206, 303–304. Available online: https://www.melkkobrew.fi/the-hops-list-by-j-healey (accessed on 1 June 2023).

21. Su, X.; Yin, Y. Aroma characterization of regional Cascade and Chinook hops (*Humulus lupulus* L.). *Food Chem.* **2021**, *364*, 130410. [CrossRef]
22. Alves, F.; Fandiño, M.; Mirás-Avalos, J.M.; Rey, B.J.; Mota, M.; Cancela, J.J. Effects of irrigation on hop (*Humulus lupulus* L.) cv. Nugget in Galicia: Yield and quality aspects. In Proceedings of the Scientific-Techinical Comission of International Hop Growers Convention, Bischoffsheim, Alsace, France, 7–11 July 2019.
23. Forteschi, M.; Porcu, M.C.; Fanari, M.; Zinellu, M.; Secchi, N.; Buiatti, S.; Passaghe, P.; Bertoli, S.; Pretti, L. Quality assessment of Cascade Hop (*Humulus lupulus* L.) grown in Sardinia. *Eur. Food Res. Technol.* **2019**, *245*, 863–871. [CrossRef]
24. Katsiotis, S.T.; Langezaal, C.R.; Scheffer, J.J.C.; Verpoorte, R. Comparative study of the essential oils from hops of various *Humulus lupulus* L. cultivars. *Flavour Fragr. J.* **1989**, *4*, 187–191. [CrossRef]
25. Adams, R.P. *Identification of Essential Oil Components by Gas Chromatography/ Mass Spectrometry*, 4th ed.; Allured Publishing Corporation: Carol Stream, IL, USA, 2007.
26. Van Den Dool, H.; Kratz, P.D. A generalization of the retention index system including linear temperature programmed gas-liquid partition chromatography. *J. Chromatogr. A* **1963**, *11*, 463–471. [CrossRef]
27. Prencipe, F.P.; Brighenti, V.; Rodolfi, M.; Mongelli, A. Development of a new high-performance liquid chromatography method with diode array and electrospray ionization-mass spectrometry detection for the metabolite fingerprinting of bioactive compounds in *Humulus lupulus* L. *J. Chromatogr. A* **2014**, *1349*, 50–59. [CrossRef]
28. Minitab, LLC. 2021. Available online: www.minitab.com (accessed on 1 June 2023).
29. Ferreira, D.F. Sisvar: A computer statistical analysis system. *Ciência E Agrotecnologia Lavras* **2011**, *35*, 1039–1042. [CrossRef]
30. Durello, R.S.; Silva, L.M.; Bogusz, S., Jr. Química do lúpulo. *Quim Nova.* **2019**, *42*, 900–919. [CrossRef]
31. Yan, D.; Wong, Y.F.; Tedone, L.; Shellie, R.; Marriott, P.; Whittock, S.; Koutoulis, A. Chemotyping of new hop (*Humulus lupulus* L.) genotypes using comprehensive two-dimensional gas chromatography with quadrupole accurate mass-time-of-flight mass spectrometry. *J. Chromatogr. A* **2017**, *1536*, 110–121. [CrossRef] [PubMed]
32. Almaguer, C.; Schönberger, C.; Gastl, M.; Arendt, E.K.; Becker, T. *Humulus lupulus*—A story that begs to be told. A review. *J. Inst. Brew.* **2014**, *120*, 289–314. [CrossRef]
33. Biendl, M.; Engelhard, B.; Forster, A.; Gahr, A.; Lutz, A.; Mitter, W.; Schönberger, C. *Hops: Their Cultivation, Composition and Usage*; Fachverlag Hans Carl: Nuremberg, Germany, 2014.
34. Variety manual. In: Hop Growers of America Website. [Yakima, WA: Hop Growers of America]. 2018. Available online: https://www.usahops.org/ (accessed on 10 August 2021).
35. Kishimoto, T. Investigations of hop-derived odor-active compounds in beer, hop flavor and aroma. In *Proceedings of the 1st International Brewers Symposium*; American Society of Brewing Chemists, Master Brewers Association of the Americas, UEA: St. Paul, Brazil, 2009.
36. Holopainen, J.K. Multiple functions of inducible plant volatiles. *Trends Plant Sci.* **2004**, *9*, 529–533. [CrossRef] [PubMed]
37. Yan, D.; Wong, Y.F.; Shellie, R.; Marriott, P.; Whittock, S.; Koutoulis, A. Assessment of the phytochemical profiles of novel hop (*Humulus lupulus* L.) cultivars: A potential route to beer crafting. *Food Chem.* **2018**, *275*, 15–23. [CrossRef]
38. Krofta, K.; Kučera, J. Mathematical model for prediction of alpha acid contents from meteorological data for 'Saaz' aroma variety. *Acta Hortic.* **2009**, *848*, 131–139. [CrossRef]
39. Sharp, D.; Townsend, M.; Qian, Y.; Shellhammer, T.H. Harvest Maturity on the Chemical Composition of Cascade and Willamette Hops. *J. Am. Soc. Brew. Chem.* **2014**, *72*, 231–238. [CrossRef]
40. Prado, M.R. *Nutrição de Plantas*; UNESP: São Paulo, Brazil, 2008.
41. De Keukeleire, J.; Janssens, I.; Heyerick, A.; Ghekiere, G.; Cambie, J.; Roldán-Ruiz, I.; De Keukeleire, D. (Relevance of organic farming and effect of climatological conditions on the formation of α-acids, β-acids, desmethylxanthohumol, and xanthohumol in hop (*Humulus lupulus* L.). *J. Agric. Food Chem.* **2007**, *55*, 61–66. [CrossRef]
42. McAdam, E.L.; Vaillancourt, R.E.; Koutoulis, A.; Whittock, S.P. Quantitative genetic parameters for yield, plant growth and cone chemical traits in hop (*Humulus lupulus* L.). *BMC Genet.* **2014**, *15*, 22. [CrossRef]
43. Sirrine, R.; Lizotte, E.; Brown, D.; O'Brien, T.; Leach, A. Estimated Costs of Producing Hops in Michigan. Available online: https://www.canr.msu.edu/uploads/resources/pdfs/estimated_costs_of_producing_hops_in_michigan_(e3236).pdf (accessed on 15 June 2023).
44. Mozny, M.; Tolasz, R.; Nekovar, J.; Sparks, T.; Trnka, M.; Zalud, Z. The impact of climate change on the yield and quality of Saaz hops in the Czech Republic. *Agric. For. Meteorol.* **2009**, *149*, 913–919. [CrossRef]

horticulturae

Article

Photosynthesis, Biochemical and Yield Performance of Grapevine Hybrids in Two Rootstock and Trellis Height

Francisco José Domingues Neto [1,*], Adilson Pimentel Junior [1], Lenon Romano Modesto [2], Mara Fernandes Moura [3], Fernando Ferrari Putti [4], Carmen Silvia Fernandes Boaro [5], Elizabeth Orika Ono [5], João Domingos Rodrigues [5] and Marco Antonio Tecchio [1]

1 School of Agriculture, São Paulo State University (UNESP), José Barbosa de Barros Street, Botucatu 18610-307, SP, Brazil
2 Plant Genetic Resources, Federal University of Santa Catarina (UFSC), Road Admar Gonzaga, Florianópolis 88040-900, SC, Brazil
3 Agronomic Institute of Campinas (IAC), Street Luiz Pereira do Santos 1500, Jundiaí 13214-820, SP, Brazil
4 School of Sciences and Engineering, São Paulo State University (UNESP), Tupã 17602-496, SP, Brazil
5 Institute of Biosciences, São Paulo State University (UNESP), Botucatu 18610-307, SP, Brazil; carmen.boaro@unesp.br (C.S.F.B.)
* Correspondence: fjdominguesneto@hotmail.com

Abstract: The interaction between variety, rootstock, and trellis height is important for grapevine management, mainly for producing new varieties of grapes for juice and wine in new wine-growing regions with high production potential. Then, this study aimed to evaluate the rootstocks and trellis height influence on photosynthesis, biochemical, and yield performance for grapevine hybrids. The experiment was carried out in a randomized block design using two factors, rootstocks ('IAC 766' and '106-8 Mgt') and trellis height (until 1.6 and 2.0 m), evaluated for two grapevine hybrids (IAC 138-22 'Maximo' and 'BRS Violeta'). During grapevine flowering, it was evaluated photosynthesis and biochemical performance, for this, the gaseous exchanges were measured using the open system photosynthesis equipment with a CO_2 analyzer and water vapor by infrared radiation, being net assimilation rate of CO_2, stomatal conductance, transpiration rate, internal CO_2 concentration, water use efficiency, carboxylation efficiency (Rubisco), and the flux density of photosynthetically active photons. At the stages of grapevine flowering and ripening berries were evaluated the antioxidant enzymes (peroxidase (POD), superoxide dismutase (SOD), and catalase (CAT)), total soluble proteins, chlorophylls, and SPAD. The interaction between rootstock and trellis heigh influenced varieties' photosynthesis, biochemical, and yield performance. In conclusion under subtropical conditions, better photosynthesis, biochemical, and yield performance were observed when both cultivars were grafted on the 'IAC 766' rootstock. The 'IAC 138-22 Maximo' was trained until 2.0 and grafted on the 'IAC 766' rootstock, increasing grape production and photosynthesis efficiency. In addition, this variety was more productive than 'BRS Violeta'.

Keywords: antioxidant enzymes; gas exchange; Rubisco enzyme; *Vitis* spp.; water use efficiency

Citation: Domingues Neto, F.J.; Pimentel Junior, A.; Modesto, L.R.; Moura, M.F.; Putti, F.F.; Boaro, C.S.F.; Ono, E.O.; Rodrigues, J.D.; Tecchio, M.A. Photosynthesis, Biochemical and Yield Performance of Grapevine Hybrids in Two Rootstock and Trellis Height. *Horticulturae* **2023**, *9*, 596. https://doi.org/10.3390/horticulturae9050596

Academic Editor: Darius Kviklys

Received: 2 May 2023
Revised: 15 May 2023
Accepted: 16 May 2023
Published: 18 May 2023

1. Introduction

Grapevine hybrids are widely used in viticulture for wine and juice production, looking for adaptation to climate change and disease resistance, resulting in lower environmental impact and food security because of the pesticides reduction [1–3]). These grapevine varieties are growing mainly for the juice industry and table grape production [4]. However, in tropical and subtropical regions grapevine hybrids are also destined to make wine [5]. Like in the case of *Vitis vinifera* (L.) cultivation, grapevine hybrids are grown on rootstocks to get resistance against Phylloxera, soil fungi, and nematodes [6,7]. In addition, the rootstock is chosen based on adaptation to environmental conditions and field management [5,8,9], in this way, new viticultural and promising areas for producing these grapes must be explored. In subtropical regions, the rootstocks are selected to be resistant to phylloxera and adapted to a humid temperate climate with dry winters and hot summers [5,6]. The interaction between cultivars and rootstocks influences photosynthesis efficiency, grapevine physiology, and production [10]. The rootstock changes the production of biochemical compounds linked with physiological stress during grapevine cultivation [11,12], occurring when grapevines suffer from stress are a decrease in photosynthesis efficiency and pigment production, modifying enzymatic activity, such as Rubisco and reactive oxygen species (ROS) [11]. Moreover, the variety decreases the stomatal conductance and increases the leaf resistance to CO_2 transport from the atmosphere to the mesophyll cells [12]; resulting in lower ATP availability that affects ribulose-1,5-bisphosphate (RuBP) regeneration, thus limiting the rate of CO_2 fixation [12]. Furthermore, the interaction between cultivar and rootstock changes ROS activity, which plays a role in the signaling route, as a key regulator of processes such as growth, development, and plant metabolism [13].

Another factor that influences grapevine adaptation in different regions is the training system [14]. The training system influences grapevine reserves, leaf development, and photosynthesis efficiency, resulting in changes in enzymatic activity [4,11,15]. Vertical shoot positioning (VSP) is the most used trellis system to grow grapevines in Brazil [4,9]. In this trellis system, the grapevine can be trained in different sizes to adapt to vegetative growth [16]. The *Vitis vinifera* cultivars are trained until 1.35 to 1.60 m from de ground [17]. The *Vitis labrusca* under subtropical conditions are trained until 1.00 to 1.80 m from the ground [12,18,19]. However, the hybrids can be more vigorous than these two species, then new training high needs to be tested (e.g., 2.00 m). In addition, vegetative growth depends on the interaction between different varieties and rootstocks. This study aimed to evaluate the physiology, fruit production, and quality of two grapevine hybrids trained in two trellis sizes and grafted onto two rootstocks under subtropical conditions looking for better adaption during grapevine cultivation.

2. Materials and Methods

2.1. Localization and Climate Conditions

The experiment was conducted at the Fruits Research Center of the Agronomic Institute (IAC), in Jundiaí, state of São Paulo, Brazil (23°06′ S, 46°55′ W, 745 m above sea level). The climate in this region is classified as Cfb, according to the Köppen, with 1400 mm of annual rainfall, 19.5 °C of average temperature, and 70.6% of relative air humidity. The soil of the experimental area was classified as dystrophic Cambisol haplic, characterized by low amounts of clay, organic matter, aluminum, and iron.

Rootstocks were planted in September 2009, at a spacing of 2.5 m between rows and 1.0 m between plants, and the cultivars were grafted in July 2010. For grape yield, the grapevines were pruned in July 2016 leaving one bud per branch and 5% hydrogen cyanamide was applied direct on the buds to homogenize sprouting. The grape harvest was conducted in December of the same year.

2.2. Treatments

The experiment was laid out in a complete randomized block using two inter-specific crossing of grapevine varieties, the 'IAC 138-22 Maximo' ('Seibel 11342' × 'Syrah') and the 'BRS Violeta' ('Niagara Rosada' × 'Bordô'), with five replicates (blocks), consisting of three plants each. The treatments were two rootstocks, 'IAC 766 Campinas' ('Ripária do Traviú' × *Vitis caribaea*) and '106-8 Mgt' [Riparia × (Cordifolia × Rupestris)] and two trellis height, until 1.6 m and 2.0 m from the ground, evaluated for each variety. The grapevines were trimmed during vegetative growth to maintain the trellis height, in both varieties.

2.3. Sampling

In the full flowering stage, the photosynthesis and biochemical assay were conducted using four complete leaves (limb and petiole) located on the opposite side of the bunch using three grapevines per block for each treatment with replicate per variety (n = 60 leaves per treatment). In addition, the SPAD index (Soil Plant Analysis Development) and biochemical assay were evaluated again in the same leaves during early berry ripening (grape maturation). The photosynthesis assay was conducted on the field using no detached leaves. For the biochemical assay, leaves were sampled, packed in foil, frozen in liquid nitrogen, and stored in an ultra-freezer at −82 °C until evaluation.

2.3.1. Photosynthesis Assay

The photosynthesis assay was conducted with an open photosynthesis system using the Infrared gas analyzer (IRGA) (model LI-64009, USA). The CO_2 assimilation rate (A), transpiration rate (E), the internal concentration of CO_2 (Ci), and stomatal conductance (gs), were calculated according to Von Caemmerer and Farquhar [20]. The water efficiency use (WUE) was determined using the ratio between CO_2 assimilation and transpiration rate. The carboxylation efficiency of the enzyme ribulose 1,5-bisphosphate carboxylase (Rubisco) (A/Ci) was calculated using the ratio between the CO_2 assimilation rate and the internal concentration of CO_2 in the leaf. For chlorophyll fluorescence, the leaves were covered with aluminum paper, kept in the dark for 30 min, and treated for 6 s with a saturation pulse of 10,000 μmol m^{-2} s^{-1} of photosynthetically active photon flux density (DFFF) to obtain the dark-adapted maximum fluorescence (Fm), the light-adapted maximum fluorescence (Fm'), the dark-adapted minimum fluorescence (Fo) and the light-adapted minimum fluorescence (Fo'). An 1150 μmol m^{-2} s^{-1} DFFF actinic light pulse of 15 s of duration was given between each saturation pulse. The maximum quantum yield (Fv/Fm) was calculated according to Kitajima and Butler [21], the effective quantum yield (ϕPSII) was calculated according to Genty et al. [22], the photochemical quenching (qP) was calculated according to Schreiber et al. [23], the non-photochemical quenching (NPQ) was calculated according to Bilger and Björkman [24], The electron transport rate (ETR) and quantum unregulated non-photochemical energy loss in photosystem II (ϕNO) were calculated according to Klughammer and Schreiber [25]. The quantum yield of non-photochemical regulated energy loss in photosystem II (ϕNPQ) was calculated according to Klughammer and Schreiber [25].

The SPAD index was evaluated using the SPAD (Model SPAD-502, Hangzhou Mindfull Technology Co.,Ltd, Tokyo, Japan) to sample three points (left, middle and right size) per leaf (n = 180).

2.3.2. Biochemical Assay

The biochemical assay was conducted using spectrophotometry (model BEL Photonics, SP UV/VIS, Brazil) and all analyses were done using technical triplicate.

The concentration of chlorophylls (a, b, and total) was determined using 100 g of fresh mass according to Sims and Gamon [26]. For protein and enzyme quantification, it was mixed 100 mg of fresh leaf and 2 mL of 0.1 mol L^{-1} potassium phosphate buffer at pH 6.8 with the addition of 100 mg polyvinylpolypyrrolidone (PVPP). Total soluble proteins were quantified according to the methodology proposed by Bradford [27] and expressed in mg of fresh mass g^{-1} protein. Superoxide dismutase (SOD) activity was determined according to the methodology proposed by Giannopolitis and Ries [28] and expressed in U/mg protein. Peroxidase (POD) activity was determined according to Teisseire and Guy [29] and expressed as μmol of purpurogaline min^{-1} mg^{-1} protein. Catalase activity (CAT) was performed according to the methodology proposed by Peixoto et al. [30] and expressed in μKatμg $prot^{-1}$. In the case of lipid peroxidation quantification, it was mixed 300 mg of fresh leaves and 5 mL of a solution containing 0.25% thiobarbituric acid (TBA) and trichloroacetic acid (TCA) at 10%. Lipid peroxidation (TBAR) was determined according to Rama Devi and Prasad [31] and expressed in μmol g^{-1} fresh mass.

2.4. Harvest, Yield, and Must Quality

Harvest was carried out when each grapevine hybrid reached its technological maturity. All bunches per plant were harvested and weighed to determine yield. In order to determine fruit quality, 10 bunches per plant were sampled (n = 1500 bunches per treatment) to evaluate the pH, soluble solids (SS), titratable acidity (TA), SS/TA ratio, and reducing sugar [32].

2.5. Statistical Analyses

The data was checked about normal distribution and homoscedasticity. Then, the data were analyzed using the variance analysis for two factors in randomized blocks. When significant in the variance analysis, the interaction between the two factors (rootstock and trellis height) or each factor separately was evaluated using Tukey's test ($p < 0.05$). In addition, principal component analysis was performed to characterize the interaction between the grapevine hybrids with each rootstock and trellis height. The variance and Tukey's tests were performed using the software SISVAR (Ferreira, 2014). The principal components analysis was performed using the software SAS [33].

3. Results

3.1. Impact of Rootstock and Trellis Height on Variety ('IAC 138-22 Maximo'): Physiological, Biochemical Parameters and Yield

During flowering, the qP increased 0.8 when the 'IAC 138-22 Maximo' was trained until 1.6 m onto 'IAC 766' and until 2.0 m onto '106-8 Mgt' then until 1.6 m onto '106-8 Mgt' (Table 1). In addition, the same combinations resulted in higher NPQ (plus 0.35) than the variety trained until 1.6 m onto 'IAC 766' and until 2.0 m onto the '106-8 Mgt'. However, the ETR presented the resulting inversely proportional, the ETR increased 25.5 μmol m^{-2} s^{-1} electrons to the 'IAC 138-22 Maximo' trained until 1.6 m onto 'IAC 766' and until 2.0 m onto the '106-8 Mgt' than the other combinations (Table 1). The 'IAC 138-22 Maximo' trained until 2.0 m onto the rootstock 'IAC 766', decreased 0.08 mol m^{-2} s^{-1} the stomatal conductance (gs) and 5.24 μmol m^{-2} s^{-1} CO_2 the assimilation (A) then trained until 1.6 m onto the '106-8 Mgt' (Table 1). In addition, the variety trained until 1.6 m onto the '106-8 Mgt' increased 1.1 the WUE compared with the other combinations. The transpiration ration (E) was 1.12 mmol m^{-2} s^{-1} water vapor bigger to this combination than the variety trained until 2.0 m onto the same rootstock, '106-8 Mgt'. However, the Ci increased 72.1 μmol mol^{-1} CO_2 when the variety was grafted onto 'IAC 766' and then trained until 2.0 m onto the '106-8 Mgt'. The lowest Fv/Fm, 0.86, was observed when this cultivar was trained until 1.6 m and grafted onto 'IAC 766', with no interaction between these two factors (Table 2). The 'IAC 138-22 Maximo' increased 9.25 mg the chlorophyll total content in 100 mg of leaves and 4.11 the SPAD index grafted onto the rootstock 'IAC 766' and then onto the '106-8 Mgt' (Table 2). During berry ripening the chlorophyll a, b, and total decreased 4.72, 3.32, and 8.05 mg, when the variety was trained until 2 m onto the '106-8 Mgt' then trained until 2.0 m onto 'IAC 766' and trained until 1.6 m onto the '106-8 Mgt', respectively (Table 1). In addition, the SPAD index during the same phenological stage increased by 2.99 when the variety 'IAC 138-22 Maximo' was grafted onto 'IAC 766' and then onto the '106-8 Mgt' (Table 2). Training the variety until 2.0 m onto the 'IAC 766' decreased 21.57 mg of CAT activity during the flowering than in other combinations (Table 1). However, the variety grafted onto this rootstock, 'IAC 766', increased 4.97 μmol of TBAR activity during flowering, and 366.53 mg of SOD and 4.75 mg of POD during berry ripening (Tables 1 and 2).

About the yield, the variety 'IAC 138-22 Maximo' trained until 2.0 m produced plus 1.6 kg of grape per plant with plus 1.06° Brix of SSC decreasing 0.8 the pH than trained until 1.6 m (Table 2 and Figure 1. Onto the rootstock 'IAC 766', the variety showed minus 0.12 percentage of tartaric acidity on must resulting in plus 3.35 to maturation index than onto '106-8 Mgt'.

Table 1. Interaction between two trellis heights and rootstock to photosynthesis and biochemical performance of grapevine 'IAC 138-22 Maximo' under subtropical conditions during flowering and berry ripening stages.

	Trellis Height	Rootstock	
		'IAC 766'	'106-8 Mgt'
Flowering			
qP	1.6 m	0.57 ± 0.02 aA	0.49 ± 0.01 bB
	2.0 m	0.54 ± 0.02 bA	0.56 ± 0.01 aA
NPQ	1.6 m	2.03 ± 0.07 bB	2.38 ± 0.01 aA
	2.0 m	2.64 ± 0.14 aA	2.08 ± 0.04 bB
ETR (μmol m^{-2} s^{-1} electrons)	1.6 m	148.58 ± 7.07 aA	133.86 ± 5.51 bB
	2.0 m	123.03 ± 3.16 bB	158.15 ± 9.93 aA
gs (mol m^{-2} s^{-1})	1.6 m	0.23 ± 0.01 aB	0.25 ± 0.01 aA
	2.0 m	0.17 ± 0.001 bB	0.22 ± 0.01 bA
E (mmol m^{-2} s^{-1} water vapor)	1.6 m	7.06 ± 0.42 aB	7.59 ± 0.30 aA
	2.0 m	6.82 ± 0.24 aA	6.47 ± 0.17 bB
WUE	1.6 m	4.53 ± 0.25 aA	4.42 ± 0.11 bA
	2.0 m	4.56 ± 0.18 aB	5.66 ± 0.15 aA
A (μmol m^{-2} s^{-1} CO_2)	1.6 m	34.75 ± 1.38 aA	36.17 ± 1.25 aA
	2.0 m	29.51 ± 1.51 bB	38.82 ± 2.48 aA
Ci (μmol mol^{-1} CO_2)	1.6 m	179.93 ± 3.74 aA	148.67 ± 5.70 aB
	2.0 m	175.31 ± 5.76 aA	103.21 ± 6.11 bB
A/Ci	1.6 m	0.19 ± 0.001 aB	0.26 ± 0.02 bA
	2.0 m	0.17 ± 0.01 bB	0.32 ± 0.01 aA
Cl b (mg 100 g^{-1} leaves)	1.6 m	16.40 ± 2.01 aA	12.22 ± 0.43 aB
	2.0 m	13.02 ± 0.66 bA	12.69 ± 2.14 aA
POD (μmol mg^{-1} min^{-1} protein)	1.6 m	36.52 ± 1.37 aA	34.83 ± 1.52 aA
	2.0 m	39.15 ± 0.83 aA	32.97 ± 2.26 aB
CAT (μg $mKat^{-1}$ protein)	1.6 m	5.68 ± 1.26 aA	6.11 ± 3.20 aA
	2.0 m	2.12 ± 0.77 bB	4.70 ± 1.69 aA
Berry ripening			
Cl a (mg 100 g^{-1} leaves)	1.6 m	35.18 ± 0.93 bA	36.03 ± 3.25 aA
	2.0 m	55.28 ± 2.60 aA	31.31 ± 3.35 bB
Cl b (mg 100 g^{-1} leaves)	1.6 m	16.03 ± 0.23 bA	17.67 ± 1.71 aA
	2.0 m	26.57 ± 0.67 aA	14.35 ± 1.98 bB
Cl total (mg 100 g^{-1} leaves)	1.6 m	51.21 ± 0.84 bA	53.71 ± 4.55 aA
	2.0 m	81.85 ± 3.20 aA	45.66 ± 5.32 bB
SOD (mg U^{-1} protein)	1.6 m	4665.88 ± 19.91 bA	3689.94 ± 34.25 aB
	2.0 m	5027.41 ± 17.90 aA	3463.00 ± 32.14 aB
CAT (μg $mKat^{-1}$ protein)	1.6 m	14.61 ± 2.61 bA	14.12 ± 0.41 bA
	2.0 m	54.62 ± 1.58 aB	86.19 ± 3.98 aA

±standard deviations. Means followed by the same lower-case letter in the column and upper-case letter in the row are not different from each other according to the Tukey's test at 5% probability. Photochemical quenching (qP), non-photochemical quenching (NPQ), electron transport rate (ETR), stomatal conductance (gs), transpiration rate (E), water use efficiency (WUE), carboxylation efficiency (A/Ci), CO_2 assimilation rate (A), internal carbon concentration (Ci), Chlorophyll b, peroxidase (POD) and catalase (CAT) enzymes activities at flowering and chlorophyll (Cl) a, b and total, activities of the superoxide dismutase (SOD) and catalase (CAT).

Table 2. Photosynthesis, biochemical and yield performance of grapevine 'IAC 138-22 Maximo' during flowering and berry ripening stages for two trellis height and rootstock under subtropical conditions.

	Trellis Height		Rootstock	
	1.6 m	**2.0 m**	**'IAC 766'**	**'106-8 Mgt'**
Flowering				
F_v/F_m	0.86 ± 0.02 b	0.90 ± 0.02 a	0.86 ± 0.01 b	0.90 ± 0.02 a
SPAD index	33.47 ± 2.79 a	32.58 ± 2.38 a	35.08 ± 1.27 a	30.97 ± 1.63 b
Cl a (mg 100 g^{-1} leaves)	33.27 ± 5.69 a	30.82 ± 4.34 a	35.55 ± 3.41 a	28.55 ± 3.88 a
Cl total (mg 100 g^{-1} leaves)	47.59 ± 8.10 a	43.68 ± 5.61 a	50.26 ± 5.54 a	41.01 ± 5.12 b
SOD (mg U^{-1} protein)	4141.42 ± 38.62 a	4331.28 ± 51.27 a	4561.96 ± 39.92 a	3910.74 ± 24.95 b
Lipid peroxidation (µmol g^{-1} leaves)	11.02 ± 1.04 a	9.59 ± 0.09 a	12.79 ± 1.06 a	7.82 ± 0.08 b
Berry ripening				
SPAD index	35.86 ± 2.93 a	37.09 ± 1.88 a	37.97 ± 2.25 a	34.98 ± 1.66 b
POD (µmol mg^{-1} min^{-1} protein)	23.45 ± 3.64 a	22.64 ± 3.18 a	25.37 ± 2.66 a	20.72 ± 2.07 b
Yield				
Yield (kg^{-1} plant)	4.16 ± 0.08 b	5.76 ± 1.04 a	5.07 ± 0.09 a	4.85 ± 0.08 a
pH	3.39 ± 0.09 a	3.31 ± 0.08 b	3.35 ± 0.08 a	3.36 ± 0.11 a
Soluble solids (°Brix)	14.19 ± 1.36 b	15.25 ± 1.55 a	15.13 ± 1.63 a	14.31 ± 1.36 a
Titratable acidity (% tartaric acid)	0.98 ± 0.16 a	0.94 ± 0.16 a	0.90 ± 0.18 b	1.02 ± 0.11 a
SS/TA ratio	14.91 ± 3.15 a	16.78 ± 4.31 a	17.52 ± 4.40 a	14.17 ± 2.25 b
Reducing sugar (%)	10.56 ± 1.89 a	10.98 ± 1.70 a	10.58 ± 1.53 a	10.96 ± 2.04 a

±standard deviations. Means followed by the same lower-case letter in the column are not different from each other according to the Tukey test at 5% probability. Quantum yield (Fv/Fm), SPAD index (Soil Plant Analysis Development), the content of chlorophyll (Cl), superoxide dismutase (SOD), peroxidase activity (POD), soluble solids/titratable acidity ratio (SS/TA).

3.2. Impact of Rootstock and Trellis Height on Variety ('BRS Violeta'): Physiological, Biochemical Parameters and Yield

During flowering, the variety 'BRS Violeta' increased 0.14 qP when trained until 1.6 m onto the 'IAC 766' than using other combinations (Table 3). In addition, the NPQ decreased by 0.71 using this combination than the variety grafted onto the '106-8 Mgt' in both trellis heights. The variety trained until 1.6 m onto the 'IAC 766' increased 10.24 µmol m^{-2} s^{-1} electrons the ETR and 4.73 µmol m^{-2} s^{-1} CO_2 the A than using the other combinations. The 'IAC 766' increased by 0.05 the Fv/Fm and 3.37 mmol m^{-2} s^{-1} water vapor the E and decreased by 0.94 the WUE on leaves of 'BRS Violeta' than the '106-8 Mgt' (Table 4). However, the Ci increased 19.16 µmol m^{-2} s^{-1} CO_2 when the variety was trained until 2.0 m then until 1.6 m.

The chlorophyll a content on leaves during flowering decreased by 5.53 mg 100 g^{-1} when the 'BRS Violeta' was trained until 1.6 m onto the '106-8 Mgt' than in the other combinations (Table 3). In addition, the SPAD index decreased by 3.59 on leaves from the variety grafted onto the '106-8 Mgt' and then trained until 2.0 m onto the 'IAC 766'. During berry ripening, the chlorophyll content, a, b, and total, increased 7.05, 3.64, and 11.86 mg 100 g^{-1} on leaves of 'BRS Violeta' trained until 1.6 m onto the 'IAC 766' than other combinations, respectively. Contributing to this result, the SPAD index in the same phenological stage increased by 3.02 when the variety was grafted onto the 'IAC 766' and then onto the '106-8 Mgt' (Table 4). Moreover, POD and SOD increased, 4.89 µmol and 1947.92 mg, during flowering when this rootstock was used compared with the '106-8 Mgt', respectively (Table 4 and Figure 2). During berry ripening, the POD activity increased by

12.64 μmol on leaves from the variety trained until 2.0 m onto the 'IAC 766' and grafted onto the '106-8 Mgt' in both trellis height then trained until 1.6 m onto the 'IAC 766' (Table 3). Moreover, the rootstock '106-8 Mgt' increased CAT and SOD activity, plus 19.05 and 682.42 mg, then the 'IAC 766' during the same phenological stage (Tables 3 and 4). However, using trellis heights until 2.0 m decreased SOD activity by 1005.63 mg than using trellis height until 1.6 m (Table 4). The variety 'BRS Violeta' produced plus 0,38 kg of grape per plant and increased by 0.1 the pH when it was grafted onto the 'IAC 766' and then onto the '106-8 Mgt'. However, higher SS contents were observed when this cultivar was trained to a high trellis system.

Table 3. Interaction between two trellis heights and rootstock to photosynthesis and biochemical performance of grapevine 'BRS Violeta' under subtropical conditions during flowering and berry ripening stages.

	Trellis Height	Rootstock	
		'IAC 766'	**'106-8 Mgt'**
Flowering			
qP	1.6 m	0.55 ± 0.02 aA	0.44 ± 0.02 aA
	2.0 m	0.45 ± 0.01 bA	0.41 ± 0.01 bB
NPQ	1.6 m	2.53 ± 0.14 bB	3.29 ± 0.09 aA
	2.0 m	3.03 ± 0.08 aB	3.24 ± 0.10 aA
ETR (μmol m^{-2} s^{-1} electrons)	1.6 m	124.38 ± 3.50 aA	114.14 ± 5.64 aB
	2.0 m	100.70 ± 5.96 bA	100.11 ± 4.87 bA
gs (mol m^{-2} s^{-1})	1.6 m	0.28 ± 0.01 aA	0.14 ± 0.001 aB
	2.0 m	0.26 ± 0.001 bA	0.13 ± 0.001 aB
A (μmol m^{-2} s^{-1} CO_2)	1.6 m	36.13 ± 1.27 aA	31.40 ± 1.06 aB
	2.0 m	31.00 ± 0.38 bA	29.74 ± 1.30 aA
Ci (μmol mol^{-1} CO_2)	1.6 m	129.29 ± 8.21 bA	132.43 ± 3.20 bA
	2.0 m	162.48 ± 3.37 aA	151.59 ± 7.21 aA
SPAD index	1.6 m	29.11 ± 2.72 aA	21.57 ± 0.57 bB
	2.0 m	30.34 ± 0.83 aA	26.75 ± 1.35 aB
Cl a (mg 100 g^{-1} leaves)	1.6 m	47.19 ± 7.00 aA	39.12 ± 3.54 bB
	2.0 m	44.65 ± 1.53 aA	48.28 ± 4.67 aA
CAT (μg $mKat^{-1}$ protein)	1.6 m	35.55 ± 3.33 aA	32.65 ± 0.98 aB
	2.0 m	19.86 ± 0.86 bA	9.41 ± 0.95 bB
Berry ripening			
Cl a (mg 100 g^{-1} leaves)	1.6 m	40.93 ± 3.44 aA	23.44 ± 2.00 bB
	2.0 m	32.98 ± 2.00 bA	29.75 ± 2.16 aA
Cl b (mg 100 g^{-1} leaves)	1.6 m	17.48 ± 2.58 aA	10.30 ± 0.79 aB
	2.0 m	13.84 ± 1.58 bA	12.68 ± 2.02 aA
Cl total (mg 100 g^{-1} leaves)	1.6 m	58.42 ± 5.86 aA	33.75 ± 2.78 bB
	2.0 m	46.82 ± 1.60 bA	42.43 ± 3.27 aA
POD (μmol mg^{-1} min^{-1} protein)	1.6 m	15.72 ± 3.36 bB	28.36 ± 2.67 aA
	2.0 m	24.68 ± 1.57 aB	30.27 ± 1.92 aA

Table 3. *Cont.*

	Trellis Height	Rootstock	
		'IAC 766'	'106-8 Mgt'
CAT (µg mKat^{-1} protein)	1.6 m	24.04 ± 4.03 aB	49.40 ± 2.93 bA
	2.0 m	29.90 ± 1.89 aB	82.95 ± 8.44 aA

±standard deviations; Means followed by the same lower-case letter in the column and upper-case letter in the row are not different from each other by Tukey's test at 5% of probability; Note: Photochemical quenching (qP), non-photochemical quenching (NPQ), electron transport rate (ETR), stomatal conductance (gs), CO_2 assimilation rate (A), internal carbon concentration (Ci), SPAD index (Soil Plant Analysis Development), chlorophyll a content and catalase activity (CAT), peroxidase (POD) and catalase (CAT) activity.

Table 4. Photosynthesis, biochemical, and yield performance of grapevine 'BRS Violeta' during flowering and berry ripening stages for two trellis heights and rootstock under subtropical conditions.

Variable	Trellis Height		Rootstock	
	1.6 m	2.0 m	'IAC 766'	'106-8 Mgt'
Flowering				
Fv/Fm	0.84 ± 0.03 a	0.86 ± 0.03 a	0.87 ± 0.01 a	0.82 ± 0.01 b
E (mmol m^{-2} s^{-1} water vapor)	7.76 ± 2.00 a	8.37 ± 1.66 a	9.75 ± 0.38 a	6.38 ± 0.56 b
WUE	4.42 ± 0.62 a	3.83 ± 0.44 a	3.66 ± 0.29 b	4.60 ± 0.44 a
A/Ci	0.19 ± 0.01 a	0.21 ± 0.01 a	0.21 ± 0.03 a	0.22 ± 0.02 a
Cl b (mg 100 g^{-1} leaves)	18.35 ± 2.43 a	20.03 ± 1.38 a	20.05 ± 2.23 a	18.33 ± 1.67 a
Cl total (mg 100 g^{-1} leaves)	61.50 ± 8.96 a	66.50 ± 4.21 a	65.97 ± 6.70 a	62.03 ± 4.21 a
POD (µmol mg^{-1} min^{-1} protein)	31.63 ± 2.20 a	32.30 ± 4.07 a	34.41 ± 2.20 a	29.52 ± 4.07 b
SOD (mg U^{-1} protein)	6685.52 ± 11.71 a	6681.96 ± 13.43 a	7657.70 ± 26.93 a	5709.78 ± 35.43 b
Lipid peroxidation (µmol g^{-1} leaves)	9.76 ± 1.04 a	9.63 ± 1.03 a	9.17 ± 2.03 a	10.22 ± 2.04 a
Berry ripening				
SPAD index	44.14 ± 1.95 a	45.12 ± 2.07 a	46.14 ± 1.66 a	43.12 ± 0.80 b
SOD (mg U^{-1} protein)	3105.15 ± 29.44 b	4110.78 ± 37.90 a	3266.75 ± 46.78 b	3949.17 ± 37.35 a
Lipid peroxidation (µmol g^{-1} leaves)	8.55 ± 0.90 a	9.44 ± 0.94 a	8.76 ± 0.94 a	9.22 ± 0.93 a
Yield				
Yield (kg^{-1} plant)	1.52 ± 0.06 a	1.55 ± 0.09 a	1.73 ± 0.09 a	1.35 ± 0.08 b
pH	3.58 ± 0.16 a	3.55 ± 0.12 a	3.62 ± 0.16 a	3.52 ± 0.10 a
Soluble solids (°Brix)	15.99 ± 0.49 a	16.32 ± 0.53 a	16.32 ± 0.51 a	15.99 ± 0.52 a
Titratable acidity (% tarctaric acid)	0.68 ± 0.09 a	0.75 ± 0.15 a	0.73 ± 0.15 a	0.70 ± 0.11 a
SS/TA	23.28 ± 2.86 a	24.50 ± 3.09 a	23.88 ± 3.46 a	23.90 ± 2.56 a
Reducing sugar (%)	12.64 ± 1.38 a	13.37 ± 1.42 a	13.27 ± 1.73 a	12.74 ± 1.04 a

±standard deviations; Means followed by the same lower-case letter in the row within the same factor are not different from each other by the test of Tukey at 5% probability. Note: Quantum yield (Fv/Fm), transpiration rate (E), water use efficiency (WUE), carboxylation efficiency (A/Ci), contents of chlorophyll (Cl) b and total, peroxidase (POD), and superoxide dismutase (SOD) activity, lipid peroxidation at flowering and SPAD index (Soil Plant Analysis Development), superoxide dismutase (SOD), soluble solids/titratable acidity ratio (SS/TA).

3.3. *Principal Component Analysis (PCA)*

The first principal component explained 45.9% of the total variation, characterizing the difference between the two grapevine hybrids using photosynthesis, biochemical, and yield performance (Figure 1A,B and Figure 2). The variables NPQ, E, SOD activity, chlorophylls content during flowering, SPAD index, POD and PER during maturation, and SSC, SSC/TA, and RS were positively correlated. However, these variables were negatively correlated with qP, ETR, gs, yield, and TA. The variety 'IAC 138-22 Maximo' presented higher qP,

ETR, and gs, and lower NPQ and E than 'BRS Violeta' (Figure 1A,B and Tables 1 and 3). However, the 'BRS Violeta' showed higher SOD activity and chlorophylls content (a, b, and total) during flowering than the 'IAC 138-22' (Figure 1A,B and Tables 1 and 3). In Addition, 'BRS Violeta' showed higher SPAD index, POD, and PER activities during maturation than 'IAC 138-22 Maximo'. About yield, the variety 'IAC 138-22 Maximo' produced more grapes than 'BRS Violeta'. Despite that, fruit from 'BRS Violeta' showed higher SSC and RS, and lower TA on must than 'IAC 138-22 Maximo' (Figure 1A,B and Tables 2 and 4).

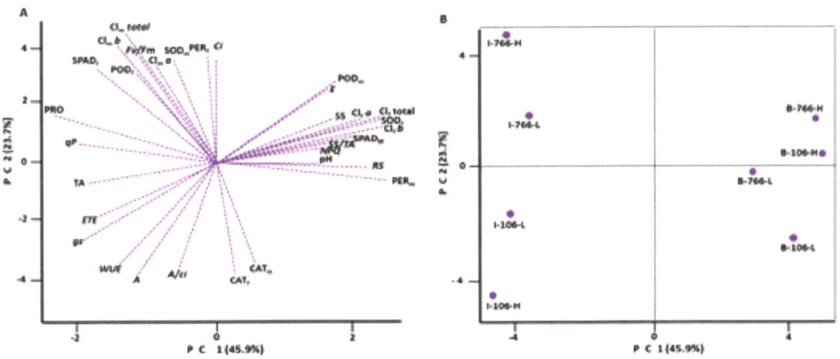

Figure 1. Photosynthesis, biochemical, and yield performance of interaction between two grapevine hybrids, 'IAC 138-22 Maximo' (I) and 'BRS Violeta' (B), trained in two trellis heights until 1.6 m (L) and 2.0 m (H) trellis and onto two rootstocks, 'IAC 766' (766) and '106-8 Mgt' (106). (**A**) Plot of the evaluated variables; (**B**) Plot the of the treatments. Note: Quantum yield (Fv/Fm), photochemical quenching (qP), Non-photochemical quenching (NPQ), electron transport rate (ETR), transpiration rate (E), water use efficiency (WUE), carboxylation efficiency (A/Ci), stomatal conductance (gs), assimilation rate (A) and internal carbon concentration (Ci), SPAD (SPADf), chlorophylls a (Clf a), b (Clf b) and total (Clf total), peroxidase (PODf) and superoxide dismutase (SODf), catalase (CATf) over flowering; SPAD (SPADm), chlorophylls a (Clm a), b (Clmb) and total (Clmtotal), peroxidase (PODm), superoxide dismutase (SODm), catalase (CATm) over-ripening phase, production (PRO), soluble solids content (SS), acidity (pH), titratable acidity (TA), reducing sugars (RA), and SS/TA relation (SS/TA).

The second principal component explained 23.7% of the total variation, characterizing the difference between the rootstocks and trellis height using photosynthesis and biochemical performance (Figure 1A,B). The variables Ci, SPAD index, PER, and POD during flowering and chlorophyll content, SOD, and Fv/Fm during berry ripening were positively correlated. However, these variables were negatively correlated with WUE, A, A/Ci, and CAT activity. The difference between the two rootstocks was observed to be 'IAC 138-22 Maximo'. When grafted on 'IAC 766', this variety showed higher SPAD index, PER, and POD during flowering and chlorophyll content, SOD, and Fv/Fm during berry ripening than onto the '106-8 Mgt'. Lower WUE, A, A/Ci, and catalase activity were observed when this genotype was grafted onto the rootstock 'IAC 766' than onto the '106-8 Mgt'. The trellis height produced less influence on the 'IAC 138-22 Maximo' performance than the rootstock. However, the difference between the two-trellis height was observed to be 'BRS Violeta'. During flowering the trellis height until 1.6 m onto '106-8 Mgt' promoted higher WUE, A, A/Ci, and catalase activity to the cultivar 'BRS Violet' than until 2.0 m onto the '106-8 Mgt' and in both trellis height onto the 'IAC 766'. However, these combinations provided higher Ci, SPAD index, PER, and POD during flowering and chlorophyll content, SOD, and Fv/Fm during berry ripening than 'BRS Violet' trained until 1.6 m onto the '106-8 Mgt'.

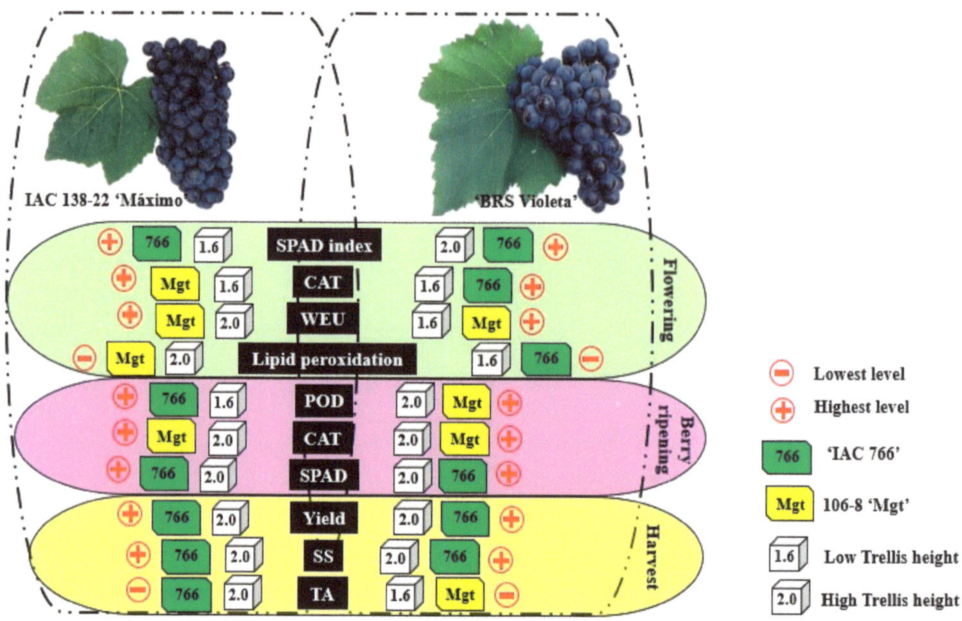

Figure 2. Graphical abstract of two grapevine hybrids, 'IAC 138-22 Maximo' and 'BRS Violeta' and their interaction with two trellis heights (low: 1.6 m and high: 2.0 m) and two rootstocks, 'IAC 766' (766) and '106-8 Mgt' (Mgt). Note: Water use efficiency (WUE), peroxidase (POD), catalase (CAT), soluble solids content (SS), and titratable (TA), SPAD index (Soil Plant Analysis Development).

4. Discussion

Maximum quantum yield (Fv/Fm) is used to detect stress during photosynthesis [34]. The varieties such as Touriga National and Chardonnay under no stress presented Fv/Fm between 0.75 and 0.83 [11], values around that were observed to both grapevine varieties in this study, independently of rootstock or trellis height used. However, the cultivar IAC 138-22 'Maximo' reduced Fv/Fm when trained until 1.6 m or grafted on the rootstock 'IAC 766'. The same was observed in 'BRS Violet' when grafted onto '106-8 Mgt'. Under unfavorable conditions, plants use absorbed light to other processes, such as thermal dissipation to protect the photosynthetic apparatus [12], decreasing the ratio Fv/Fm and the capacity of the primary acceptor to reduce the QA (quinone A) at photosystem II [11]. In this study, the photosynthesis and biochemical activity on leaves of IAC 138-22 'Maximo' and 'BRS Violeta' were highly influenced by the interaction between the rootstock and trellis height.

The use of the rootstock '106-8 Mgt' improve the water efficiency use (WUE) compared to 'IAC 766' for both varieties, IAC 138-22 'Máximo' and 'BRS Violeta'. The rootstock '106-8 Mgt' is recommended for grapevine regions with less water available because this rootstock is more efficient during carbohydrates synthesis [35]. However, for the variety IAC Maximo the highest Ci was observed onto IAC 766. In addition, IAC 138-22 'Máximo' onto this rootstock increased the chlorophyll during flowering and SPAD index during berry ripening to than onto '106-8 Mgt'. The 'IAC 766' increased the TBAR activity during flowering and SOD and POD during berry to IAC 138-22 'Máximo'. In addition, the IAC 766' increased POD and SOD activity during flowering to 'BRS Violeta'. However, 'BRS Violeta' onto this rootstock decreased CAT and SOD during berry ripening than onto '106-8 Mgt'. The increase of ROS production is correlated with abiotic stress [13,36]. The two studies cultivars produced more grapes when grafted onto the rootstock 'IAC 766' than onto '106-8 MGT'. The ideal interaction between canopy and rootstock results in high photochemical efficiency, gas exchange, and fruit yield with less ROS activity [11,37–39].

Under temperate climate conditions, the 'IAC 766' provided better yield performance to 'IAC 138-22 Máximo', 'BRS Lorena', and 'Bordô' than the rootstock '106-8 Mgt' [12]. In addition, the rootstock '106-8 Mgt' and 'IAC 766' resulted in the same yield and fruit quality as 'BRS Violeta' and 'IAC 21-14 Madalena' [12].

About the trellis height, the variety IAC 138-22 'Máximo' trained until 1.6 m improved the WUE. Despite that, the 'BRS Violeta' trained until 2.0 m increased the assimilation of CO_2 (Ci). Both trellis heights did not cause physiologic or biochemical stress to the studied varieties. However, the variety IAC 138-22 'Máximo' trained until 2.0 m increased the CAT activity during berry ripening. On the other hand, 'BRS Violeta' trained until 2.0 m decreased CAT activity during flowering and SOD activity during berry ripening. ROS activity increases in response to abiotic stress because cellular protection breaks down H_2O_2 [13]. The differences in plant morphology trained using different trellis systems or heights influence the microclimate and light interception [40]. The trellis height until 1.6 m decreased the grape yield and SSC of 'IAC 138-22 Maximo', increasing pH. Canopy size is correlated with canopy light environment, increasing leaves area, photosynthesis efficiency, and fruit yield [18,41]. For the variety Chardonnay, training until 1.65 m was recommended for low to moderate-sized canopies, and training until 1.35 m was suited for moderate to large canopies [18]. However, under subtropical conditions, the Niagara Rosada was trained until 1.8 m onto the rootstock 'IAC 766' [18]. In addition, under temperate conditions, the hybrids 'BRS Carmem', 'BRS Cora', and IAC 138-22 'Máximo' were trained until 1.0 m onto the 'IAC 766' [12].

Productivity and grape quality are the most important factors in selecting grapevine varieties for growing in a given region [5,41]. The cultivar 'IAC 138-22 Maximo' yielded more than 'BRS Violeta' in the study region (subtropical condition). However, 'BRS Violeta' present better photosynthesis and biochemical performance under subtropical condition than 'IAC 138-22 Maximo'. The better combination between canopy, training system, and rootstock, improves variety performance resulting in climate adaptation, high yield, and fruit quality [35,37–39,41,42]. In conclusion, IAC 138-22 'Maximo' presented better photosynthesis, biochemical, and yield performance trained until 2.0 m onto the rootstock 'IAC 766'. In addition, 'BRS Violeta' grafted onto 'IAC 766' showed better yield performance.

Author Contributions: F.J.D.N., J.D.R. and M.A.T. planned and designed the experiment. F.J.D.N. and A.P.J. performed plant physiological analyses, chemical, biochemical and enzyme analyses. F.J.D.N., L.R.M. and F.F.P. performed data analyses. F.J.D.N. and L.R.M. created the tables and figures. F.J.D.N., L.R.M., M.F.M., E.O.O., J.D.R. and C.S.F.B. wrote and revised the manuscript. All authors have read and agreed to the published version of the manuscript.

Funding: To Fundação de Amparo à Pesquisa do Estado de São Paulo (FAPESP—Research Support Foundation of the State of São Paulo, process no. 2016/07510-2) for granting the scholarship for the first author, for the programs that supported the study for setting the experimental area (2011/03440-6, 2013/8915-5 and 2015/16440-5) and to CNPq for the Research Productivity Scholarship (process no. 307377/2021-0).

Data Availability Statement: Not applicable.

Conflicts of Interest: The authors declare no conflict of interest.

References

1. Bellée, A.; Comont, G.; Nivault, A.; Abou-Mansour, E.; Coppin, C.; Dufour, M.C.; Corio-Costet, M.F. Life traits of four Botryosphaeriaceae species and molecular responses of different grapevine cultivars or hybrids. *Plant Pathol.* **2017**, *66*, 763–776. [CrossRef]
2. Nikolić, D.; Ranković-Vasić, Z.; Jovanović-Cvetković, T.; Petrović, A.; Radojević, I.; Sivčev, B. Effect of genotype x environment interactions of grapevine hybrids characteristics. *BIO Web Conf.* **2017**, *9*, 1018. [CrossRef]
3. Marín, D.; Armengol, J.; Carbonell-Bejerano, P.; Escalona, J.; Gramaje, D.; Hernández-Montes, E.; Intrigliolo, D.; Martínez-Zapater, J.; Medrano, H.; Mirás-Avalos, J.; et al. Challenges of viticulture adaptation to global change: Tackling the issue from the roots. *Aust. J. Grape Wine Res.* **2021**, *27*, 8–25. [CrossRef]
4. Júnior, M.J.P.; Hernandes, J.L.; Moura, M.F. Performance of juice and wine grape cultivars in different training systems. *Rev. Bras. Frutic.* **2018**, *40*, 1–8. [CrossRef]

5. Mattar, G.S.; Modesto, L.R.; Hernandes, J.L.; Tecchio, M.A.; Moura, M.F. Cycle, physicochemical characterization and climatic adaptation of a white hybrid grape on different rootstocks. *Bragantia* **2019**, *78*, 470–478. [CrossRef]
6. Smith, H.M.; Clarke, C.W.; Smith, B.P.; Carmody, B.M.; Thomas, M.R.; Clingeleffer, P.R.; Powell, K.S. Genetic identification of SNP markers linked to a new grape phylloxera resistant locus in Vitis cinerea for marker-assisted selection. *BMC Plant Biol.* **2018**, *18*, 360. [CrossRef]
7. Divers, M.; Gomes, C.B.; Menezes-Netto, A.C.; Lima-Medina, I.; Nondillo, A.; Bellé, C.; Filho, J.V.D.A. Diversity of plant-parasitic nematodes parasitising grapes in Southern Brazil. *Trop. Plant Pathol.* **2019**, *44*, 401–408. [CrossRef]
8. Zhang, L.; Marguerit, E.; Rossdeutsch, L.; Ollat, N.; Gambetta, G.A. The influence of grapevine rootstocks on scion growth and drought resistance. *Theor. Exp. Plant Physiol.* **2016**, *28*, 143–157. [CrossRef]
9. Tecchio, M.A.; da Silva, M.J.R.; Callili, D.; Hernandes, J.L.; Moura, M. Yield of white and red grapes, in terms of quality, from hybrids and Vitis labrusca grafted on different rootstocks. *Sci. Hortic.* **2020**, *259*, 108846. [CrossRef]
10. Nowicka, B.; Ciura, J.; Szymańska, R.; Kruk, J. Improving photosynthesis, plant productivity and abiotic stress tolerance—Current trends and future perspectives. *J. Plant Physiol.* **2018**, *231*, 415–433. [CrossRef]
11. Dinis, L.T.; Malheiro, A.; Luzio, A.; Fraga, H.; Ferreira, H.; Gonçalves, I.; Pinto, G.; Correia, C.; Moutinho-Pereira, J. Improvement of grapevine physiology and yield under summer stress by kaolin-foliar application: Water relations, photosynthesis and oxidative damage. *Photosynthetica* **2018**, *56*, 641–651. [CrossRef]
12. Silva, J.R.; Rodrigues, W.P.; Ferreira, L.S.; de Paula Bernado, W.; Paixão, J.S.; Patterson, A.E.; Poni, S. Deficit irrigation and transparent plastic covers can save water and improve grapevine cultivation in the tropics. *Agri. Water Manag.* **2018**, *202*, 66–80. [CrossRef]
13. Tombesi, S.; Cincera, I.; Frioni, T.; Ughini, V.; Gatti, M.; Palliotti, A.; Poni, S. Relationship among night temperature, carbohydrate translocation and inhibition of grapevine leaf photosynthesis. *Environ. Exp. Bot.* **2019**, *157*, 293–298. [CrossRef]
14. Noctor, G.; Reichheld, J.-P.; Foyer, C.H. ROS-related redox regulation and signaling in plants. *Semin. Cell Dev. Biol.* **2018**, *80*, 3–12. [CrossRef]
15. Jimenes, I.M.; Mayer, N.A.; Dias, C.T.D.S.; Filho, J.A.S.; da Silva, S.R. Influence of clonal rootstocks on leaf nutrient content, vigor and productivity of young 'Sunraycer' nectarine trees. *Sci. Hortic.* **2018**, *235*, 279–285. [CrossRef]
16. Ferreira, T.D.O.; da Costa, R.R.; Félix, D.T.; Neto, E.R.D.A.; Cruz, M.D.M.; de Lima, M.A.C. Quality and antioxidant potential of 'BRS Magna' grapes harvested in the first half of the year under different training systems and rootstocks in a tropical region. *Cien. Agrot.* **2019**, *43*, e029518. [CrossRef]
17. Leão, P.C.D.S.; Chaves, A.R.D.M. Training systems and rootstocks on yield and agronomic performance of 'Syrah' grapevine in the Brazilian semiarid. *Ciência Agrotecnol.* **2019**, *43*, 1–9. [CrossRef]
18. Gladstone, E.A.; Dokoozlian, N.K. Influence of leaf area density and 487 trellis/training system on the light micro-climate within grapevine canopies. *Vitis* **2003**, *42*, 123–132.
19. Sanchez-Rodriguez, L.A.; Spósito, M.B. Influence of the trellis/training system on the physiology and production of *Vitis labrusca* cv. Niagara Rosada in Brazil. *Sci. Hortic.* **2020**, *261*, 109043. [CrossRef]
20. Von Caemmerer, S.; Farquhar, G.D. Some relationships between the biochemistry of photosynthesis and the gas exchange of leaves. *Planta* **1981**, *153*, 376–387. [CrossRef]
21. Kitajima, M.; Butler, W. Quenching of chlorophyll fluorescence and primary photochemistry in chloroplasts by dibromothymoquinone. *Biochim. Biophys. Acta (BBA) Bioenerg.* **1975**, *376*, 105–115. [CrossRef]
22. Genty, B.; Briantais, J.-M.; Baker, N.R. The relationship between the quantum yield of photosynthetic electron transport and quenching of chlorophyll fluorescence. *Biochim. Biophys. Acta (BBA) Gen. Subj.* **1989**, *990*, 87–92. [CrossRef]
23. Schreiber, U.; Schliwa, U.; Bilger, W. Continuous recording of photochemical and non-photochemical chlorophyll fluorescence quenching with a new type of modulation fluorometer. *Photosynth. Res.* **1986**, *10*, 51–62. [CrossRef] [PubMed]
24. Bilger, W. Role of the xanthophyll cycle in photoprotection elucidated by measurements of light-induced absorbance changes, fluorescence and photosynthesis in leaves of *Hedera canariensis*. *Photosynth. Res.* **1990**, *25*, 173–185. [CrossRef]
25. Klughammer, C.; Schreiber, U. Complementary PS II quantum yields calculated from simple fluorescence parameters measured by PAM fluorometry and the Saturation Pulse method. *PAM Appl. Not.* **2008**, *1*, 27–35.
26. Sims, D.A.; Gamon, J.A. Relationships between leaf pigment content and spectral reflectance across a wide range of species, leaf structures and developmental stages. *Remote Sens. Environ.* **2002**, *81*, 337–354. [CrossRef]
27. Bradford, M.M. A rapid and sensitive method for the quantitation of microgram quantities of protein utilizing the principle of protein-dye binding. *Anal. Biochem.* **1976**, *72*, 248–254. [CrossRef]
28. Giannopolotis, C.N.; Ries, S.K. Superoxide dismutases: I. Occurrence in higher plants. *Plant Physiol.* **1977**, *59*, 309–314. [CrossRef]
29. Teisseire, H.; Guy, V. Copper-induced changes in antioxidant enzymes activities in fronds of duckweed (*Lemna minor*). *Plant Sci.* **2000**, *153*, 65–72. [CrossRef]
30. Peixoto, P.H.P.; Cambraia, J.; Sant'anna, R.; Mosquim, P.R.; Moreira, M.A. Aluminium effects on lipid peroxidation and the activities of enzymes of oxidative metabolism in sorghum. *Rev. Brasil Fis. Veg.* **1999**, *11*, 37–143.
31. Devi, S.R.; Prasad, M.N.V. Copper toxicity in *Ceratophyllum demersum* L. (Coontail), a free floating macrophyte: Response of antioxidant enzymes and antioxidants. *Plant Sci.* **1986**, *138*, 157–165. [CrossRef]
32. Nelson, N.A. A photometric adaptation of Somogy method for the determination of glucose. *J. Biol. Chem.* **1944**, *135*, 136–175.
33. SAS Institute. *SAS Version 9 (Vol. 2)*; SAS Institute Incorporated: Cary, NC, USA, 2000.

34. Bertamini, M.; Grando, M.S.; Zocca, P.; Pedrotti, M.; Lorenzi, S.; Cappellin, L. Linking monoterpenes and abiotic stress resistance in grapevines. *BIO Web Conf.* **2019**, *13*, 01003. [CrossRef]
35. Sucu, S.; Yağcı, A.; Yıldırım, K. Changes in Morphological, Physiological Traits and Enzyme Activity of Grafted and Ungrafted Grapevine Rootstocks under Drought Stress. *Erwerbs-Obstbau* **2018**, *60*, 127–136. [CrossRef]
36. Miller, G.; Coutu, J.; Shulaev, V.; Mittler, R. Reactive Oxygen Signaling in Plants. In *Annual Plant Reviews Online*; Wiley: Hoboken, NJ, USA, 2018; pp. 189–201. [CrossRef]
37. Ollat, N.; Bordenave, L.; Tandonnet, J.P.; Boursiquot, J.M.; Marguerit, E. Grapevine rootstocks: Origins and perspectives. *Acta Hortic.* **2016**, *1136*, 11–22. [CrossRef]
38. Ollat, N.; Cookson, S.J.; Lauvergeat, V.; Marguerit, E.; Barrieu, F.; Gambetta, G.A.; Goutouly, J.P.; Tandonnet, J.P.; Vivin, P.; Delrot, S. Grapevine roots: The dark side. *Acta Hortic.* **2017**, *1188*, 213–226. [CrossRef]
39. Ollat, N.; Cookson, S.; Destrac-Irvine, A.; Lauvergeat, V.; Ouaked-Lecourieux, F.; Marguerit, E.; Barrieu, F.; Dai, Z.; Duchêne, E.; Gambetta, G.; et al. Grapevine adaptation to abiotic stress: An overview. *Acta Hortic.* **2018**, *1248*, 497–512. [CrossRef]
40. Kraus, C.; Pennington, T.; Herzog, K.; Hecht, A.; Fischer, M.; Voegele, R.T.; Kicherer, A. Effects of canopy architecture and microclimate on grapevine health in two training systems. *Vitis* **2018**, *57*. [CrossRef]
41. Ferreira, D.F. Sisvar: A Guide for its Bootstrap procedures in multiple comparisons. *Ciência Agrotecnol.* **2014**, *38*, 109–112. [CrossRef]
42. Reynolds, A.G.; Heuvel, J.E.V. Influence of Grapevine Training Systems on Vine Growth and Fruit Composition: A Review. *Am. J. Enol. Vitic.* **2009**, *60*, 251–268. [CrossRef]

horticulturae

MDPI

Article

Profile of Bioactive Compounds in Orange Juice Related to the Combination of Different Scion/Rootstocks, Packaging and Storage

Rafaelly Calsavara Martins [1], Sarita Leonel [1,*], Jackson Mirellys Azevedo Souza [2], Giuseppina Pace Pereira Lima [3], Magali Leonel [4], Fernando Ferrari Putti [5], Gean Charles Monteiro [1], Patrícia Graosque Ülguim Züge [1], Gabriel Maluf Napoleão [1], Ricardo Figueira [1] and Jaime Duarte Filho [6]

1 Department of Horticulture, School of Agriculture (FCA), São Paulo State University (UNESP), Botucatu 18610-034, SP, Brazil
2 Department of Agronomy, Federal University of Viçosa (UFV), Viçosa 36570-900, MG, Brazil
3 Department of Chemistry and Biochemistry, Institute of Biosciences (IBB), São Paulo State University (UNESP), Botucatu 18618-970, SP, Brazil
4 Center for Tropical Root and Starches (CERAT), São Paulo State University (UNESP), Botucatu 18610-034, SP, Brazil
5 Faculty of Science and Engineering, São Paulo State University (UNESP), Tupã 17602-496, SP, Brazil
6 Regional Development Coordination, Department of Agriculture and Supply of the State of São Paulo, Botucatu 18607-050, SP, Brazil
* Correspondence: sarita.leonel@unesp.br

Abstract: Citrus scion/rootstock combinations alter the concentration of bioactive compounds in orange juice. The shelf life of freshly squeezed juice can be maximized through packaging and storage. The profiles of ascorbic (AA), dehydroascorbic acid (DHAA), and phenolic compounds were analyzed in juices of four sweet orange scions, Sanguínea de Mombuca (SM), Rubi (R), Lue Gin Gong (LGG), and Valência Delta Seedless (VDS), grafted onto 'Rangpur' lime (RL) and 'Swingle' citrumelo (SC) rootstocks. The juices obtained from the combination of the 'Rubi' orange in both rootstocks stood out by their higher concentration of ascorbic acid (AA) and dehydroascorbic acid (DHAA). Overall, all SC-grafted scions showed higher AA and DHAA and some phenolic compound concentrations. In all combinations, phenolic compounds showed the highest concentrations in the juices at the time of fruit extraction and decreased during storage. Dark packaging provided higher bioactive compounds in juices stored for longer periods. These findings can contribute to the diversification of scion/rootstock cultivars in order to increase the variety of orchards by choosing the best combinations for pasteurized orange juice with higher nutritional value.

Keywords: *Citrus* spp.; ascorbic acid; cultivar diversification; phenolic compounds

Citation: Martins, R.C.; Leonel, S.; Souza, J.M.A.; Lima, G.P.P.; Leonel, M.; Putti, F.F.; Monteiro, G.C.; Züge, P.G.Ü.; Napoleão, G.M.; Figueira, R.; et al. Profile of Bioactive Compounds in Orange Juice Related to the Combination of Different Scion/Rootstocks, Packaging and Storage. *Horticulturae* **2023**, *9*, 347. https://doi.org/10.3390/horticulturae9030347

Academic Editor: Charalampos Proestos

Received: 26 January 2023
Revised: 2 March 2023
Accepted: 4 March 2023
Published: 6 March 2023

1. Introduction

Brazil is the world's largest producer of orange juice. The Brazilian fruit stands out for presenting characteristics such as color, aroma, flavor, and nutritional value, which provide high quality juice [1], and mainly for presenting low cost due to the large scale production process [2]. The orange juice, besides being pleasant to the palate, presents benefits to health because it is a source of bioactive compounds, such as vitamin C and phenolic compounds, among others [3].

The sustainability of citrus crops is a global concern [4,5]. Brazil is among the vulnerable regions that cultivate a reduced number of citrus genotypes, which leads to greater susceptibility to pests and diseases, as well as less economic competitiveness among growers [6]. Diversification of citrus cultivars is an approach adopted by growers to increase the variety and profitability of orchards and to adapt to climate changes. New scion/rootstock cultivar combinations are a constant need of the citrus growers and also aim to satisfy the preferences of consumers, who are becoming increasingly demanding in relation to the quality attributes of the fruit and the orange juice consumed.

The cultivar Sanguínea de Mombuca (*Citrus sinensis* (L.) Osbeck) arose from spontaneous mutation and was subsequently selected and released by the Agronomic Institute of Campinas. This cultivar is a rich source of nutraceuticals, including the carotenoids β-carotene and lycopene that are responsible for the red color of the pulp [7,8]. The tree has early-ripening spherical-shaped medium-sized (140 g) fruits that contain 55% juice on average [8].

The cultivar Rubi originated in the active germplasm bank located in the city of Araras, São Paulo State, Brazil. The cultivar exhibits early fruit ripening, an average size of 172 g, annual production of up to 40 t/ha, and a juice yield of 49% [9]. Lue Gi Gong is a cold-tolerant Valencia-type cultivar, which is tolerant to citrus canker (*Xanthomonas axonopodis* Starr and Garces emend; Vauterin et al. pv. *citri* (Hasse) Dye) and shows late maturation of the fruit, which, under refrigeration conditions after harvest, can be preserved for longer than one month. The main limitations of this cultivar are its propensity to produce small fruits and its alternate bearing [10]. The Valência Delta Seedless cultivar originated from spontaneous bud mutation of the Valencia cultivar or by nucellar seedlings, with the Valencia cultivar as the genitor. It has tolerance to citrus canker, late ripening, and seedless fruits, but is also alternate bearing [11].

Grafting is a widely used technique in citrus farming [12] and, when performed with the use of proper rootstock, it can provide important improvements for the scion [13], such as juvenile period reduction, homogeneous tree architecture, pest and disease protection, water and nutrient absorption, tolerance to abiotic stress [14], and increased yield and fruit quality. Fruit size, juice quality, sugar and acid content, fruit skin color and thickness, and fruit ripening and production duration are also influenced by rootstock [15].

The 'Rangpur' lime (*Citrus limonia* Osbeck) tree is a natural hybrid of *Citrus medica* L. and mandarin (*Citrus reticulata* Blanco) and is suggested to be native to India [16]. In Brazil, the rootstock of 'Rangpur' lime has previously been used in citrus orchards due to its vigor, drought tolerance, high yield, precocity, and early fruit maturation [17]. Although it is tolerant to *Citrus* tristeza virus (CTV), it is susceptible to *Citrus exocortis* viroid (CEVd) and *Citrus* sudden death-associated virus (SCDaV) [18].

'Swingle' (*P. trifoliata* (L.) Raf × *C. paradisi* Macf.) is the most cultivated citrumelo in the world. It is among the main rootstocks used for diversification of orange groves, providing scions with high quality fruits, high juice yield, greater soluble solids content and yield, and lower scion vigor. This cultivar is ideal for semi-dense planting in cooler locations [1]. It is resistant to *Citrus* sudden death-associated virus and decline [19].

Citrus is one of the most important fruit crops widely investigated for its bioactive composition and its health benefits [20]. The bioactive compounds present in the fruits prevent the oxidative damage of cells by detoxifying the free radicals, thus minimizing the incidence of various diseases [21].

Significant advancements have been made to study the composition, content, and health-promoting activities of citrus fruits' bioactive compounds [22]. However, new studies should be addressed to identify the traditional and new cultivar variations and contents of bioactive compounds. This information can help to select bioactive-rich cultivars for food formulations. Moreover, precise identification of the bioactive-rich growth stage of citrus fruits processing into juice suitable for consumption is necessary [3]. Post-harvest processing can induce changes in the levels of several primary and secondary metabolites, including storage of fruits and derivatives. In the world, there is a continuous increase in the search for packaged foods and beverages that maintain the nutritional and phytochemical characteristics, including the composition of compounds with antioxidant properties. The food industry has been looking for ways to preserve quality, because it is known that during processing and storage, nutritional and sensory changes can occur, which are a limiting factor in determining the shelf life of the juice [23,24].

Nowadays, there is an increasing demand for nutritious food and many attempts have been made to maximize the retention of nutrients during storage as much as during the processing [24]. The shelf life of freshly squeezed juice can be maximized through packaging and storage.

Vitamin C or L-ascorbic acid is a water-soluble unstable vitamin that has been used as an important marker or indicator of fruit juice quality [23]. The vitamin C content in orange juice can be different depending on the raw material and the processing conditions. Vitamin C and bioactive flavonoids play an important role in oranges to scavenge free radicals and to prevent some diseases [24]. The study was performed by UHPLC (ultra-high performance liquid chromatography) and aimed to evaluate the degradation of vitamin C, as well as the content of phenolic compounds in orange juice stored in different packages for longer periods and from different combinations of scion/rootstock cultivar combinations.

2. Materials and Methods

2.1. Experimental Area Characterization

The experiment was conducted at the São Manuel Experimental Farm, School of Agriculture, São Paulo State University (FCA UNESP), Brazil (22°44′28″ S, 48°34′37″ W) located at an altitude of 740 m a.s.l. According to the Köppen–Geiger climate classification, the climate of the area is Cwa, or warm temperate (mesothermal) and humid, and the average temperature of the warmest month is approximately 22 °C [25]. The soil is classified as a sandy-textured Latossolo Vermelho distroférrico according to the Brazilian system of soil classification [26], that is, a dystrophic Typic Hapludox [27].

2.2. Plant Material and Crop Management

A replicated trial was performed in two consecutive harvest seasons (2019–2020) in a non-irrigated orchard of trees of three and four years of age, respectively. The trees were planted with 6 m spacing between rows and 4 m spacing between trees (i.e., 416 trees/ha).

The sweet orange scion cultivars Sanguinea de Mombuca (SM), Rubi (R), Lue Gin Gong (LGG), and Valencia Delta Seedless (VDS) were used, grafted on the rootstocks of 'Rangpur' lime (RL) and 'Swingle' citrumelo (SC) trees (Figure 1).

Figure 1. Fruits of sweet orange scion cultivars grafted onto two rootstocks: SM/RL (**A**); SM/SC (**B**); R/RL (**C**); R/SC (**D**), LGG/RL (**E**); LGG/SC (**F**); VDS/RL (**G**); VDS/SC (**H**).

The experimental area was prepared based on soil analysis and orange crop recommendations, using ploughing, sorting, and liming. The trees received the standard management practices recommended for citrus orchards.

2.3. Treatments and Experimental Design

The treatments consisted of eight scion/rootstock combinations: 'Sanguínea de Mombuca'/'Rangpur' lime (SM/RL); 'Sanguínea de Mombuca'/'Swingle' citrumelo (SM/SC); 'Rubi'/'Rangpur' lime (R/RL); 'Rubi'/'Swingle' citrumelo (R/SC); 'Lue Gim Gong'/ 'Rangpur' lime (LGG/RL); 'Lue Gim Gong'/'Swingle' citrumelo (LGG/SC); 'Valência Delta Seedless'/'Rangpur' lime (VDS/RL); 'Valência Delta Seedless'/'Swingle' citrumelo (VDS/SC).

The experimental design was completely randomized, considered separately for each scion/rootstock combination, using five replicates. Each replicate consisted of three trees per experimental plot, with guard trees external to the trial.

A split-plot design, with rootstocks (RL and SC) considered as plots and storage (transparent and dark bottles) as subplots, was used to determine the contents of phenolic compounds and ascorbic and dehydroascorbic acids in juices.

2.4. Fruit Harvesting and Sample Preparation

The harvest was performed when the fruits reached the ripeness index or *ratio* (soluble solids/titratable acidity) between 8.5 and 10. The Rubi and SM cultivars are early ripening and their fruits were harvested at 246 and 244 days after anthesis, respectively. The late ripening cultivars LGG and VDS were harvested at 402 and 406 days after anthesis, respectively.

After harvesting, the preparation of whole juices took place in the Beverage Laboratory of the Horticulture Department of FCA/UNESP. Juices were extracted using a semi-industrial juicer and after were pasteurized. Then, the pasteurization units (PU) were calculated. The calculation for PU units was performed using the method described by Peña et al. [28].

$$TL = (T_{obs} - T_{ref})/Z$$

where

TL is lethal rate;
T_{obs} is observed temperature;
T_{ref} is reference temperature;
Z is interval of temperature (causes a variation of 10 times in the speed of destruction).

To reach the desired PU, the counting of the PU was initiated when the juice samples (scion/rootstock combinations) reached 70 °C (reference temperature). At 70 °C, the temperature was recorded every minute using a digital thermometer and the number of PU was calculated.

During this process, the temperature was constantly balanced at around 75 °C. Upon reaching the desired PU (50 PU), cooling took place using a serpentine heat exchanger (chiller). The cooling lasted approximately 30 s, at which time the juice samples reached a temperature below 70 °C (reference temperature).

The juices were packaged in 300 mL transparent and dark polyethylene terephthalate bottles, covered with aluminium foil, and sealed by inverting the rotation cap. Subsequently, the samples were stored in a refrigerator at 4 °C for 0, 7, 14, 21, 28, and 35 days.

2.5. The Content of Ascorbic Acid (AA) and Dehydroascorbic Acid (DHA) in Juices

The determination of ascorbic and dehydroascorbic acid in juices was performed by UHPLC (ultra-high performance liquid chromatography). The methodology for extracting and quantifying ascorbic acid and dehydroascorbic acid in the samples was performed according to the methodology described by Spínola et al. [29], adapted.

Weekly (0, 7, 14, 21, 28, and 35 days of storage), juice samples stored in transparent bottles were evaluated. On day thirty-five the analysis of the juice stored in dark bottles was also performed. The evaluation of the ascorbic acid and dehydroascorbic acid content during the whole storage period was carried out only for juices packaged in transparent bottles, since this is the predominant packaging in the commercialization of this type of juice in Brazil.

Samples were diluted in Milli-Q water (1:9), filtered (PTFE, 0.45 μm, Hydrophilic, MA, USA), and injected (20 μL) into a CLUE system (Ultimate 3000 BioRS, Dionex-Thermo Fisher Scientific Inc., Fair Lawn, NJ, USA) equipped with a diode array detector (DAD), Ace 5 C18 4.6 mm × 250 mm column (Advanced Chromatography Technologies, Aberdeen, UK) at 25 °C. The flow was 0.8 mL min^{-1} for 17 min and the reading was performed at 245 nm. The substances were identified by comparing their retention times and the areas under the curves were determined and compared with standard curves of ascorbic acid ($y = 2300.3x - 1.1376$ $r^2 = 0.99$) and dehydroascorbic acid ($y = 9.1003x - 1.0026$ $r^2 = 0.99$), with purity ≥ 95% (Sigma-Aldrich, Saint Louis, MO, USA). Results were expressed as mg of ascorbic acid 100 mL^{-1} of juice.

2.6. Profile of Phenolic Compounds

The profile of the phenolic compounds in juices from the scion/rootstock combinations was carried out in UHPLC (ultra-high performance liquid chromatography, Sigma-Aldrich, São Paulo, Brazil). The separation, identification, and quantification of these compounds was according to the method described by Natividade et al. [30], adapted. The analysis was performed on the juice preparation day (day 0), and after thirty-five (35) days of storage.

The juices were filtered (PTFE, 0.45 μm, Millipore, MA, USA) and injected (20 μL) into a CLUE system (Ultimate 3000 BioRS, Dionex-Thermo Fisher Scientific Inc., USA) equipped with a cluster array detector diode (DAD), Luna® 2.5 μm C18 column (2) HST 2.0 × 50 mm (Phenomenex®, Torrance, CA, USA). The run temperature was 39 °C and the flow rate was 0.6 mL/min. The mobile phase consisted of 0.85% phosphoric acid solution (solvent A) and 100% acetonitrile (solvent B). The gradient used was: 0–2.5 min: 4% B; 2.5–7.5 min: 8% B; 7.5–15 min: 12% B; 15–18 min: 15% B; 18–20 min: 20% B; 20–21 min: 25% B; 21–22 min: 35% B; 22–24 min: 65% B; 24–25 min: 65% B; 25–25.5 min 35% B; 25.5–26 min: 0%; 26–27 min: 0% B. The absorbance was measured at 280 nm, 320 nm, 360 nm, and 520 nm using a UV-Vis spectrophotometer. Calibration curves were prepared with commercial standards (hesperidin, naringerin, caffeic acid, chlorogenic acid, *p*-coumaric acid, trans-ferulic acid, and synaptic acid, Sigma-Aldrich) and, based on their retention times, compound quantification was performed. Data were expressed as mg/L. All analysis was performed in triplicate.

2.7. Statistical Analysis

Two-year data of the evaluated variables were analysed as repeated measures. Analysis of variance (ANOVA) was performed with a significance level of 5% and differences between means were determined by Tukey's test, using the Sisvar program (Lavras, MG, Brazil). Analyses were performed in triplicate. Regression analysis was used for weekly assessments of ascorbic and dehydroascorbic acid content in juices stored in transparent bottles.

Principal component analysis (PCA) was performed [31] using XLSTAT version 2019.4.1 (Addinsoft, New York, NY, USA) to obtain a better visualization and explanation of the variability in the evaluated variables.

3. Results and Discussion

3.1. Ascorbic Acid (AA) and Dehydroascorbic Acid (DHAA) Concentration of Orange Juice According to Rootstock, Storage Time, and Packaging

There was a significant effect of the rootstock (R) and storage (S) interaction on DHAA concentration for the SM cultivar. The cultivar R did not have any interaction and the cultivars LGG and VDS showed interaction on AA and DHAA concentration (Table 1).

Table 1. F-values, degree of freedom (DF), and coefficient of variation (CV) values of AA and DHAA concentrations in juices from different scion/rootstock combinations after 35 days of storage.

		SM		R		LGG		VDS	
	DF	**AA**	**DHAA**	**AA**	**DHAA**	**AA**	**DHAA**	**AA**	**DHAA**
Block		0.228 ns	2.552 ns	4.663 ns	0.659 ns	12.102 ns	3.642 ns	1.441 ns	0.027 ns
R	1	1414.012 **	329.793 **	15.067 ns	173.077 **	805.399 **	1072.488 **	51.630 *	637.588 **
S	2	1324.995 **	294.266 **	2745.036 **	45.851 **	5833.517 **	213.374 **	4512.925 **	256.593 **
R x S	2	0.245 ns	14.201 **	0.103 ns	2.397 ns	61.421 **	48.364 **	518.471 **	24.911 **
CV (%)		2.13	3.76	1.33	3.23	1.57	2.45	0.93	2.72
CV (%)		2.36	2.73	2.5	3.05	1.41	2.71	1.87	2.46
Mean		375.35	155.58	368.32	428.96	326.35	189.65	280.6	242.31

** = statistically different at 1%; * = statistically different at 5%; ns = do not differ statistically by the F test $p < 0.05$. R—rootstock; S—storage; AA—ascorbic acid; DHAA—dehydroascorbic acid; SM—Sanguínea de Mombuca; R—Rubi; LGG—Lue Gin Gong; VDS—Valência Delta Seedless.

The cultivar R showed no difference between rootstocks in AA concentration. The highest concentrations were found in juices at the time of fruit extraction; for stored juices there was no difference. The R/SC combination provided the highest concentrations of DHAA. Juices at the time of fruit extraction and stored in transparent bottles had the highest concentrations (Table 2).

In the SM/SC combination, the highest concentrations were found in juices stored in transparent bottles, followed by dark bottles and juices at the time of fruit extraction. Regarding the rootstock, the SM/SC combination showed the highest concentrations compared with the SM/RL combination, with no difference between the juices at the time of extraction and storage (Table 2).

There was no difference between rootstocks for AA concentration in the LGG cultivar. The highest concentrations were obtained in juices at the time of fruit extraction, regardless of packaging (Table 2). Evaluation of AA concentration in the rootstocks indicated that the LGG/RL combination showed the highest concentrations in non-stored and stored juice (Table 2).

The VDS/RL and VDS/SC combinations showed the highest concentrations in juices at the time of fruit extraction, followed by those stored in dark and transparent bottles (Table 2). The VDS/SC combination showed the highest concentrations in juices stored in transparent bottles. The VDS/RL combination presented the highest AA concentrations in juices at the time of fruit extraction and stored in dark bottles. In the VDS/SC combination, the highest concentrations were obtained in juices stored in transparent bottles (Table 2).

The harvest quality and optimal citrus harvest time are based on the SS concentration and TA and their relationship (the ripeness index (RI) or ratio (SS:AT)). The RI or ratio represents the balance between the sugar and the organic acid concentration in the fruit; it is associated with juice taste and is widely used in the orange juice industry as an indicator of ripening and fruit quality [1].

AA concentration varied according to scion and harvest season, although the average data from the two harvest seasons were evaluated together. This may have been influenced by weather conditions, since rainfall occurred from March onwards and temperatures remained high, resulting in an increase in concentration due to water loss in the fruits. There were periods of severe drought during the assessment period. Environmental factors, such as irradiation and stress, can stimulate the expression of genes involved in AA production [32].

Table 2. Ascorbic acid and dehydroascorbic acid concentrations in juices from Rubi, Sanguínea de Mombuca, Lue Gin Gong, and Valencia Delta Seedless cultivars with different rootstock combinations submitted to 35 days of storage.

		Rubi		
		AA (mg/L)		
RL	**SC**	**0**	**35**	**35D**
363.83 a *	372.80 a	595.47 a	253.33 b	356.14 b
		DHAA (mg/L)		
RL	**SC**	**0**	**35**	**35D**
386.06 b	471.87 a	458.59 a	439.58 a	388.71 b
		Sanguínea de Mombuca		
		AA (mg/L)		
RL	**SC**	**0**	**35**	**35D**
304.21 b	446.37 a	527.61 a	296.35 b	302.07 b
		DHAA (mg/L)		
	RL	SC		
0	101.94 Bb *	147.12 Ac		
35	151.38 Ab	216.28 Aa		
35D	138.28 Ab	178.46 Ab		
		Lue Gin Gong		
	AA (mg/L)		**DHAA (mg/L)**	
	RL	**SC**	**RL**	**SC**
0	516.74 Aa *	463.41 Ba	137.66 Bb	299.11 Ab
35	244.34 Ac	194.70 Bc	184.82 Ba	319.07 Aa
35D	320.72 Ab	218.18 Bb	139.10 Bb	181.09 Ac
		Valencia Delta Seedless		
	AA (mg/L)		**DHAA (mg/L)**	
	RL	**SC**	**RL**	**SC**
0	460.31 Aa *	410.04 Ba	266.89 Ba	299.11 Ab
35	98.32 Bc	201.07 Ac	213.99 Ba	319.07 Aa
35D	296.49 Ab	217.35 Bb	168.54 Bb	226.27 Ac

* Means followed by the same letter, lower case letter in the column (storage) and upper case letter in the row (rootstock), do not differ statistically, Tukey test at 5% probability level. AA—ascorbic acid; DHAA—dehydroascorbic acid; RL—'Rangpur' lime; SC—'Swingle' citrumelo; 0 = juice not stored; 35 = juice stored in transparent bottle for 35 days; 35D = juice stored in dark bottle for 35 days.

The sum of AA and DHAA concentrations in descending order for the scions evaluated were: R (797.28 mg/L), SM (530.93 mg/L), VDS (522.91 mg/L), and LGG (516.00 mg/L) (Table 1). The 'Rubi' orange stood out for presenting the highest concentrations of ascorbic acid (AA) in both rootstocks and the SM/SC combination for the highest DHAA concentration. The main hypothesis for these results is that AA and DHAA acids are genotype dependent variables. Moreover, the cultivars R and SM are classified as early maturing, in which less climatic variations occurred during the fruit ripening. In the late ripening cultivars LGG and VDS, the fruit remained longer in the field, having the influence of greater climatic changes that had an effect on the ripening period of the fruit.

In these cultivars, the fruit harvest occurred after a long dry season and higher temperatures, favoring a higher concentration of AA and DHAA. On the contrary, the harvest of the early maturing cultivars occurred after a period of high rainfall, which favored an increase in the mass of the fruits, as well as the dilution of the organic acids, decreasing the concentration of AA and DHAA. Acid concentration can be influenced by growing conditions, climate, and even fruit size, which is influenced by the characteristics that the rootstock and scion can play in the fruit mass. The higher the fruit yield, the greater the dilution of organic acids, carbohydrates, and vitamins, thus decreasing the concentration of AA and DHAA [3,4]. Orange is a rich source of AA, which has several biological functions related to the immune system, collagen formation, iron absorption, nitrosamine inhibition, and antioxidant activity, therefore the SM and R scions, which have higher AA concentrations, are relevant [33]. The vitamin C concentrations in citrus juices are different from each other depending on processing conditions and raw material, such as 38 mg 100 g^{-1} for grapefruit juice, 46 mg 100 g^{-1} for lemon juice, 50 mg 100 g^{-1} for orange juice, and 31 mg 100 g^{-1} for mandarin juice [24]. The results obtained in this experiment confirm this report, as there were variations in AA and DHAA concentrations in each scion/rootstock cultivar combination evaluated.

In general, all SC-grafted scions showed higher AA and DHAA concentrations. The effect of rootstock on fruit quality can be attributed to several factors such as nutrient absorption and transport, compatibility, hormonal signaling, and gene expression [4]. The rootstock plays an important role in fruit ripening because it can accelerate or delay citrus tree development [1].

'Swingle' citrumelo is among the main rootstocks used in the diversification of orange groves because it provides scions with high-quality fruits with high juice yield and SS concentrations [34,35]. Scions on trifoliate orange rootstocks and their hybrids, as SC, produced better-quality fruits than those on other commonly used rootstocks. This has been well documented, but the genetic factors affecting fruit quality through the interaction between the scion and rootstock remain unclear. The results obtained by Hu et al. [15] demonstrated consistent correlations with the fruit quality of four 'Daya' mandarin cultivars grafted onto *Poncirus trifoliata* rootstocks related to the differential gene expression of small RNAs.

AA biosynthesis is generated from d-glucose, with nucleotides and sugars as intermediates, and the SC rootstock presents lower tolerance to water deficit, limiting the photosynthetic capacity of the tree, which may explain these results [32]. Photosynthesis, temperature, and light exposure can affect AA synthesis and production [36].

The concentration of acids can be influenced by growing conditions, climatic changes, and the size of the fruit, which is influenced by the characteristics that the rootstock and scion can play in the mass of the fruit [37]. The higher the fruit yield, the greater the dilution of organic acids, carbohydrates, and vitamins [38], thus decreasing the concentration of these compounds. The highest concentrations were observed in the juices at the time of fruit extraction, differing from those subjected to storage (Table 2). Similar results were reported by Nakilcioğlutaş and Ötleş [24], who, when evaluating the degradation of vitamin C in citrus juices, observed that the lowest vitamin C losses were found in non-stored juices. The lowest losses were recorded in juices stored at 4 °C.

Vitamin C (L-ascorbic acid) is a water-soluble and highly unstable vitamin. Vitamin C is often considered a nutrient quality indicator undergoing the processing and storage of foods, since it is seen that other nutrients are well preserved [39].

Data showed that the AA concentration in all scion/rootstock combinations had a negative linear effect, i.e., the concentrations decreased throughout the storage period. Differently, for DHAA concentrations, for SM and R cultivars, both the scion/rootstock combinations presented a positive quadratic effect. The cultivars LGG and VDS, grafted onto two rootstocks, showed a cubic effect during storage. This means that, for these combinations, there was no regular pattern of response (Figure 2). This result can be attributed to the late ripening of these cultivars, with physiological responses under constraining environments.

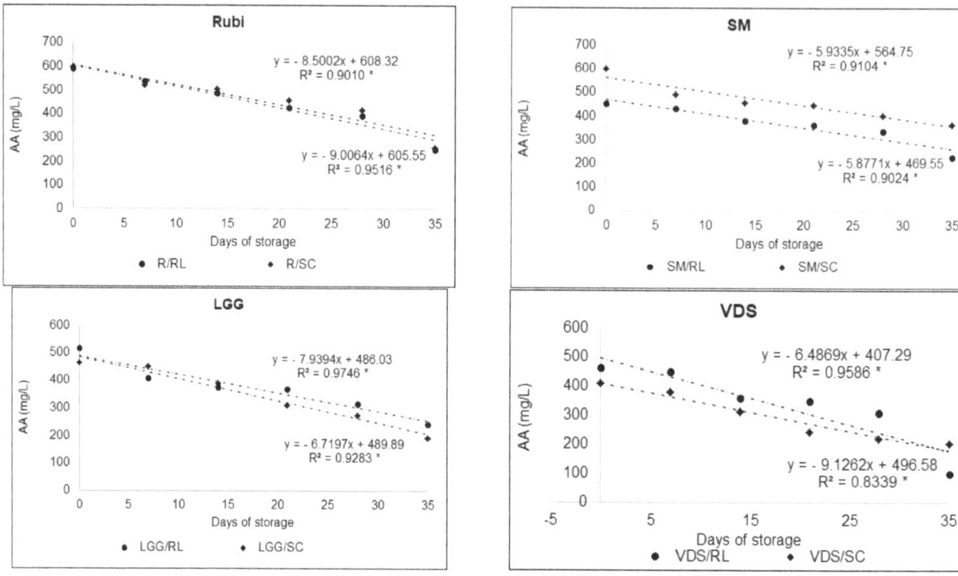

Figure 2. Ascorbic acid concentration (mg/L) in juices of Sanguínea de Mombuca (SM), Rubi (R), Lue Gin Gong (LGG), and Valencia Delta Seedless (VDS) sweet orange cultivars with different rootstock combinations subjected to 0, 7, 14, 21, 28, and 35 days of storage. RL—'Ranpur' lime; SC—'Swingle' citrumelo.

The decrease in AA concentrations occurs from the moment of processing and continues during the storage period of the juices. During storage, numerous deteriorating reactions occur, causing the degradation of AA, consequently leading to changes in the taste, color, texture, and appearance of the juice [23]. These results were confirmed in all scion/rootstock combinations evaluated, in which the concentrations of flavonoids, hesperidin, and naringerin decreased during the storage period.

Vitamin C in citrus juices is generally easily oxidized and therefore is lost in storage. There are many variables that affect this oxidation process such as light exposure, dissolved oxygen level, storage temperature, and presence of sugar and metal ions [39]. During storage, L-ascorbic acid oxidizes to dehydroascorbic acid (DHAA). This does not cause the loss of vitamin C, because DHAA can be converted back to ascorbic acid [40]. However, DHAA is easily hydrolyzed to 2,3-diketogulonic acid (DKGA) due to being highly unstable. DKGA has no biological activity [23]. These oxidation stages have been found to be particularly sensitive to oxygen availability, long-term heat treatment in the presence of oxygen, and exposure to light [40]. These reports were the main hypothesis to explain why, in some raw materials, juices stored in transparent bottles had higher concentrations of AA.

The variation in DHAA concentration observed in this experiment can occur in response to AA oxidation (Figure 3). AA degradation occurs during the storage of citrus juices and is mainly due to oxidation caused by storage temperature, light, and the presence of oxygen [24]. Due to this oxidation, AA is transformed into DHAA, which may explain the higher concentration of DHAA in stored juices. Wibowo et al. [23] studied the changes in acids, sugars, oxygen, and vitamin C due to the storage of orange juice and also observed that during storage, mainly due to the presence of oxygen, there were variations in the concentration of DHAA.

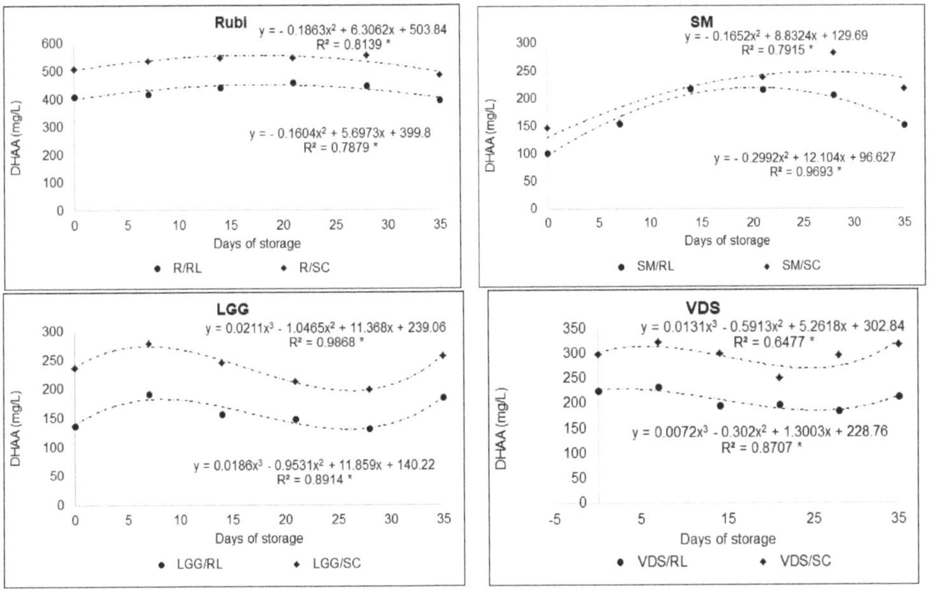

Figure 3. Dehydroascorbic acid concentration (mg/L) in juices from Sanguínea de Mombuca (SM), Rubi (Rubi), Lue Gin Gong (LGG), and Valencia Delta Seedless (VDS) sweet orange cultivars with different rootstock combinations submitted to 0, 7, 14, 21, 28, and 35 days of storage. RL—'Ranpur' lime; SC—'Swingle' citrumelo.

Principal component analysis (PCA) allowed demonstrating the concentration of AA and DHAA for each cultivar combined with the RL and SC rootstocks. In this study, PCA was applied to evaluate the concentration of AA and DHAA in response to juice storage and packaging. Variability was explained by two principal components, PC1 and PC2, for all cultivars. The cv. SM, accounting for 50.95% and 49.05%; cv. Rubi, responsible for 69.62% and 30.08%; cv. LGG, accounting for 50.52% and 49.42%; cv. VDS, responsible for 60.25% and 30.34% of the variation in the data (Figure 4).

The PCA demonstrated that the SM, LGG, and VDS sweet orange cultivars showed similar results in the first (PC1-DHAA) and second principal component (PC2-AA) and the cultivar R presented AA and DHAA in the first principal component (PC1).

Overall, the PCA allowed concluding that during juice storage there was an inverse relationship between AA and DHAA. It was also possible to identify that the highest AA concentrations occurred in juices at the time of fruit extraction, as well as the highest DHAA concentrations occurred in stored juices.

Despite the difference in the response of the cultivars, it is important to point out that all scion/rootstock combinations showed optimal levels of vitamin C (AA + DHAA) (Table 2), which is, according to Stinko et al. [37], around 529 mg L^{-1}. The degradation of AA and DHAA is more associated with the presence of oxygen than with light, since oxygen determines the rate of oxidative degradation of the compounds [24]. Oxygen is usually incorporated into the juice during preparation, processing, and storage and can pass through the package by diffusion process [41].

Studies that aim to quantify the vitamin C concentration before and after degradation are important, since vitamin C is a reliable indicator of the nutritional value and quality deterioration of the processed juice [42].

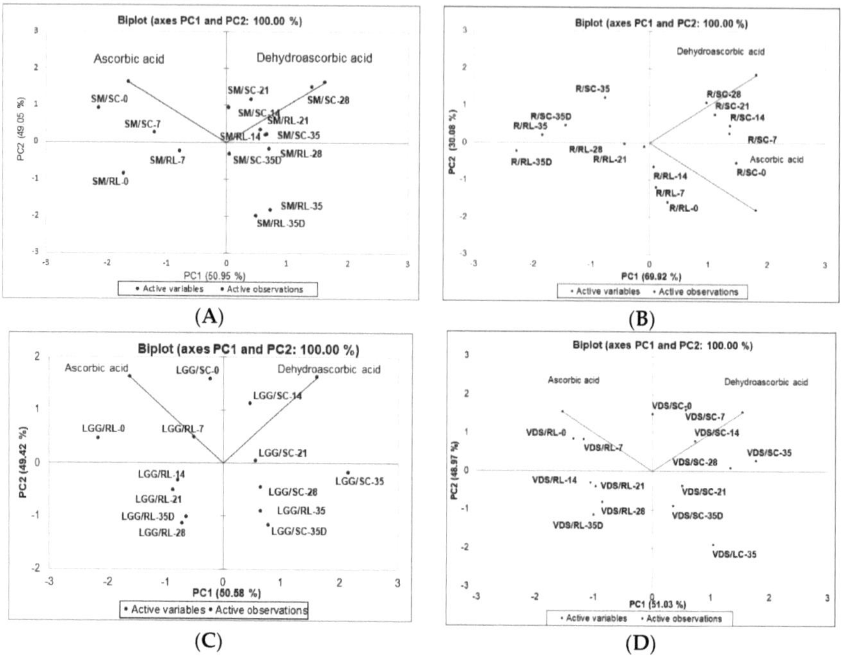

Figure 4. Principal component analysis (PCA) of AA and DHAA in juices from different scion/rootstock combinations subjected to 0, 7, 14, 21, 28, and 35 days of storage in transparent bottle and dark bottle (35D). (**A**) SM—Sanguínea de Mombuca; (**B**) R—Rubi; (**C**) LGG—Lue Gin Gong; (**D**) VDS—Valencia Delta Seedless; RL—'Rangpur' lime; CS—'Swingle' citrumelo.

3.2. Phenolic Compound Concentration of Orange Juice According to Rootstock, Storage Time, and Packaging

There was a significant effect of the rootstock and storage interaction for hesperidin, naringerin, caffeic acid, *p*-coumaric acid, and synaptic acid in the SM cultivar. The same result was observed for hesperidin, caffeic acid, chlorogenic acid, trans-ferulic acid, and synaptic acid in cultivar R and for hesperidin, naringerin, caffeic acid, and chlorogenic acid in 'LGG' orange and in cultivar VDS for hesperidin and synaptic acid. The results observed for each of the cultivars showed that there were variations in the concentrations of these compounds over the storage period of the orange juice. These variables are also genotype dependent (Table 3).

Regardless of the rootstock combination, SM scion had the highest concentrations of hesperidin, naringerin, caffeic acid, and *p*-coumaric acid in juices at the time of fruit extraction, except for synaptic acid, which had the highest concentrations in juices stored in dark bottles (Table 4). The degradation of phenolic compounds over time may occur due to storage conditions, processing, bottling, temperature, and exposure to light [43]. These variables can lead to the oxidation process and consequently the loss of the compounds [44], which may explain the higher concentration of the compounds at the time of juice extraction.

The hesperidin and caffeic acid had similar performances; for the combinations SM/RL and SM/SC, the highest losses occurred when the juice was stored in transparent bottles (Table 4). Naringerin showed different concentrations between the scion/rootstock combinations. The scion/rootstock combinations did not have differences in coumaric acid concentration between stored juices. Caro et al. [45], studying the concentrations of flavonoids in stored citrus juices, also reported a significant increase in some bioactive compounds during storage.

Table 3. F-values, degree of freedom (DF), and coefficient of variation (CV) of phenolic compounds in juices from different scion/rootstock combinations subjected to 35 days of storage.

					Sanguínea de Mombuca			
FV	GL	Hesperidin	Naringerin	Cafffeic Acid	Chlorogenic Acid	*p*-Coumaric	Ferulic Acid	Sinaptic Acid
Block	2	5.295 ns	0.47 ns	61.00 **	0.717 ns	1.00 ns	0.33 ns	2.71 ns
RS	1	1357.05 **	27.84 *	7921.00 **	167.84 **	90.75 **	12.00 ns	43.75 *
S	2	3628.31 **	2238.00 **	569.71 **	492.23 **	650.67 **	470.86 **	132.72 **
RS x S	2	56.87 **	23.23 **	208.00 **	2.08 ns	32.00 **	3.43 ns	108.14 **
CV (%)		0.36	3.48	0.51	3.3	4.58	4.56	1.53
CV (%)		1.61	2.04	1.36	3.41	2.8	2.46	2.78
Mean		46.82	0.3	0.5	0.81	0.18	0.18	0.82
					Rubi			
Block	2	0.21 ns	21.00 *	7.00 ns	10.34 ns	1.00 ns	19.00 *	0.18 ns
RS	1	6637.85 **	484.00 **	240.25 **	26.23 *	6.25 ns	484.00 **	6.49 ns
S	2	21,648.56 **	468.12 **	403.15 **	1549.72 **	522.47 **	490.75 **	409.30 **
RS x S	2	98.210 **	1.65 ns	37.25 **	8.56 **	0.12 ns	4.75 *	214.25 **
CV (%)		0.26	0.65	0.84	2.04	2.58	1.26	3.49
CV (%)		0.51	3.31	2.27	2.32	3.76	2.51	3.73
Mean		40.17	0.36	0.56	0.9	0.18	0.19	0.82
					Lue Gin Gong			
Block	2	0.10 ns	0.08 ns	3.86 ns	2.71 ns	1.00 ns	0.11 ns	3.00 ns
RS	1	34.83 *	9.31 ns	96.57 **	57.14 **	42.25 *	25.00 *	32.00 **
S	2	20.356.46 **	271.05 **	0.273 ns	43.19 **	76.00 **	7.54 **	2.17 ns
RS x S	2	77.03 **	18.85 **	17.55 **	78.23 **	4.00 ns	1.39 ns	0.66 ns
CV (%)		0.87	1.76	1.01	1.29	2.17	4.7	0.26
CV (%)		0.67	2.65	1.25	1.36	1.09	3.99	0.69
Mean		67.71	0.48	0.63	0.97	0.22	0.15	1.85
					Valência Delta Seedless			
Block	2	3.75 ns	1.00 ns	4.33 ns	0.01 ns	0.33 ns	1.00 ns	0.16 ns
RS	1	0.02 ns	2.15 ns	75.00 **	2.81 ns	6.75 ns	1.00 ns	565.12 **
S	2	15,922.90 **	340.98 **	7.72 **	73.32 **	6.32 *	1.00 ns	13.23 **
RS x S	2	135.27 **	0.12 ns	2.91 ns	0.95 ns	3.12 ns	1.00 ns	5.05 *
CV (%)		0.71	4.12	0.57	4.52	3.7	1.58	1.33
CV (%)		1.01	3.29	1.87	1.35	3.78	1.58	0.95
Mean		79.49	0.55	0.71	0.99	0.22	0.15	2.01

** = statistically different at 1%; * = statistically different at 5%; ns = not statistically different by F-test at <0.05. RS—rootstock; S—storage; CV—coefficient of variation.

The evaluation of the rootstocks showed that the highest hesperidin concentrations were observed in the non-stored juice in the VDS/SC combination. In stored juices, regardless of the storage, the highest concentrations were recorded in the VDS/RL combination. The highest concentrations of synaptic acid were observed in the VDS/RL combination, regardless of the storage. There was no statistical difference between scion/rootstock combinations for naringerin and, caffeic, *p*-coumaric and *trans*- ferulic acids concentrations (Table 5).

Juices stored in dark bottles had higher concentrations of caffeic acid and juices at the time of fruit extraction had higher concentrations of *p*-coumaric acid. *Trans*-ferulic acid showed no difference for storage (Table 5).

Table 4. Concentration of phenolic compounds (mg/L) in 'Rubi' and 'Sanguinea de Mombuca' juices with different combinations of rootstock subjected to 35 days of storage in transparent bottles and dark bottles (D).

					Rubi					
	Hesperidin		**Caffeic Acid**		**Chlorogenic Acid**		**Ferulic Acid**		**Sinaptic Acid**	
	RL	**SC**	**RL**	**SC**	**RL**	**SC**	**RL**	**SC**	**RL**	**SC**
0	54.51 aB *	57.41 aA	0.69 aA	0.66 aB	0.62 cA	0.55 cB	0.22 aB	0.25 aA	0.76 bA	0.30 bB
35	31.77 bB	35.47 bA	0.42 cB	0.52 cA	0.90 bA	0.83 bB	0.15 bB	0.17 cA	0.89 aB	1.04 aA
35D	31.11 cB	33.74 cA	0.52 bB	0.56 bA	1.25 aA	1.26 aA	0.16 bB	0.18 bA	0.87 aB	1.07 aA
			Naringerin					**p-Coumaric**		
	RL	**SC**	**0**	**35**	**35D**	**RL**	**SC**	**0**	**35**	**35D**
	0.35 b	0.38 a	0.48 a	0.33 b	0.28 c	0.19 a	0.18 a	0.26 a	0.14 b	0.15 b
					Sanguinea de Mombuca					
	Hesperidin		**Naringerin**		**Caffeic Acid**		***p*-Coumaric**		**Sinaptic Acid**	
	RL	**SC**	**RL**	**SC**	**RL**	**SC**	**RL**	**SC**	**RL**	**SC**
0	72.35 aA *	64.06 aB	0.41 aB	0.44 aA	0.43 aB	0.62 aA	0.20 aB	0.27 aA	0.82 bA	0.57 bB
35	34.92 cA	35.02 cA	0.19 cB	0.24 bA	0.38 cB	0.43 cA	0.13 bB	0.16 bA	0.78 bB	0.92 aA
35D	37.61 bA	36.97 bA	0.24 bA	0.24 bA	0.41 bB	0.48 bA	0.14 bB	0.16 bA	0.90 aA	0.90 aA
			Chlorog					**Ferul**		
	RL	**SC**	**0**	**35**	**35D**	**RL**	**SC**	**0**	**35**	**35D**
	0.72 b	0.89 a	0.54 c	0.86 b	1.03 a	0.17 a	0.19 a	0.22 a	0.15 c	0.16 b

* Means followed by the same letter, lower case in the column (storage) and upper case in the row (rootstock), do not differ statistically, Tukey test at 5% probability level. RL—'Rangpur' lime; SC—'Swingle' citrumelo; 0 = non-stored juice; 35 = juice stored in transparent bottle for 35 days; 35D = juice stored in dark bottle for 35 days.

Table 5. Concentration of phenolic compounds (mg/L) in 'Lue Gin Gong' and 'Valencia Delta Seedless' juices with different combinations of rootstock submitted to 35 days of storage.

							Lue Gin Gong							
				Hesperidin		**Naringerin**		**Caffeic Acid**		**Chlorogenic Acid**				
				RL	**SC**	**RL**	**SC**	**RL**	**SC**	**RL**	**SC**			
			0	99.24 aA	97.39 aB	0.61 aA	0.55 aB	0.62 aA	0.63 bA	1.00 aA	0.85 bB			
			35	48.83 cB	53.44 bA	0.43 bB	0.45 bA	0.61 bB	0.64 bA	0.96 bB	1.01 aA			
			35 E	52.61 bB	54.75 bA	0.42 bA	0.42 cA	0.61 bB	0.65 aA	0.99 aA	0.98 aA			
		***p*-Coumaric Acid**					**Ferulic Acid**					**Sinaptic Acid**		
RL	**SC**	**0**	**35**	**35D**	**RL**	**SC**	**0**	**35**	**35D**	**RL**	**SC**	**0**	**35**	**35D**
0.21 a *	0.22 a	0.23 a	0.22 b	0.21 c	0.14 b	0.16 a	0.16 a	0.15 b	0.15 b	1.65 b	2.04 a	1.84 a	1.85 a	1.84 a

			Valencia Delta Seedless						
	Hesperidin		**Sinaptic Acid**				**Chlorogenic Acid**		
	RL	**SC**	**RL**	**SC**	**RL**	**SC**	**0**	**35**	**35D**
0 **35** **35E**	122.80 aB 55.88 cA 59.84 bA	131.42 aA 52.83 bB 54.17 bB	2.20 aA 2.17 bA 2.17 bA	1.87 aB 1.85 aB 1.87 aB	1.01 a	0.98 b	0.94 b	1.01 a	1.03 a
		Naringerin					**Caffeic Acid**		
RL	**SC**	**0**	**35**	**35D**	**RL**	**SC**	**0**	**35**	**35D**
0.55 a *	0.54 a	0.70 a	0.46 b	0.47 b	0.72 a	0.71 a	0.73 a	0.70 b	0.72 a
		***p*-Coumaric Acid**					**Ferulic Acid**		
RL	**SC**	**0**	**35**	**35D**	**RL**	**SC**	**0**	**35**	**35D**
0.23 a	0.22 a	0.20 a	0.21 b	0.22 b	0.15 a	0.15 a	0.15 a	0.15 a	0.15 a

* Means followed by the same letter, lower case in the column (storage) and upper case in the row (rootstock), do not differ statistically, Tukey test at 5% probability level. RL—'Rangpur' lime; SC—'Swingle' citrumelo; 0 = non-stored juice; 35 = juice stored in transparent bottle for 35 days; 35D = juice stored in dark bottle for 35 days.

Differences in the concentrations of phenolic compounds may occur according to the raw material, pasteurization methods, storage conditions, and temperature, as well as the specific compound. In a comparative study, in the juice sacs of ripened fruits, flavonone hesperidin was the dominant phenolic compound in lemon (2213 mg/kg DW) and oranges (1957 and 1975 mg/kg DW in Washington Navel and Tarocco, respectively), whereas flavonone narirutin was the most prevalent in grapefruit (292 mg/kg DW) [20].

The reactions of the degradation of phenolic compounds occur through hydroxylation, methylation, isoprenylation, dimerization, and glycosylation effects. The enzymes polyphenol oxidase (PPO), peroxidase (POD), pectinamethylesterase (PME), and phenylalanine ammonia lyase (PAL) can also catalyze the oxidation of phenolic compounds in the presence of oxygen, causing the formation of dark compounds and consequently contributing to the loss of juice quality [46].

The phenolic compound concentrations are highly influenced by the ripening stage and the cropping system, genotype and environmental, since they are secondary compounds that are produced by the plant under stress conditions [47].

Overall, scions grafted on SC rootstocks had a higher concentration of some phenolic compounds due to SC showing less tolerance to water deficit [48]. Trees under water stress showed increased secondary metabolism, mainly phenolics, terpenes, alkaloids, and cyanogenic glycosides [49] (Table 4). Some compounds and/or scion/rootstock combinations in the stored juices presented higher concentrations in transparent bottles and others in dark bottles. Similarly, Giuffrè et al. [50] reported that hesperidin is the main flavonoid in orange juice and that storage was responsible in decreasing the flavonoid concentration. Chlorogenic acid is one of the main components present in citrus fruits and it usually occurs in larger amounts, as in this study [51]. The authors also reported that, during juice storage, the concentrations of these acids may change and, consequently, there may be a decrease in flavonoids.

The result of the principal component analysis (PCA) allowed an overview of the phenolic compounds for each cultivar combined with the RL and SC rootstocks. PCA was applied in order to evaluate the performance of phenolic compounds in response to juice storage and packaging. For all cultivars, the variability was explained by two principal components, PC1 and PC2; the cv. SM accounting for 79.68% and 13.96%, cv. R accounting for 87.33% and 9.00%, cv. LGG accounting for 54.14% and 29.10%, and 'VDS' accounting for 60.25% and 30.34% of the variation in the data (Figure 5).

The PCA showed that all scion cultivars presented similar performances. The juices of the 'SM' and 'Rubi' oranges presented in the first principal component (PC1) the flavonoids hesperidin and naringerin and the phenolic acids caffeic, *p*-coumaric, and trans-ferulic and in the second principal component (PC2), the phenolic acids chlorogenic and synaptic. The juices of the cultivars LGG and VDS presented in the first principal component (PC1) the flavonoids hesperidin and naringerin and the phenolic acids caffeic, *p*-coumaric, trans-ferulic, and synaptic (Figure 5). For all cultivars, chlorogenic acid was inversely proportional to the other compounds present in PC1. This same compound was observed in larger concentrations in stored juices, transparent and dark bottles, and in practically all the scion/rootstock combinations. Chlorogenic acid is derived from most phenolic acids, especially caffeic, *p*-coumaric, and trans-ferulic acids [20].

Most scion/rootstock combinations showed the highest concentrations of these phenolic acids at the time of juice extraction, which can explain the higher concentrations of chlorogenic acid in stored juices. These acids are biosynthesized by hydroxylation of the coumaroyl ester of the chemical acid. This hydroxylation produces the ester of shikimic acid, which is converted to chlorogenic acid [52].

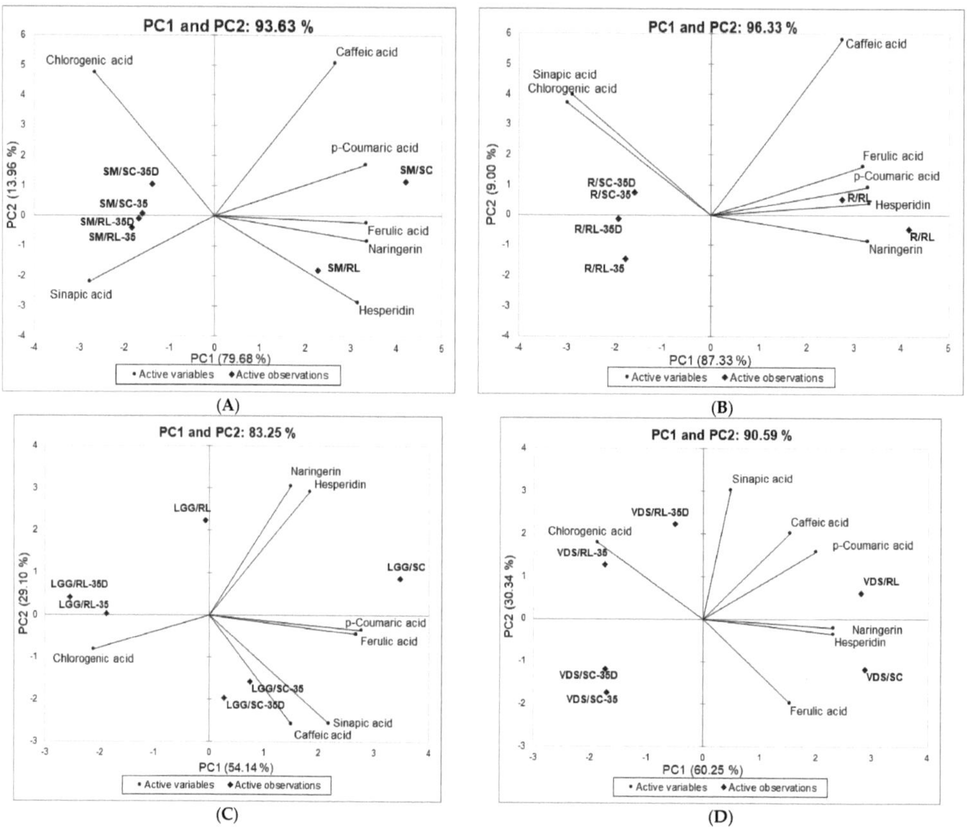

Figure 5. Principal component analysis (PCA) of phenolic compounds in juices from different scion/rootstock combinations subjected to 35 days of storage. (**A**) SM—Sanguínea de Mombuca; (**B**) R—Rubi; (**C**) LGG—Lue Gin Gong; (**D**) VDS— Valencia Delta Seedless; RL—'Rangpur' lime; CS—'Swingle' citrumelo; 0 = juice at the time of fruit extraction, 35 = juices stored for 35 days in transparent bottles; 35D = juices stored for 35 days in dark bottles.

The flavonoids hesperidin and naringerin also presented similar performances for all scion/rootstock combinations. The highest concentrations of these compounds were obtained in juices at the time of extraction, i.e., there was a decrease during storage (Figure 3). Zhang et al. [22] also concluded that flavonoid concentration decreased after juice storage. These authors hypothesized that the decrease in concentrations during storage is associated with the degradation of vitamin C and hydroxycinnamic acids, corroborating the results of the present study.

Summarizing, in all scion/rootstock combinations, the phenolic compounds hesperidin, naringerin, caffeic acid, *p*-coumaric acid, and trans-ferulic acid had the highest concentrations in the juices at the time of fruit extraction, so they decreased during storage. Hesperidin was the main flavonoid, as chlorogenic and synaptic acids were the main hydroxycinnamic acids. For stored juices, chlorogenic acid and synaptic acid were higher for most scion/rootstock combinations.

Both ascorbic acid and phenolic compounds are antioxidant substances that play an important role as indicators of juice quality. Information about their contents helps to add commercial and industrial value to orange juice. In general, the data showed that the contents varied with the scion/rootstock combinations and decreased during storage.

Significant studies have been made on the composition, content, and health-promoting activities of citrus juice bioactive compounds [20]. However, further investigations with new cultivars are needed to identify genetic variation and composition, due to these data may contribute to the selection of bioactive-rich citrus cultivars suitable for natural consumption and for processing into juice. These findings may also be useful in planning diversification of scion/rootstock combinations for new orchards by identifying genotypes best adapted to undesirable climatic conditions.

During the processing and storage of orange juice, nutritional and sensory changes may occur, which is a limiting factor for determining the shelf life of the juice. Studies that improve the processing and storage of orange juice, allowing the reduction of the degradation of bioactive compounds, have been increasingly requested. Choosing the best type of packaging and storage time to minimize nutritional losses in pasteurized orange juice is very important and should be the subject of future research considering the diversity of the raw material.

4. Conclusions

The juices obtained from the combination of the 'Rubi' orange in both rootstocks stood out by their higher concentration of AA and DHAA. However, all juice combinations evaluated showed optimal AA and DHAA concentrations in their composition and the highest AA concentrations were obtained in the juices at the time of extracting the fruit, without any storage, and were inversely proportional to the DHAA concentrations. With regard to combinations, the dark packaging provided a higher concentration of bioactive compounds in juices stored for longer periods. The data obtained from this study may provide additional contributions for choosing the best scion/rootstock combinations for processing orange juice with higher nutritional values and how to package and store orange juice.

Author Contributions: Conceptualization, G.P.P.L., S.L., R.C.M. and J.M.A.S.; methodology, G.P.P.L., G.C.M., R.F. and R.C.M.; validation, S.L., M.L., P.G.Ü.Z., G.M.N., R.F., R.C.M. and J.M.A.S.; formal analysis, F.F.P.; investigation, G.M.N., P.G.Ü.Z., J.D.F., R.F., J.M.A.S., G.C.M. and R.C.M.; resources, S.L. and M.L.; data curation, F.F.P. and R.C.M.; writing— original draft preparation, R.C.M. and S.L.; writing—review and editing, S.L.; supervision, S.L.; project administration, S.L.; funding acquisition, S.L. and J.D.F. All authors have read and agreed to the published version of the manuscript.

Funding: This work was partially supported by the Brazilian National Council for Scientific and Technological Development (CNPq), grant numbers 302611/2021-5 and 302848/2021-5.

Institutional Review Board Statement: Not applicable.

Informed Consent Statement: Not applicable.

Data Availability Statement: Data are contained within the article.

Acknowledgments: The authors thank the Secretary of Agriculture and Supply of the State of São Paulo for the donation of seedlings through the UAT/CDR agreement.

Conflicts of Interest: The authors declare no conflict of interest. The funders had no role in the design of the study; in the collection, analyses, or interpretation of data; in the writing of the manuscript; or in the decision to publish the results.

References

1. Domingues, A.R.; Marcolini, C.D.M.; Gonçalves, C.H.S.; Gonçalves, L.S.A.; Roberto, S.R.; Carlos, E.F. Fruit ripening development of 'Valência' orange trees grafted on different 'Trifoliata' hybrid rootstocks. *Horticulturae* **2021**, *7*, 3. [CrossRef]
2. Citrusbr. Associação Nacional dos Exportadores de Sucos Cítricos. 2020. Available online: http://www.citrusbr.com/mercadoexterno/?me=01\T1\textgreater{} (accessed on 2 September 2021).
3. Lu, X.; Zhao, C.; Shi, H.; Liao, Y.; Xu, F.; Du, H.; Xiao, H.; Zheng, J. Nutrients and bioactives in citrus fruits: Different citrus varieties, fruit parts, and growth stages. *Crit. Ver. Food Sci. Nutr.* **2021**, *4*, 1–24. [CrossRef] [PubMed]
4. Tietel, Z.; Srivastava, S.; Fait, A.; Tel-Zur, N.; Carmi, N.; Raveh, E. Impact of scion/rootstock reciprocal effects on metabolomics of fruit juice and phloem sap in grafted Citrus reticulata. *PLoS ONE* **2020**, *15*, e0227192. [CrossRef] [PubMed]

5. Carvalho, E.V.; Cifuentes-Arenas, J.C.; Raiol-Junior, L.L.; Stuchi, E.S.; Girardi, E.A.; Lopes, S.A. Modeling seasonal flushing and shoot growth on different citrus scion-rootstock combinations. *Sci. Hortic.* **2021**, *288*, 110358. [CrossRef]
6. Carvalho, L.M.; Carvalho, H.W.L.; Barros, I.; Martins, C.R.; Soares Filho, W.S.; Girard, E.A.; Passos, O.S. New scion-rootstock combinations for diversification of sweet orange orchards in tropical hardsetting soils. *Sci. Hortic.* **2019**, *243*, 169–176. [CrossRef]
7. Hamedani, M.; Rabiel, V.; Moradi, H.; Ghanbari, A.; Azimi, R.M. Determination of storage duration and temperature effects on fruit quality parameters of blood orange (*Citrus sinensis* cv. Torocco). *Biharean Biol.* **2012**, *6*, 10–13. Available online http://biozoojournals.ro/bihbiol/cont/v6n1/bb.111129.Ghambari.pdf (accessed on 10 October 2020).
8. Latado, R.R. Laranja Mombuca é Rica em Licopeno Potente Agente Antioxidante Natural Que Ajuda Prevenir o Câncer 2021. Available online: https://www.agricultura.sp.gov.br/noticias/laranja-mombuca-e-rica-em-licopeno-potente-agente-antioxidante-natural-que-ajuda-prevenir-o-cancer (accessed on 20 November 2021).
9. Pio, R.M.; Figueiredo, J.O.; Stuchi, E.S.; Cardoso, S.A.B. Variedade de Copas. In *Citros*; Mattos Júnior, D., Pio, R.M., De Negri, J.D. Pompeu Júnior, J., Eds.; Instituto Agronômico e Fundag: Campinas, Brazil, 2005; 929p.
10. Oliveira, R.P.; Scivitaro, W.B. *Lue Gim Gong: Laranja Tipo Valência Mais Tardia e Tolerante Ao Frio e Cancro Cítrico*; Embrapa Clima Temperado: Pelotas, Brazil, 2008; Available online: http://www.infoteca.cnptia.embrapa.br/infoteca/handle/doc/746632 (accessed on 14 September 2021).
11. Oliveira, R.P.; Scivitaro, W.B.; Gomes da Rocha, P.S. *Cultivares de Laranjeira Resistentes ao Cancro Cítrico*; Embrapa Clima Temperado: Pelotas, Brazil, 2008; 31p. Available online: https://www.infoteca.cnptia.embrapa.br/bitstream/doc/746885/1/documento249.pdf (accessed on 12 January 2020).
12. Bowman, K.D.; Joubert, J. Citrus rootstocks. In *The Genus Citrus*; Talon, M., Caruso, M., Gmitter, F.G., Jr., Eds.; Elsevier: Amsterdam, The Netherlands, 2020; pp. 105–127.
13. Balfagón, D.; Rambla, J.L.; Granell, A.; Arbona, V.; Gómez-Cadenas, A. Grafting improves tolerance to combined drought and heat stresses by modifying metabolism in citrus scion. *Environ. Exp. Bot.* **2022**, *195*, 104793. [CrossRef]
14. Rasool, A.; Mansoor, S.; Bhat, K.M.; Hassan, G.I.; Baba, T.R.; Alyemeni, M.N.; Alsahli, A.A.; El-Serehy, H.Á.; Paray, B.A.; Ahmad, P. Mechanisms underlying graft union formation and rootstock scion interaction in horticultural plants. *Front. Plant Sci.* **2020**, *11*, 1778. [CrossRef]
15. Hu, Z.; Wang, F.; Yu, H.; Zhang, M.; Jiang, D.; Huang, T.; Xiang, J.; Zhu, S.; Zhao, X. Effects of scion-rootstock interaction on citrus fruit quality related to differentially expressed small RNAs. *Sci. Hortic.* **2022**, *298*, 110974. [CrossRef]
16. Curk, F.; Olitrautt, F.; Garcia-Lor, A.; Luro, F.; Navarro, L.; Ollitraut, P. Phylogenetic origin of limes and lemons revealed by cytoplasmic and nuclear markers. *Ann. Bot.* **2016**, *117*, 565–583. [CrossRef]
17. Oliveira, C.R.M.; Mello-Farias, P.C.; Oliveira, D.S.C.; Chaves, A.L.S.; Herter, F.G. Water availability effect on gas exchanges and on phenology of 'Cabula' orange. *Acta Hortic.* **2017**, *1150*, 133–138. [CrossRef]
18. Fadel, A.L.; Mourão Filho, F.A.A.; Stuchi, E.S.; Wulff, N.A.; Couto, H.T.Z. Citrus sudden death-associated vírus (CSDaV) and citrus tristeza vírus (CTV) in eleven rootstocks for 'Valência' sweet orange. *Rev. Bras. Frutic.* **2018**, *40*, e-788. [CrossRef]
19. Ribeiro, R.V.; Núñez, E.E.; Júnior, J.P.; Mourão Filho, F.A.A. Citrus rootstocks for improving the horticultural performance and physiological responses under constraining environments. *Improv. Crops Era Clim. Changes* **2014**, *1*, 1–37. [CrossRef]
20. Saini, R.K.; Ranjit, A.; Sharma, K.; Prassad, P.; Shang, X.; Gowda, K.G.M.; Keum, Y.S. Bioactive compound of Citrus fruits: A review of composition and health benefits of carontenoids, flavonoids, limonoids, and terpenes. *Antioxidants* **2022**, *11*, 239. [CrossRef] [PubMed]
21. Zou, Z.; Xi, W.; Hu, Y.; Nie, C.; Zhou, Z. Antioxidant activity of Citrus fruits. *Food Chem.* **2016**, *196*, 885–896. [CrossRef]
22. Zhang, M.; Zhu, S.Y.; Yang, W.J.; Huang, Q.R.; Ho, C.T. The biological fate and bioefficacy of citrus flavonoids: Bioavailability, biotransformation, and delivery systems. *Food Funct.* **2021**, *12*, 3307–3323. [CrossRef] [PubMed]
23. Wibowo, S.; Grauwet, T.; Santiago, J.S.; Tomic, J.; Vervoort, L.; Hendrickx, M.; Van Loey, A. Quality changes of pasteurised orange juice during storage: A kinetic study of specific parameters and their relation to colour instability. *Food Chem.* **2015**, *187*, 140–151. [CrossRef] [PubMed]
24. Nakilcioglu-tas, E.; Ötles, S. Kinetic modelling of vitamin C losses in fresh citrus juices under different storage conditions. *An. Acad. Bras. Cienc.* **2020**, *92*, 1–9. [CrossRef] [PubMed]
25. Cunha, A.R.; Martins, D. Classificação climática para os municípios de Botucatu e São Manuel, SP. *Irriga* **2009**, *14*, 1–11. [CrossRef]
26. Santos, H.G.; Jacomine, P.K.T.; Anjos, L.H.C.; Oliveira, V.A.; Lumbreras, J.F.; Coelho, M.R.; de Almeida, J.A.; Cunha, T.J.F.; de Oliveira, J.B. *Sistema Brasileiro de Classificação de Solos*, 3rd ed.; Embrapa: Brasília, Brazil, 2013; 353p, Available online: https://www.embrapa.br/solos/sibcs (accessed on 11 February 2021).
27. Soil Survey Staff. *Soil Taxonomy: A Basic System of Soil Classification for Making and Interpreting Soil Surveys*, 2nd ed.; USDA, NRCS: Washington, DC, USA, 1999. Available online: https://www.nrcs.usda.gov/wps/portal/nrcs/main/soils/survey/class/taxonomy/ (accessed on 27 July 2021).
28. Peña, W.E.L.; Massaguer, P.R.; Teixeira, L.Q. Microbial modeling of Alicyclobacillus acidoterrestris CRA7152 growth in orange juice with nisin addition. *J. Food Prot.* **2006**, *69*, 7. [CrossRef]
29. Spínola, V.; Mendes, B.; Câmara, J.S.; Castilho, P.C. An improved and fast UHPLC-PDA methodology for determination of L-ascorbic and dehydroascorbic acids in fruits and vegetables. Evaluation of degradation rate during storage. *Anal. Bioanal. Chem.* **2012**, *403*, 1049–1058. [CrossRef] [PubMed]

30. Natividade, M.M.P.; Corrêa, L.C.; Souza, S.V.C.; Pereira, G.E.; Lima, L.C.O. Simultaneous analysis of 25 phenolic compounds in grape juice for HPLC: Method validation and characterization of São Francisco Valley samples. *Microchem. J.* **2013**, *110*, 665–674. [CrossRef]
31. Rencher, A.C.; Christensen, W.F. *Methods of Multivariate Analysis*, 3rd ed.; John Wiley & Sons: New York, NY, USA, 2012; 758p. [CrossRef]
32. Loannidi, E.; Kalamaki, M.S.; Enginner, C.; Pateraki, I.; Alexandrou, D.; Mellidou, I.; Giovannonni, J.; Kanellis, A.K. Expression profiling of ascorbic acid-related genes during tomato fruit development and ripening and in response to stress conditions. *J. Exp. Bot.* **2009**, *60*, 663–678. [CrossRef]
33. Castillo-Velarde, E.R. Vitamin C in health and disease. *Rev. Fac. Med. Hum.* **2019**, *19*, 95–100. [CrossRef]
34. Castle, W.S. A career perspective on Citrus rootstocks, their development, and commercialization. *Hortscience* **2010**, *45*, 11–15. [CrossRef]
35. Pompeu Júnior, J.; Blumer, S. Citrumelos como porta-enxertos para laranja 'Valência'. *Pesqui. Agropecu. Bras.* **2011**, *46*, 105–107. [CrossRef]
36. Lee, S.K.; Kader, A.A. Preharvest and postharvest factors influencing vitamin C content of horticultural crops. *Postharvest Biol. Technol.* **2020**, *20*, 207–220. [CrossRef]
37. Stinco, C.M.; Fernández-Vásquez, R.; Hernanz, D.; Heredia, F.J.; Meléndez-Martínez, A.J.; Vicario, I.M. Industrial orange juice debittering: Impact on bioactive compounds and nutritional value. *J. Food Eng.* **2013**, *116*, 155–161. [CrossRef]
38. Beber, P.M.; Álvares, V.d.S.; Kusdra, J.F. Qualidade industrial e maturação de frutos de laranjeiras-doce em Rio Branco, Acre. *Citrus Res. Technol.* **2018**, *39*, e1030. [CrossRef]
39. Sapei, L.; Hwa, L. Study on the kinetics of vitamin C degradation in fresh orange juices. *Procedia Chem.* **2014**, *9*, 62–68. [CrossRef]
40. Cruz, R.M.S.; Vieira, M.C.; Silva, C.L.M.; Effect of heat and thermosonication treatments on watercress (*Nasturticum officinale*) vitamin C degradation kinetics. Innov. *Food Sci. Emerg. Technol.* **2008**, *9*, 483–488. [CrossRef]
41. Bacigalupi, C.; Lemaistre, M.H.; Boutroy, N.; Bunel, C.; Peyron, S.; Guillard, V. Changes in nutritional and sensory properties of orange juice packed in PET bottles: An experimental and modelling approach. *Food Chem.* **2013**, *141*, 3827–3838. [CrossRef]
42. Klimczak, I.; Malecka, M.; Szlachta, M.; Gliszcynska-Świglo, A. Effect of storage on the content of polyphenols, vitamin C and the antioxidant activity of orange juices. *J. Food Compos. Anal.* **2007**, *20*, 313–322. [CrossRef]
43. Teixeira, M.; Monteiro, M. Degradação da vitamina C em suco de fruta. *Aliment. e Nutr.* **2006**, *17*, 219–227. Available online: https://www.academia.edu/63811880/Degrada%C3%A7%C3%A3o_da_vitamina_C_em_suco_de_fruta (accessed on 8 September 2022).
44. Laorko, A.; Tongchitpackdee, S.; Youravong, W. Storage quality of pineapple juice non-thermally pasteurized and clarified by microfiltration. *J. Food Eng.* **2013**, *116*, 554–561. [CrossRef]
45. Caro, A.D.; Piga, A.; Vacca, V.; Agabbio, M. Changes of flavonoids, vitamin C and antioxidant capacity in minimally processed citrus segments and juices during storage. *Food Chem.* **2004**, *84*, 99–105. [CrossRef]
46. Lago-Vanzela, E.S. Influência de reações químicas e enzimáticas sobre a cor e o aroma de vinhos. In *Uvas e Vinhos: Química, Bioquímica e Microbiologia*, 1st ed.; Silva, R., Lago-Vanzela, E.S., Baffi, M.A., Eds.; Editora Senac, Editora Unesp: São Paulo, Brazil, 2015; pp. 107–132. Available online: https://editoraunesp.com.br/catalogo/9788539305643,uvas-e-vinhos (accessed on 20 December 2021).
47. La Peña, M.; Morales, D.E.; Salvia-Trujillo, L.; Rojas-Graü, M.A.; Martín-Belloso, O. Changes on phenolic and carotenoid composition of high intensity pulsed electric field and thermally treated fruit juice-soymilk beverages during refrigerated storage. *Food Chem.* **2011**, *129*, 982–990. [CrossRef] [PubMed]
48. Bastos, D.C.; Ferreira, E.A.; Passos, O.S.; Sá, J.F.; Ataíde, E.M.; Calgaro, M. Cultivares copa e porta-enxertos para a citricultura brasileira. *Inf. Agrop.* **2014**, *35*, 36–45. Available online: https://www.embrapa.br/busca-de-publicacoes/-/publicacao/1007492/cultivares-copa-e-porta-enxertos-para-a-citricultura-brasileira (accessed on 30 March 2022).
49. Palacios, C.; Serra, D.; Torres, P. Papel Ecológico Dos Metabólitos Secundários Frente ao Estresse Abiótico. In *Botânica No Inverno*; Lopez, A.M., Nagai, A., Furlan, C.M., Eds.; Instituto de Biociências da Universidade de São Paulo: São Paulo, Brazil, 2013; pp. 52–59. Available online: https://www.academia.edu/42993539/Papel_ecol%C3%B3gico_dos_metab%C3%B3litos_secund%C3%A1rios_frente_ao_estresse_abi%C3%B3tico_1 (accessed on 2 August 2020).
50. Giuffré, A.M.; Zappia, C.; Capocasale, M. Physicochemical stability of blood orange juice during frozen storage. *Int. J. Food Prop.* **2017**, *20*, 1930–1943. [CrossRef]
51. Agcam, E.; Akydlddz, A.; Evrendilek, G.; Akdemir, E. Comparison of phenolic compounds of orange juice processed by pulsed electric fields (PEF) and conventional thermal pasteurisation. *Food Chem.* **2014**, *143*, 354–361. [CrossRef]
52. Boerjan, W.; Ralph, J.; Baucher, M. Lignin biosynthesis. *Ann. Rev. Plant Biol.* **2003**, *54*, 519–546. [CrossRef] [PubMed]

horticulturae

MDPI

Review

Achievements of Banana (*Musa* sp.)-Based Intercropping Systems in Improving Crop Sustainability

Sarita Leonel [1,*], Magali Leonel [2], Paulo Ricardo Rodrigues de Jesus [1], Marco Antonio Tecchio [1], Marcelo de Souza Silva [1], Hebert Teixeira Cândido [1], Nicholas Zanette Molha [1] and Lucas Felipe dos Ouros [2]

1 Department of Crop Sciences, School of Agriculture (FCA), São Paulo State University (UNESP), Botucatu 18610-034, SP, Brazil; pr.jesus@unesp.br (P.R.R.d.J.); marco.a.tecchio@unesp.br (M.A.T.); marcelo.s.silva@unesp.br (M.d.S.S.); hebert.candido@gmail.com (H.T.C.); nz.molha@unesp.br (N.Z.M.)

2 Center for Tropical Root and Starches (CERAT), São Paulo State University (UNESP), Botucatu 18610-034, SP, Brazil; magali.leonel@unesp.br (M.L.); lucas.ouros@unesp.br (L.F.d.O.)

* Correspondence: sarita.leonel@unesp.br

Abstract: Sustainable agricultural practices need to be continuously sought after so that a greater number of producers can adopt them, taking into account, above all, the food security scenario, land use efficiency, and climate change. Intercropping—a cultivation system in which two or more species are grown in close proximity in the same field—is one strategy to increase diversity in the agroecosystem. However, for intercropping systems to be adopted, their productive and economic advantages over monoculture must be clearly demonstrated. Banana (*Musa* sp.) growers are interested in crop diversification as a potential strategy to increase production yields and, consequently, economic income. The management of banana crops can be facilitated by intercropping, as this system plays an important role in increasing biodiversity and reducing the need for weed control in the crop rows, promoting better land use efficiency. However, this system should be evaluated alongside other indicators. Banana intercropping has significant potential and many benefits, but success depends on the interaction between the component species, appropriate management practices, and favorable environmental conditions. This review aims to provide an overview of recent studies on banana intercropping systems, focusing on the contextualization of land use, monoculture and intercropping, and evaluating intercropping indicators, as well as the benefits, risks, and disadvantages discussed in the literature, and the main outcomes of banana-based intercropping systems. The main findings relate to the possibility of using intercrops with aromatic species and the preliminary reports on the contributions of intercrops to the suppression of Fusarium wilt disease.

Keywords: environmental resources; climate changes; competitive indices; plantation management; sustainable agriculture

Citation: Leonel, S.; Leonel, M.; Jesus, P.R.R.d.; Tecchio, M.A.; Silva, M.d.S.; Cândido, H.T.; Molha, N.Z.; Ouros, L.F.d. Achievements of Banana (*Musa* sp.)-Based Intercropping Systems in Improving Crop Sustainability. *Horticulturae* **2024**, *10*, 956. https://doi.org/10.3390/horticulturae10090956

Academic Editor: Esmaeil Fallahi

Received: 4 August 2024
Revised: 2 September 2024
Accepted: 5 September 2024
Published: 6 September 2024

1. Introduction

The banana (*Musa* sp.) is a tropical fruit with a high economic and nutritional value. The crop occupies a prominent position in world agricultural production, as it is the most widely produced fruit globally and is grown in more than 125 countries [1], predominantly in Asia, Latin America, and Africa. India is the world's largest banana producer, representing 26.4% of total production. China is the second largest producer, representing 9.3%, followed by Indonesia with 7.0%, and Brazil with about 5.4% of world banana production [2]. Bananas are an important food crop and trade product for many developing countries, where their role in food security and income generation has been widely recognized [3]. As one of the most popular and commonly consumed fruits in the world, bananas are appreciated for their sweet taste and soft texture. In addition, the fruit is rich in carbohydrates, vitamins, and minerals—such as potassium and magnesium, vitamin C, bioactive compounds, and resistant starch—making it highly nutritious [4,5]. Bananas are also economically and socioeconomically important, as they have a continuous production cycle and relatively rapid economic returns [6] and their cultivation provides an excellent source of income for small, medium, and large-scale producers [7].

The expansion of banana plantations can help to increase food production and, consequently, reduce food insecurity and hunger around the world [5], as well as generate employment and income throughout the production chain. Currently, there is a growing concern regarding the ecological and social implications of horticultural crops. The preservation of biodiversity and the promotion of agricultural practices that respect the environment are essential to ensuring the continuity of food production in the long term [1]. Adverse climatic environments are leading to a decline in soil fertility and an increase in the incidence of pests and diseases in banana fields, reducing crop yields [8].

Intercropping—i.e., planting two or more crops on the same land over a full or partial harvest—makes it possible to obtain a higher yield from the same or a smaller area. It is also considered a sustainable management strategy [9]. Intercropping can reduce management factors and result in sustainable systems that more effectively utilize and even potentially replenish the natural resources used during crop production in the context of the long-term management of agricultural land. Some of the benefits of intercropping for the farmer are risk minimization, effective use of available resources, efficient use of labor, increased production per unit area, erosion control, and food security [9].

The banana is a perennial crop that grows for long periods in the same fields and is predominantly grown as a monoculture [6]. In view of the vulnerability of banana monoculture to the combined effects of climate change, pests, and diseases, the diversification of cropping systems should be a recognized priority. Bananas can be intercropped with other species, both as a main and a secondary crop [10]. As the main crop, bananas can be grown in consortium with annual food crops, such as beans (*Phaseolus vulgaris*), maize (*Zea mays*), rice (*Oryza sativa*), and cassava (*Manihot esculenta*), among others, or with cover crops, which provide benefits to the soil and to the plant. As a secondary crop, bananas can be combined with perennial trees, such as coffee (*Coffea arabica*), the oil palm (*Elaeis gineensis*), cupuaçu (*Theobroma grandiflorum*), and cocoa (*Cocos nucifera*) [11]. It can also be used in diverse and sustainable agroforestry systems. Banana plants not only provide fruit of high nutritional and economic value, but also contribute to environmental benefits, such as soil conservation, microclimate regulation, and increased biodiversity [12].

This literature review aimed to provide a comprehensive discussion of recent studies discussed in the literature, focusing on the contextualization of land use, monoculture, and intercropping, as well as on evaluating intercropping in terms of metrics, benefits, risks, and disadvantages, and the main outcomes from the adoption of intercropping systems in banana plantations. The purpose is to demonstrate the application of a cultivation technology in banana plantations and provide its indicators for improving banana farming in a sustainable way.

2. Growing Bananas Sustainably Requires Changes in Land Use and Crop Management

Agriculture is one of the most important economic activities worldwide, and requires strategic planning to support social, political, and cultural development [13,14]. In addition, agriculture plays a prominent role in developing countries. However, due to population growth and the conversion of agricultural land through urbanization and industrial growth, productivity must be increased to meet the needs of a growing population [15].

Sustainable agriculture is a type of farming that uses resources more efficiently than conventional agriculture, benefits human beings, and is in balance with the environment [13]. The main approaches to implementing sustainable agriculture are the restoration of agricultural ecosystem diversity and effective management [16]. Sustainable agriculture must be ecologically sound, economically viable, and socially desirable [8].

Kaliz et al. [17] defined sustainable development as one that seeks to fulfill the needs and aspirations of the present without compromising the ability to satisfy those of the future. Sustainable development emphasizes, among other issues, the efficient use of the resources on Earth.

Changes in land use and occupation are associated with alterations to the world's surface. Land use includes the ways in which land is utilized, including as pasture, arable land, and forest, among others. Land cover refers to the coverage of the land surface with a certain type of vegetation, bare soil, infrastructure, or water, but does not describe land use, which can differ for the same type of land cover [18]. Alterations in land use are considered fundamental to sustainable development. With rapid urbanization, rural transformation, and the development of modern agriculture, land is becoming fragmented and changes in land use threaten sustainable development [19]. The banana growing and trading system has been characterized by unequal positions of control between the international corporations that own plantations and supply the market and the farmers who grow and harvest the fruit [8].

Changes in land use must also be associated with the implementation of agricultural practices to mitigate the impact on biomes, mainly by reducing nitrogen and phosphorus loads to the environment and conserving biodiversity [20]. In view of the vulnerability of monocultures to the combined threats of climate change, pests, and diseases, the diversification of plantations must be a prioritized area. Biodiversity performance is very poor for banana producers due to its intensive monoculture production system. Climate change is increasingly threatening economic sustainability in several important producing regions, requiring responses in terms of management and cropping systems. Sustainability in banana plantations is a worldwide concern. Banana growers are interested in maintaining or increasing production gains and preserving environmental resources for the continuity of their plantations. In addition, there is a consensus that fruit produced in sustainable agricultural systems tends to have a higher market value. Similarly, banana consumers are interested in purchasing nutritious, high-quality fruit to support a healthy lifestyle.

3. Monoculture and Intercropping Systems

Monoculture refers to the cultivation of a single crop in a given area [21]. It can also refer to the practices of large-scale agriculture in which a single crop is grown over a wide area. The practice of monoculture can have a significant impact on biodiversity, as it often involves the removal of natural vegetation and the planting of large expanses of the same crop [14]. This cropping system can deplete soil nutrients, increase susceptibility to pests and diseases, and lead to a reliance on chemical inputs [22].

This practice can have negative environmental and agricultural consequences [14]. For large-scale banana farmers, intensive monoculture is easier to implement as it enables the use of machinery. However, for small-scale farmers, it is more difficult, as they have limited access to the market and to marketing information, and often grow crops for their family's subsistence [23].

In conventional agriculture and monoculture systems, the high yield per unit area may be enough to meet the nutritional needs of growing populations in some areas; however, these systems require investments of inputs and energy that come from fossil fuels [13]. In these systems—based on conventional assessments of agricultural productivity—growth can be achieved by improving production factors, provided that the increase in output is even greater [24].

Monocultures are prevalent in most tropical countries. Bybee-Filey and Ryan [25] reported that, in Australia, landowners have adopted the broadacre system [9,26], which refers to farms using large-scale crop production. The authors emphasized that farming systems in Australia are dominated by intensive monocultures managed through crop rotation and the integration of livestock, when possible, as mixed farming enterprises. These practices are based on the economic view of specialization and economies of scale, which occur when a farmer increases the scale of production, thus distributing fixed costs over many production units and reducing the cost of production per unit [9].

Bananas are perennial crops that grow for long periods in the same fields and are predominantly grown as monocultures [6,27]. Banana monoculture is widely used in intensive agriculture [28]. The system of growing a single crop, such as bananas, repeatedly on the same piece of land was invented to increase food supply and to fight hunger. Unfortunately, its unintended consequences threaten greater global insecurity and exacerbate climate change [14].

Intercropping is an ancient farming practice [29]. Mousavi and Eskandari [13] pointed out that there is evidence that planting crops in consortium has a long history. This agricultural system consists of growing two or more crops simultaneously in the same field for all or part of the growing season [30]. It is important to note that in intercropping systems, the plants do not have to be planted at the same time; the aim is for two or more crops to grow together in one area during part of or the entire crop cycle [13,31].

Crop diversification through intercropping can improve the results and stability of agricultural production in the face of seasonal variability and climate change. Different species react differently to environmental conditions, so if one species is negatively affected by adverse weather conditions, another intercropped species can still produce a feasible yield.

Intercropping is a cultural practice in which two or more crops are grown on the same field in a year with different cropping patterns. In this multiple cropping system, biodiversity and pest suppression are enhanced. Biodiversity can restore the natural elements of the agricultural ecosystem because almost all elements favorable to the natural enemies of pests are available in a diversified agricultural ecosystem. Modern energy-intensive technology used in agriculture is one of the vital causes of biodiversity loss. With the intercropping system, enhanced biological pest control can be ensured with a higher level of crop diversity, as opposed to energy-intensive farming [30,31].

One of the biggest challenges of intercropping with two or more crops is maintaining the productivity of each crop [32]. Yet, intercropping can help achieve a higher yield than planting just one crop at a time [22]. Therefore, it is important to choose a combination of crops that grow well together in order to use environmental resources more efficiently—such as solar energy and regarding water per unit area per unit time—and to maintain soil health while improving yield [33]. Intercropping is widely practiced by small-scale farmers, as it supports their livelihood by producing a diverse range of food crops [34].

The economic logic of intercropping is based on the theory of economies of scope, which arises when a farmer can use the same inputs to produce two or more products, thus reducing the cost of producing them separately [9]. Table 1 summarizes the main characteristics of monoculture and intercropping systems.

Table 1. Summarized contextualization of monoculture and intercropping systems in banana plantations.

Monoculture	Reference	Intercropping	Reference
Single crop	[6,21,27]	Two or more crops simultaneously	[13,29]
Large scale	[23]	Small scale	[34]
Impact on biodiversity	[14]	Increased biodiversity	[30,31]
Deplete soil nutrients	[22]	Stability environmental resources	[14]
Increase susceptibility to pest and diseases	[22]	Increased pests and diseases suppression	[22]
Reliance on chemical inputs	[14]	Less reliance on chemical inputs	[14]
Negative environmental and agriculture consequences with greater impact from climate change	[14]	Stability agricultural production due to seasonal variability with less impact on climate change	[14,30,31]
Higher yield per unit area	[13]	Can achieve higher yield per unit area with two or more component crops	[22]
Specialization of economies of scale when increasing the scale of production leads to a reduction in production costs per unit	[9]	Economies of scope when the same inputs are used to produce two or more products	[9]

4. Evaluating Intercropping Indicators

There are several indicators used to compare intercropping and monocropping systems. In most research projects, all intercropped species are also tested as monocultures, so the advantages and disadvantages of intercropping can be compared to the practice of monoculture [9].

The agronomic viability of intercropping can be evaluated using productivity and competitive indices (Table 2), including land use efficiency (LUE), the land equivalent ratio (LER), the area-time equivalent ratio (ATER), the land equivalent coefficient (LEC), the relative density coefficient (RDC), aggressivity (A), the competitive ratio (C), the system productivity index (SPI), the intercropping advantage (IA), gross income (GI), net income (NI), rate of return (RR), profit margin (PM), and actual yield loss (YL), among others (Table 2) [9,35,36]. These indices are used to not only estimate the effects of competition among different crops, but also to assess which system is most effective in managing environmental resources to provide greater productivity and sustainability [22,33].

LUE considers crop yields in intercropping and monoculture systems and relates them to land use equivalence; it is one of the most widely used indices to evaluate intercropping [31]. LER provides a rough estimate of the area of land needed to obtain the same yields as an intercropping system. ATER is an alternative index to LUE because the latter does not consider time. As such, LUE can overestimate the advantage of intercropping, especially when the crops differ significantly in crop cycle duration [22,33].

A systematic assessment of LUE needs to support decision-making in land use management and to promote its use in a better and more efficient way [37]. Lin and Hülsbergen [38] presented a new method for calculating LUE, starting with an overview of the different approaches to assessing agricultural LUE. This method takes into account the quality and function of agricultural products and the relationship between the yield of the farm assessed and the average yield of the reference region with comparable soils, climate, and socio-economic conditions. The main conclusion is that LUE should be used in combination with agri-environmental indicators to ensure efficient and sustainable land use. The methods used to quantify the effects of changes in land use are still the subject of intense research, stimulating much scientific discussion [17]. This study presents research on land use indicators in the context of land use efficiency. The overall aim is to fill the knowledge gap on responsible and sustainable land use management.

According to Ferreira and Féres [39], the relationship between property size and land use efficiency in the Brazilian Amazon was negative; the authors concluded that the current process of land concentration observed in this region would result in an increase in land use inefficiency.

The LEC is obtained from the product of the LER of each individual crop in the intercropping system, and must be at least 25% for intercropping to have a productive advantage [35].

The RDC index represents a measure of the dominance of one crop over another. The A index indicates the extent to which the relative increase in the yield of one crop is greater than that of the other crop in an intercropping system. This index measures the dominance among the intercropped species [15]. The C index represents the number of times one crop is more competitive than the other [22]. C offers an alternative to assess competition between different crops and provides a more accurate measure of the competitive capacity of the crops [40]. C represents the proportions of individual LUEs of the two component crops and considers the proportion in which they are initially planted [22].

The SPI normalizes the yield of the secondary crop in relation to the main crop [35]. IA represents the real income losses related to the prices of the intercropped species [40].

The GI is obtained by multiplying the crop yield of each component species of the intercrop by the price (P) paid to the producer in the regional market. Net income (NI) is calculated by subtracting the total costs (TC) of production for inputs and services from the GI. The RR is the ratio between the GI and the TC, which corresponds to the amount of revenue obtained in relation to that invested. The PM is obtained from the ratio between NI and GI, expressed as a percentage [40].

Table 2. Competitive and productivity indicators used to assess the efficiency of intercropping in relation to monoculture.

Indicator	Formula	Criteria for Decisions	Reference
LUE	LUE = (Yai)/(Ybm) + (Ybi)/Yam)	LUE > 1 indicates a productive advantage of intercropping; LUE = 1 no productive advantage; LUE < 1 productive disadvantage	[9,22,33,35,40]
LER	LER = Yam/Ybm + Yai/Ybi	LER > 1 intercropping is most effective; LER < 1 intercropping has a negative effect on the yield	[22]
ATER	ATER = [(LUEa × ta) + (LUEb × tb)]/Tbi	ATER > 1 productive advantage; ATER = 1 no productive advantage; ATER < 1 productive disadvantage	[40]
RDC	RDC = {(Yai × Zb)/[(Yam − Yai × Za)]} × {(Ybi × Za)/[(Ybm − Ybi) × Za]}	RDC > 1 productive advantage; RDC = 1 no productive advantage; RDC < 1 productive disadvantage; RDC_{ai} > RDC_{bi} indicates that the main crop presents strong interspecific competition	[35,40]
A	Aa = [Ybi/(Yam × Za)] − [Ybi/(Ybi × Zb)] and Ab = [Ybi/(Ybm × Zb)] − [Ybi/(Ybm × Zb]	Both crops are equally competitive when A = 0. When A is +, the culture with a + sign is dominant and the culture with a—sign is dominated	[35,40]
C	C = Cb + Cl Cb = $(LUE_a/LUE_b) \times (Z_a/Z_b)$ Cb = $(LUE_b/LUE_a) \times (Z_b/Z_a)$		[22]
SPI	SPI = $[(Y_{am}/Y_{bm}) \times Y_{bai}] + Y_{abi}$		[40]
IA	IA = $AY_{at} \times P_{at} + AY_{bc} \times P_{bt}$	IA > 0 intercropping advantage; IA ≤ 0 intercropping disadvantage	[36,40]
GI	$CP_a \times P_a$; $CP_b \times P_b$		[40]
NI	NI = GI − TC		[40]
RR	RR = GI/TC		[40]
PM	PM = (NI/GI) × 100%		[40]
YL	YL = (WL/WI) × 100%		[40]

LUE = land use efficiency; ATER = area-time equivalent ratio; RDC = relative density coefficient; A = aggressivity; C = competitive ratio; SPI = system productivity index; IA = intercropping advantage; GI = gross income; NI = net income; RR = rate of return; PM = profit margin; Y_{ai} = yield of main crop in the intercropping; Y_{bi} = yield of second crop in the intercropping; Y_{am} = yield of main crop in the monocropping; Y_{bm} = yield of second crop in the monocropping; t_a = duration of main crop cycle; t_b = duration of second crop cycle; t_{ab} = duration of the total time of intercropping system; Z_a = proportion of main crop intercropping with second crop; Z_b = proportion of the second crop intercropping with main crop; A_a = aggressivity of main crop; A_b = aggressivity of second crop; C = ratio for intercropping; C_a = ratio for intercropping main crop; C_b = ratio for intercropping second crop; AY_a = yield losses of main crop; AY_b = yield losses of second crop; P_a = price of main crop; P_b = price of second crop; WL = weight loss; WI = weight inputs.

Khanal et al. [9] proposed the use of total economic value (TEV) generated by the cropping system (intercropping or monoculture). TEV can be classified into use values, i.e., values that people obtain from the use of services, and non-use values, i.e., values that people place on the existence of resources and the opportunity to pass them on intact to the next generation. However, the authors reinforced the idea that the main challenge is quantifying the non-use value of the benefits generated by intercropping systems.

Appropriate profitability and risk metrics need to be used when evaluating intercropping. The metrics consider all possible differences in costs and benefits between intercropping and monoculture systems [9].

Ditzler et al. [41] suggested a FarmDESIGN model to quantify the profitability, sustainability, and nutritional yield of current banana-based intercropping systems. The farms' levels of agroecological intensification were grouped according to variables such as farm size, number of crops, cover, agroforestry, shade- and drought-tolerant species, and production constraints and orientations. The authors noted the disparities in agroecological practices and socioeconomic constraints among farmers, and that the FarmDESIGN model was a valuable tool for assessing farm performance and could help reduce costs and time-consuming trials.

An evaluation of the metrics of the banana and bean intercropping system [42] concluded that bananas appeared to be more competitive than beans in the intercropping system [42]. The yield of beans in the intercropping system was 52 percent of the yield of beans in the monocropping system, due to shading and nutritional effects. The LER of the banana and bean intercrop during the three seasons was 1.60. The results obtained by [43] showed that vigorous intercropping with climbing beans (*Phaseolus coccineus*) and soya (*Glycine max*) often reduced banana growth and yield. The greater economic efficiency in banana monocrop plots suggests that reliance on the LER alone may be insufficient to inform intercropping decisions.

The evaluation of the agronomic and economic benefits of coffee-banana intercropping has shown that this system is advantageous for NI when compared to banana or coffee monocultures [44]. Sonavane et al. [45] evaluated several scenarios related to the percentage of area allocated for banana–onion (*Allium cepa*) intercropping. The highest net revenue was recorded with 58% of the area allocated along the row and 60% the area allocated between the rows for intercropping.

According to Almeida et al. [6], the intercropping of banana plants with other crops is a common practice in agroforestry systems, with the aim of optimizing LUE, diversifying production, and increasing GI. The intercropping of sweet gourd (*Momordica cochinchinensis*), bitter gourd (*Momordica charantia*), red amaranth (*Amaranthus cruentus*), and radish (*Raphanus sativus*) with banana showed a lower yield when compared to banana monocropping [46]. However, their economic analysis indicated that banana intercropped with the evaluated species showed the maximum cost-benefit ratio compared to banana as a monocrop.

Siqueira et al. [47] conducted an economic analysis of 'Conilon' coffee intercropping with perennial forest species in Brazil and found that intercropping with banana plants was economically viable. The authors also emphasized that this type of intercropping is more efficient in terms of LUE than monoculture coffee. In addition, banana–coffee intercropping can provide food security, an important factor for family coffee growers, who can consume or sell the fruit. The intercropping of banana plants with coffee is beneficial and can increase the revenue of an area by more than 50% [48].

Intercropping banana and yacon (*Smallanthus sonchifolius*), considered a functional food, optimizes the use of the area and is profitable for the farmer [49]. The study's economic analysis found that this system had a higher GI than banana monoculture due to the market value of yacon.

Dissanayake and Palihakkara [50] reported a yield percentage of a banana intercrop of 60.64% compared to monoculture, and their results can help inform sustainable LUE on oil palm plantations.

5. Intercropping Benefits

Intercropping is positioned as a potential cropping system that is environmentally friendly and can help address the challenge of increasing production with less or equivalent amounts of land, thereby improving food security [26].

Intercropping is especially advantageous when the associated crops exhibit some complementarity, which can depend on the management of the system. However, the bio-agro-economic efficiency of such systems is directly linked to crop species, production factors, spatial arrangement, and growing seasons [40]. It is therefore important to choose a combination of crops that grow well together in order to efficiently use environmental resources, as noted above, and support soil health, while also improving yield [33].

There are several benefits to adopting intercropping systems, especially for perennial crops [51] such as bananas. The main advantages include increasing or maintaining productivity and profitability, minimized losses in productivity and profitability, effective use of natural resources, weed control, pest and disease reduction, nutrients cycling, and improving nutritional management and crop resilience (Figure 1).

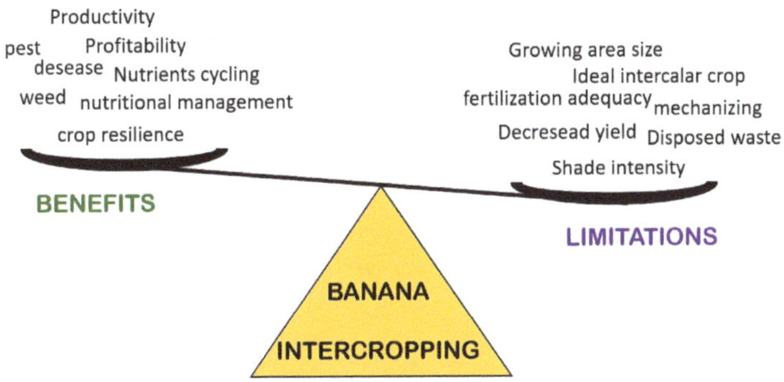

Figure 1. Overview of the benefits and limitations of intercropping in banana plantations.

5.1. *Increasing or Maintaining Productivity and Profitability*

Banana and other fruit farmers implement crop diversification strategies for a variety of reasons, including maximizing yields [51] and supporting the establishment of a permanent intercropping system. In this case, both crops are cultivated over several years and a temporary intercrop is used to improve the economic viability of implementing a banana plantation [11]. The main reason for intercropping bananas/plantains is to obtain both additional food and a cash return, as well as to reduce the cost of establishing the plantation [27,52].

Intercropping can help achieve a higher yield than planting just one crop at a time [22,34,53]. Increased yield is important, especially for small-scale farmers and in areas where the growing season is short [13,34]. A higher yield in intercropping can be due to more effective use of resources—such as nutrients, solar radiation, and water [24]—and more effective and complementary interaction between the component species [32]. An increase in productivity can lead to greater profitability. However, extra labor, material, and financial resources were not considered in the profitability metrics.

Yogendra et al. [15] emphasized that intercropping offers a viable solution to achieve greater productivity within the constraints of limited available land and provides an increase in yield. Intercropping can reduce production costs and diversify and stabilize farm income [31].

Field research conducted at scale in Costa Rica indicated that the conventional coffee–banana intercropping system could be scaled up to achieve a productive and profitable system that produces high-quality bananas [54].

The use of aromatic species in intercropping can provide farmers with additional income, contribute to the qualitative and quantitative diagnosis of plant formations and entomofauna balance of crops, and reduce costs and environmental damage caused by the excessive use of pesticides [55]. Income generated from aromatic species can be more profitable from an economic point of view than subsistence crops often used in association with banana plantations [56]. Lemongrass (*Cymbopogon citratus*), which has various medicinal properties, is widely grown for commercial essential oil extraction [57]. It is commonly consumed as a tea, but also has uses in the pharmaceutical, food, cosmetics, and perfume industries [58]. Furthermore, intercropping can facilitate entry into consumer markets for crops like lemongrass, taking advantage of the demand for other, better-known crops such as bananas [30].

Oil palm is the main edible, oil-producing plant in the world, and in tropical countries it is well established as a perennial plantation crop. However, during the oil palm juvenile phase, there is almost no income for the producer, so intercropping could provide an opportunity to obtain revenue before the oil palm's first harvest [50].

Aromatic plants are a source of essential oils, cosmetics, and biocides, and in intercropping systems they play a positive role in increasing farmers' additional income due to the greater added commercial value of their essential oils [59].

Maintaining productivity can be considered as an advantage of the intercropping system, as it can lead to maintaining or increasing profitability. Almeida et al. [6] reported that no difference was found in banana productivity when intercropped with acai (*Euterpe oleracea*). Rodrigues de Jesus et al. [60] concluded that the growth and yield of banana cultivars exhibited similar performance in both monoculture and intercropping with lemongrass.

Intercropping can minimize the risk of losses for producers [30]. This is related to the fact that different component species respond differently to seasonal variations in the climate [61,62]. Consequently, if losses occur for one component, they may be compensated by the other. Minimizing productivity losses can mitigate profitability losses [63]. Production failure risk in intercropping systems is lower than in monoculture and monocropping systems [40].

5.2. Promoting the Effective Use of Natural Resources

Intercropping can help control erosion due to providing increased soil cover that reduces surface runoff [23]. In addition, intercropping can improve the physical, chemical, and biological properties of the soil [64], as well as increase the circulation and efficiency of nutrient use and the recovery of degraded areas [16]. Intercropping is proposed as a potential cropping system that is environmentally sound in the current climate change scenario, due to its ability to enhance radiation and water use efficiency [60]. Other reported benefits are an increase in organic matter, earthworm and soil microbial activity, and improvement in soil structure [59].

The components of intercrops do not compete for the same ecological niche due to morphological and physiological differences, and competition between species is less prominent than competition within species [13].

5.3. Weed Control

Intercropping, when well-managed, offers advantages over monocultures, including in weed control [31]. Intercropping is more effective than monocropping in suppressing weeds, but its effectiveness varies widely [21]. Banana–bean intercropping systems common in East Africa are characterized by low banana productivity. In these systems, the soil is manually ploughed twice a year before the beans are planted, with potentially detrimental effects on the banana plant's shallow root system [65].

The advantages of weed control are twofold: usurping weed resources and suppressing weed growth through allelopathy [13]. Controlling weeds is one of the main reasons for establishing a banana-based intercropping system [27,52]. According to Concenço et al. [66], the shade provided by banana plants proved to be an efficient management strategy for weed suppression in the coffee–banana intercrop.

Intercropping can facilitate banana crop management by reducing the need to control weeds in the crop rows due to the cultivation of another component species in the rows [67]. Rodrigues de Jesus et al. [60] reported that the banana–lemongrass intercropping system facilitated banana crop management by reducing weed control in the crop rows.

5.4. Pest and Disease Reduction

Pests and diseases are a major risk to the sustainability of banana production, through the direct impact of agrochemicals on the environment, the loss of income that increases the area needed for production, and the associated health risks for workers in the sector. Intercropping can reduce damage from pests and disease [31]. A greater diversity of species in agricultural ecosystems can help to mitigate the spread of plant pathogens [6,13,68]. Intercropped systems provide different benefits in pest management according to two hypotheses. One is the concentration of resources hypothesis, and the other is the natural enemies hypothesis. Intercropping directly and indirectly influences the increase in biodiversity, which results in a reduction in the density of pests in crop fields. Consequently, less expenditure on the use of pesticides is required and, ultimately, a higher yield also brings some financial benefits. The intercropping system uses the plant's inherent ability to protect against pests. Therefore, more knowledge is needed about crop genotypic diversity, plant diversity, and the plant's ability to protect itself from pests [6,31].

Improving the microecological environment of soil for banana roots is crucial to promote the stable and sustainable development of banana farming. Intercropping banana and sweet potato (*Ipomoea batatas*) had a significant effect on regulating the structure and composition of the soil microbial population and improving the abundance and diversity of the microbial population [69].

With regard to pest management in banana intercropping, it was reported that the number of banana weevils (*Cosmopolites sordidus* Germar), was lower in banana intercropped with millet (*Panicum miliaceum*) [52,70]. The probable reason for this is that the root exudates from the millet, which may inhibit the presence of the weevil. Leguminous crops such as *Canavalia muzzina* and *Tephrosia vogelli* have been reported to have repellent or insecticidal properties against the banana weevil [52].

Intercropping is a useful strategy for providing food and alternative habitats for arthropods, including generalist predators. In sustainable agriculture, ants are important predators and have complex and often strong effects on pests [71]. With the aim of optimizing control of the banana weevil, *Cosmopolites sordidus*, Dassou et al. [71] studied maize, taro (*Xanthosoma sagittifolium*), and gourd (*Lagenaria siceraria*) as intercrops in a banana field. The effects of the intercropping on the abundance of ants and the damage caused by *C. sordidus* larvae to the banana stalks were assessed. Intercropping had significant effects on ant abundance, which was negatively correlated with the damage caused by *C. sordidus* to the ants. Intercropping in banana plantations has the potential to alter the structure of the ant community, which contributes to the control of weevils, but the effect of the intercropped plant species remains unclear.

Banana production faces significant challenges due to Fusarium wilt, a destructive disease caused by the soil-borne fungus *Fusarium oxysporum* f.sp. *cubense* [72]. Fusarium wilt in bananas is managed by planting disease-resistant cultivars, using appropriate cultural practices, biological control agents, and intercropping [73,74].

Several studies have shown that intercropping can lead to the recruitment of beneficial indigenous soil microbial taxa via root exudates, leading to increased host protection against pathogens [75].

Intercropping contributes to the suppression of Fusarium wilt disease (*Fusarium oxysporum* f.sp. *cubense*) in bananas. In one study, intercropping with Chinese chives (*Allium tuberosum* Rottler) showed the potential to reduce the incidence of the disease in bananas [76]. Meanwhile, Yang et al. [77] developed an approach to reduce Fusarium wilt disease by rebuilding the soil microbiome through intercropping with green manure. Intercropping bananas and green manure demonstrated the biological basis of the disease-suppressing microbiome in terms of agricultural practices and soil management [78]. *Trifolium repens* effectively reduced the incidence of banana wilt disease by regulating soil microorganisms and enriching beneficial bacterial and fungal microorganisms. However, it remains to be determined whether protist communities, important soil microbial components, contribute to disease suppression in intercropping management systems [77].

Ren et al. [68] also explored intercropping as a strategy to manage Fusarium wilt disease by remodeling soil protist communities. Protists are particularly important predators that feed on microbes [79] and are important consumers of bacteria and fungi in the soil food system. Thus, they affect soil microbial composition and function. They can feed selectively on microbial prey, leading to differential impacts on soil microbial communities [80]. Through this selective predation and induction of activity, protists can promote some resistant bacteria to increase pathogen-suppressive secondary metabolites [81]. Predatory protists can potentially reduce the effect of Fusarium wilt on bananas by promoting the expression of disease-suppressive secondary metabolite synthesis genes, including, for example, the non-ribosomal peptide synthetase gene in pathogen-suppressive bacteria [82].

Ren et al. [68] assessed changes in the microbiome with a focus on protists in a banana–legume consortium using *Trifolium repens*. Their results highlighted that predatory protists are important agents underlying disease suppression in the consortium system, which may offer new avenues to promote plant health in sustainable agriculture. These effects have been associated with intercrop-induced changes in soil microbial composition and function, as well as modulation of the microbial community composition to increase the host plant's functional resilience and stress tolerance [68]. They also propose that predatory protists could be the advantage controller of the soil microbiome, contributing to the suppression of soil-borne diseases.

5.5. Nutrient Cycling

Intercropping systems improve both soil nutrient cycling through the activity of the microbial community and, consequently, land productivity. However, the mechanism of interactions between the soil microbiome and nutrient cycling in the perennial orchard has yet to be identified [59]. The authors reported that, in orchards intercropped with aromatic plant species, the chemical diversity of the mixed aromatic plant species led to increased diversity, complexity, and stability of the soil microbial community and, consequently, to nutrient cycling. In addition, it has been assumed that the composition and quantity of exudates from intercrops play an important role in regulating the microbial community and nutrient availability. In general, the introduction of functional plants, such as aromatic plants, can increase primary production, alter the chemical characteristics of crop residues—e.g., N and P concentrations [83]—and alter the characteristics of roots, such as nitrogen-fixing bacteria and mycorrhizal fungi [84].

Soil microbes are associated with a variety of ecosystem processes, including the decomposition of plant residues, the degradation of organic matter, and the cycling of C and N, through their interactions with plants in the soil [85]. These processes are affected by local biotic and abiotic conditions, local vegetation patterns, and their intra-species and inter-species interactions, as well as by the introduction of new plant species, which destabilize microbial communities and their function in the rhizosphere and soil due to changes in vegetation composition and the composition and decomposition of plant residues. By decomposing the soil and mediating the biogeochemical cycles of C and N, soil microbes have the ability to adapt to the composition of different resources and can thus alter their nutrient utilization efficiencies [59].

Many of the practical benefits of nutrients cycling are related to the extensive root systems of the component species of the intercrop, which contribute to the cycling of nutrients from the deep layers of the soil and the storage of carbon in the soil, thus improving the fertility and quality of the soil [86].

According to Lin and Hülsberger [38], nitrogen-use efficiency is correlated with land degradation, which damages soil health. Nitrogen losses can be useful for analyzing the effects of various factors related to soil, climate, and cropping systems. Researchers [17] have proposed an environmental indicator to assess the sustainability of farming and cropping systems based on a simple model that simulates nitrate leaching and gaseous emissions of nitrogen, NH_3, and N_2O, in a quantitative way.

5.6. Improving Nutritional Management

The main purposes of research into banana intercropping have been to improve the use of land cultivated with bananas through more intensive cropping and to increase the nutritional base of banana plantations [27].

Bananas require a large amount of nutrients [28] and are highly efficient in phytomass production [87]. The obtaining of a high banana yield requires that nutrients are in adequate quantities and proportions in the plant [28]. Mazzafera et al. [8] emphasized the importance of adequate nitrogen requirements and the need for nitrogen fertilization to guarantee the yields and profitability of intercropping systems. Maia et al. [88] evaluated the initial growth of banana trees intercropped with green manure and concluded that the *Cajanus cajan* and *Crotalaria juncea* species provided greater banana growth. Grevillea (*Grevillea robusta*)–banana agroforestry systems were evaluated in central Kenya, Africa, by Musongora et al. [89], who found that low soil fertility continually restricts production and that emerging technologies are needed to address this challenge.

5.7. Crop Resilience

Due to global warming, high temperatures and resulting droughts are having a particularly damaging impact worldwide [61,90]. To combat climate change, farmers must innovate through ecological intensification to increase food production, increase resilience to extreme weather events, and reduce the carbon burden of agriculture. Intercropping can strengthen and stabilize agroecosystems in the context of climate change by improving resource use efficiency, increasing soil water retention capacity, and increasing the diversity and quality of habitats for beneficial insects that ensure pollination and natural pest control [91].

Intercropping is related to climate regulation through carbon balance. It is also receiving increased global attention as a sustainable agricultural practice, as farmers strive to improve sustainability and maintain soil health.

6. Intercropping Limitations and Risks

Intercrop systems are complex, with non-uniform competition between the component species during the growth cycle. It generally leads to unequal relative yields, making them difficult to evaluate. The direct benefits of intercropping, such as increased yields and reduced inputs, can be quantified using productivity indices. However, the environmental benefits have a long-term impact on the cropping system and cannot be measured directly by competitive indices [9].

Although intercropping has numerous benefits, it also has limitations and risks, including those detailed in Figure 1.

6.1. Size of the Growing Area

Banana plantations around the world are cultivated in two different land use contexts: in the first, by the largest and most specialized growers, they are cultivated on a large scale and are predominantly monocultures, with their production sold on different markets. In the second, crops are produced by small farms and are used almost exclusively for subsistence or, when commercialized, are sold in small markets [10,42,52]. Smallholders adopt banana-based intercropping systems in East and Central Africa, where fruit production is their livelihood and contributes to food and economic security [43,92].

Intercropping is an option to better utilize cultivated areas. However, these systems increase the complexity of management and require a steep learning curve for successful management, so they are not adopted in large areas of cultivation [91].

6.2. Decreased Crop Yield

Despite numerous benefits, intercropping has yet to be widely adopted on a large scale due to perceived risks and challenges, including decreased crop yield for both or for just one of the crop components of the systems. In addition, relatively few studies document ecosystem services conferred by intercrops alongside labor costs, which are key to economic sustainability for small, medium, and large-scale farmers [91].

6.3. Appropriate Choice of Component Crops

The use of appropriate cultivars for intercropping is one of the challenges for banana growers who adopt these systems. This is because intraspecific competition tends to be greater than competition between species [52]. Competition and dominance between species in the intercropping context must be continually assessed so that the system does not lead to losses for producers [10].

6.4. Proper Fertilization and Nutritional Status

Fertilization in the intercropping system still needs to be better evaluated and requires further study [53]. The fertilizer used for the banana plantation can serve as a residual fertilizer for the other component crop. Rodrigues de Jesus et al. [60] pointed out that in the banana–lemongrass intercropping system the inputs used for fertilization were directed towards the banana crop, without the need for the selection of specific fertilizers for lemongrass. As a result, with the amount of inputs directed exclusively to banana production, the intercropping produced more per unit area than monocropping banana. Authors also reported that the inputs used for fertilization can be directed to the banana crop without the need to select specific fertilizers for lemongrass. In addition, there were no differences in the average macronutrient contents in banana leaves, both in monocropping and in intercropping with lemongrass. These results show that lemongrass did not interfere with nutrient absorption by banana plants, which indicates that the nutritional management of the bananas did not need to be modified as a result of intercropping.

However, Rao and Edmunds [27] emphasized the need for more inputs in banana-based intercropping systems. Banana and beans were grown in monocropping and intercropping systems to evaluate the effects on the nutritional requirements. Nutrient concentration levels in the foliar tissues indicated that low potassium and high manganese availability constrained intercrop bean yield, while banana yields were associated with potassium levels in the soil [42]. Investments in external inputs are crucial for a sustainable banana intercrop system [43,93].

6.5. Use of Machinery

Banana growing is affected by numerous pests and pathogens. The yellow sigatoka (*Mycosphaerella musicola*) and Black leaf spot—Black Sigatoka (*Pseudocercospora fijiensis*)—diseases, caused by fungus, are the most important diseases of banana leaves [4,7]. These diseases are controlled through frequent application of fungicides or mineral oil sprays, which requires the use of machinery [6].

Intercropping can limit the use of machinery from planting through to harvest if a component crop is planted between the crop rows. However, in intercropping with banana plants, the other component crop can be grown in the crop rows. This is the main limitation and advantage for large-scale farmers in adopting an intercropping system, as the use of machinery over large areas is necessary during crop management [54].

Furthermore, the use of machinery in banana fields is less intensive compared to other fruit crops. This is due to the frequent spraying of pests and diseases by airplane or drone on large farms. For smallholders, the use of machinery is fairly low, due to the cost of purchasing machinery [30,54].

6.6. Shade Intensity

The main risk in adopting intercropping in orchards is related to the influence of shade intensity on the growth, biomass allocation, yield, and quality of fruit trees as the main crop in the system [16]. In their study, Kishore et al. [16] emphasized that the physiological functions of plants change with the level of irradiation; therefore, limiting light intensity is vital to guarantee production.

In the case of banana cultivation, the selection of suitable crops for intercropping depends on the relevant cultivation restrictions, namely the availability of light under the banana canopy. This acts as a limiting factor for growing annual crops, with the availability of light depending on the spatial distribution of the banana plants and the density of cultivation. As the fruit ripens, the canopy becomes larger, reducing the light in the area. Therefore, to correctly use intercropping in banana plantations, a succession of short-cycle annual crops and more shade-tolerant species is recommended at the ripening stage [10].

In East and Central Africa, Ntamwira et al. [10] noted that small-scale farmers' banana fields are often intercropped with various annual crops to optimize land use, a practice limited by the availability of light under the banana canopy. The bananas produced by small-scale farmers in the African Great Lakes region are often pruned to provide more light to shorter intercrops, reducing the overall profitability of the farm [43]. Banana–legume intercropping is important in several countries in Africa's Great Lakes region. This practice is widely used because of the high population pressure on the land. In this region, banana leaf pruning facilitates annual legume intercropping [94].

6.7. Disposed Waste

Growing bananas generates a large amount of waste that is discarded as a result of cultural practices. The thinning of the tillers, the removal of the male inflorescence and old leaves, and the removal of the pseudostem after harvesting result in organic material returned to the soil between the rows of banana trees, which can make intercropping more complex. Although nutrient cycling in banana plantations is very important, this means that this banana crop waste needs to be disposed of in another possible manner in order to make the banana-based intercropping system feasible [52].

7. Banana Planting Design

The spacing between banana plants in the planting rows varies according to the cultivar, the location, and the level of technology used. The most commonly used spacings are 3.0 × 2.0 m and 3.0 × 2.5 m in single rows and 4.0 × 2.0 × 2.0 m and 4.0 × 2.0 × 3.0 m in double rows, which allow for better use of the land [6].

Dense banana cultivation, an agronomic practice widely used by producers, makes better use of the land, labor, and inputs while increasing productivity. However, this strategy is only recommended for farms with favorable soil and climate conditions to support full development [7]. In general, a higher plant density can reduce the mass of the bunch and the length and diameter of the fruit. Nevertheless, productivity increases due to the greater number of plants per area [6]. The competitiveness of a crop is proportional to the increase in plant density [67].

To achieve cropping system production, many farmers are increasingly using dense cultivation [6], while intercropping has been the subject of research with the aim of developing recommendations for producers. Leaf area index was linearly correlated to yield in the intercrop system, suggesting that a higher plant density may result in higher yields [42]. The relationship between planting density and the growth and development of component crops in rubber–banana intercropping systems was evaluated by Rodrigo et al. [95], who concluded that increasing the density of banana plants from one to three rows increased biomass productivity per unit area, with no adverse effect on the growth and yield of the rubber or banana component crops. The increase in the growth of intercropped rubber trees was attributed to better crop management, since farmers tend to pay more attention to intercropped crops than to single crops, due to the additional initial income they provide.

Intercropping, according to Yogendra et al. [15], allows for better utilization of the available space, since the slower initial spacing prevents competition between crops and weeds. Banana–yacon (*Smallanthus sonchifolius*) intercropping optimizes the use of the area, especially when yacon is planted in double rows alternating between the banana rows [49].

Bananas intercropped with coffee is common in the Democratic Republic of Congo, Africa, but it is predominantly implemented by large-scale farmers who use a wide spacing between banana plants [10]. Studies conducted at this location concluded that climbing beans, bush beans, and vegetable amaranth had reasonable yields when intercropped with new banana plantations, regardless of plant density. However, in the second annual harvest season, a decline in yield was observed with increased banana canopy formation [10].

Cocoa (*Cocos nucifera*) and coffee are the crops most intercropped with bananas. However, farmers who intercrop bananas and cocoa tend to reduce the density of the banana rows, eventually replacing bananas with cocoa trees [43].

8. Banana-Based Intercropping Outcomes

Diversifying banana production systems is an important strategy to improve food and nutritional security, improve ecosystem health, and strengthen the resilience of small-scale agricultural systems [10]. Banana producers are concerned about increasing the profitability and sustainability of their plantations. The main challenges identified by growers are related to adapting cropping systems to current needs to reduce losses in the production and marketing process and, above all, to improve the final product quality to increase consumption [96].

Intercropping is becoming increasingly important in areas of the world where land is increasingly scarce, such as East and Central Africa. Bananas are a staple food for millions of people in low and middle income countries. The banana export trade that supplies North America, Europe, and other wealthy nations has a history of exploitation and conflict. The price of cheap bananas has been environmental degradation, violence, and poverty. Only recently have efforts been made to address the power imbalances in this trade. Voluntary certification schemes, in addition to the implementation of different cropping systems and management, aim to address various sustainability issues, while research into biological control has accelerated plant breeding and efficient irrigation will help prepare the sector for pest, disease, and climate change risks [97].

The growing social and environmental concerns of producers and consumers has led to the establishment of changes in the banana production sector. These modifications establish a series of criteria around social, economic, and environmental sustainability, according to which producers are classified by registered certification agencies. Producers pay a fee for the certification process and receive higher prices and, potentially, market access as rewards for complying with sustainability standards [97].

Alongside clearer demonstrations of the economic viability of intercropping, banana farmers also need technical support during the adoption process to help them resolve the complexities and location-specific challenges of managing polycultures. The environmentally friendly intensification of banana plantations requires a strategic approach than simplifies production systems, which is not without its inherent risks and challenges. Banana plants can be intercropped with various species, with different aims and outcomes. The main intercropping systems with banana plants are summarized in Table 3.

The use of intercropping in banana plantations has its benefits and drawbacks. However, there are still some questions that require further analysis, evaluation, and research results over more harvest seasons, so that the outcomes can be applied to a greater number of producers. Thus, research in this area should increase, especially considering efforts to achieve more efficient and sustainable banana plantations. The most significant results concern weed control and the potential use of intercropping with aromatic species, as well as the contribution of intercropping to suppressing the Fusarium wilt disease. Intercropping can also have an important impact on regulating the community and structure of soil bacteria and fungi and improve the diversity and abundance of soil microbes, which can act as an insect repellent or have insecticidal properties. The central element of intercropping is the use of natural resources. However, there is a long way to go before intercropping systems are more widely adopted in the predominant monoculture system due to their productive and economic advantages over banana monoculture. These indicators need to be proven in order to convince a large number of farmers.

Table 3. Main outcomes reported in banana-based intercropping systems and a model to quantify the profitability and sustainability of current systems.

Component Crop	Outcomes	Reference
Green Manures: *Cajanus cajans* and *Crotalaria juncea*	Greater banana growth	[88]
Coffee (*Coffea arabica*)	Increase economic viability; advantageous for NI, better LUE efficiency, increase the revenue, high quality bananas and weed supression	[11,44,47,48,54,66]
Bean (*Phaseolus vulgaris*)	Bananas appeared more competitive, low banana productivity and the need for investment in external inputs	[42,43,65]
Climbing beans (*Phaseolus coccineus*) and soya (*Glycine max*)	Reduced banana growth and yield	[43]
Onion (*Allium cepa*)	Highest net revenue	[45]
Sweet goud (*Momordica cochinchinensis*), Bitter gourd (*Momordica charantia*), red amaranth (*Amaranthus cruentus*) and radish (*Raphanus sativus*)	Lower yield and economic analysis with maximum cost-benefit ratio	[46]
Yacon (*Smallanthus sonchifolius*)	Higher GI and optimizes the use of the area	[49]
Aromatic species	Additional income, reduce costs and environmental damage	[55,56,59]
Lemongrass (*Cymbopogon citratus*)	Entry into consumer markets, similar performance compared with monocropping, reduced weed control and without the need to select specific fertilizers for lemongrass	[30,60]
Sweet potato (*Ipomoea batatas*)	Regulating the structure and compositions and improving the abundance and diversity of soil microbial population	[69]
Millet (*Panicum miliaceum*)	Lower number of banana weevil	[52,70]
Leguminosae (*Canavalia muzzina* and *Tephrosia vogelli*)	Repellent or insecticidal properties	[52]
Maize (*Zea mays*), taro (*Xanthosoma sagittifolium*) and gourd (*Lagenaria siceraria*)	Alter the structure of ant community which contributes to the control of weevil (*Cosmopolites sordidus*)	[71]
Chinese chives (*Allium tuberosum*)	Potential to reduce Fusarium wilt disease	[76]
Leguminosae White clover (*Trifolium repens*)	Reduced the incidence of Fusarium wilt disease	[68,77,78]
Oil palm (*Elaeis gineensis*)	Sustainable LUE and revenue	[50]
Grevillea (*Grevillea robusta*)	Low soil fertility continually restricts production	[89]
Rubber (*Hevea brasiliensis*)	Increase in the growth of rubber	[95]
Cocoa (*Cocos nucifera*)	Necessity to reduce the density of banana plantation	[43]
Agroforestry systems	Optimizzing LUE, diversifying production and increasing GI	[6]
FARMdesign model	Disparity in agroecological practices and socioeconomic constraints	[41]

Author Contributions: S.L., M.L. and P.R.R.d.J.: writing—original draft; S.L., M.L., P.R.R.d.J., N.Z.M., L.F.d.O., M.A.T., M.d.S.S. and H.T.C.: assembly of previous research articles on this topic; S.L., M.L. and P.R.R.d.J.: writing—review and editing; S.L. and M.L.: funding acquisition, supervision. All authors have read and agreed to the published version of the manuscript.

Funding: The authors acknowledge the financial support of the National Council for Scientific and Technological Development (CNPq) (Grant numbers 302848/2021-5, 302611/2021-5).

Data Availability Statement: The authors are unable or have chosen not to specify which data have been used.

Conflicts of Interest: The authors declare that they have no known competing financial interests or personal relationships that could have appeared to influence the work reported in this paper.

References

1. FAO. Food and Agriculture Research of United Nations. FAOSTAT: Food and Agriculture Data. 2023. Available online: https://www.fao.org/faostat/en/#home (accessed on 15 December 2023).
2. FAOSTAT. Food and Agriculture Research of United Nations. Food and Agriculture Data. 2024. Available online: https://www.fao.org/faostat/en/#data/QCL (accessed on 2 June 2024).
3. Petsakos, A.; Prager, S.; Gonzalez, C.E.; Gama, A.C.; Sulser, T.B.; Gbegbelegbe, S.; Kikulwe, E.M.; Hareau, G. Understanding the Consequences of Changes in the Production Frontiers for Roots, Tubers, and Bananas. *Glob. Food Secur.* **2019**, *20*, 180–188. [CrossRef]
4. Bolfarini, A.C.B.; Putti, F.F.; Souza, J.M.A.; Silva, M.D.S.; Ferreira, R.B.; Leonel, M.; Tecchio, M.A.; Leonel, S. Application of Phosphate Fertilization on Banana Hybrid 'FHIA 18' and Its Impact on Production Performance. *Aust. J. Crop Sci.* **2020**, *14*, 744–750. [CrossRef]
5. Cândido, H.T.; Marzullo, Y.O.T.; Leonel, M. Green Banana Flour Technology: From Raw Material to Sensory Acceptance of Products Made with Green Banana Flour in the Brazilian Scenario. *Braz. Arch. Biol. Technol.* **2023**, *66*, e23210543. [CrossRef]
6. Almeida, U.O.; Andrade Neto, R.C.; Lunz, A.M.P.; Cades, M.; Costa, D.A.; Araújo, J.M.; Teixeira Júnior, D.L.; Rodrigues, M.J.S. Production of Banana, Cultivar D'Angola, Intercropped with Açai Single in Different Planting Arrangements. *Rev. Bras. Agropecuária Sustentável* **2019**, *9*, 80–89.
7. Napoleão, G.M.; Rodrigues de Jesus, P.R.; Leonel, S. Cultivar Diversification of Banana Production in Brazil. *Agron. Sci. Biotechnol.* **2021**, *7*, 1–14. [CrossRef]
8. Mazzafera, P.; Favarin, J.L.; de Andrade, S.A.L. Editorial: Intercropping Systems in Sustainable Agriculture. *Front. Sustain. Food Syst.* **2021**, *5*, 634361. [CrossRef]
9. Khanal, U.; Stott, K.J.; Armstrong, R.; Nuttal, J.G.; Henry, F.; Christy, B.P.; Mirchell, M.; Riffkin, P.A.; Wallace, A.J.; McCaskill, M.; et al. Intercropping—Evaluating the Advantages to Broadacre Systems. *Agriculture* **2021**, *11*, 453. [CrossRef]
10. Ntamwira, J.; Ocimati, W.; Kearsley, E.; Safari, N.; Bahati, L.; Amini, D.; Lubobo, A.K.; Waswa, B.; Blomme, G. The Integration of Shade-Sensitive Annual Crops in Musa spp. Plantations in South Kivu, Democratic Republic of Congo. *Agronomy* **2021**, *11*, 368. [CrossRef]
11. Alves, E.P.; Silva, M.L.; Oliveira Neto, S.N.; Barrela, T.P.; Santos, R.H.S. Economic Analysis of a Coffee-Banana System of a Family-Based Agriculture at the Atlantic Forest Zone, Brazil. *Cienc. Agrotec.* **2015**, *39*, 232–239. [CrossRef]
12. Ferreira, T.M.C.; Vasconcelos, M.; Cantão, B.P.; da Silva, J.L.; Aguiar, W.K. Land Use Based on Agroforestry System: A Study at São Domingos do Capim County, Pará. *Rev. Ciênc. Agroamb.* **2016**, *14*, 93–99.
13. Mousavi, S.R.; Eskandari, H. A General Overview on Intercropping and Its Advantage in Sustainable Agriculture. *J. Appl. Environ. Biol. Sci.* **2011**, *1*, 482–486.
14. Nwaogu, C.; Cherubin, M.R. Integrated Agricultural Systems: The 21st Century Nature-Based Solution for Resolving the FEEEs Challenges. *Adv. Agron.* **2024**, *185*, 1–73. [CrossRef]
15. Yogendra, N.D.; Keerhi, P.E.; Nazeer, M.; Jnanesha, A.C.; Verma, R.S.; Sundaresan, V. Livelihood Enhancement and Resource Use Efficiency under Lemongrass Intercropping with Food Crops. *Ecol. Front.* **2024**, *44*, 266–274. [CrossRef]
16. Kishore, K.; Rupa, T.R.; Samant, D. Influence of Shade Intensity on Growth, Biomass Allocation, Yield and Quality of Pineapple in Mango-Based Intercropping System. *Sci. Hortic.* **2021**, *278*, 109868. [CrossRef]
17. Kalisz, B.; Zuk-Golaszewska, K.; Radawiec, W.; Golaszewski, J. Land Use Indicators in the Context of Land Use Efficiency. *Sustainability* **2023**, *15*, 1106. [CrossRef]
18. Boehmel, C.; Lewandowski, I.; Claupein, W. Comparing Annual and Perennial Energy Cropping Systems with Different Management Intensities. *Agric. Syst.* **2008**, *96*, 224–236. [CrossRef]
19. Bockstaller, C.; Guichard, L.; Makowski, D.; Aveline, A.; Girardin, P.; Plantureux, S. Agri-Environmental Indicators to Assess Cropping and Farming Systems: A Review. *Agron. Sustain. Dev.* **2008**, *28*, 139–149. [CrossRef]
20. Cassidy, E.S.; West, P.C.; Gerber, J.S.; Foley, J.A. Redefining Agricultural Yields: From Tonnes to People Nourished per Hectare. *Environ. Res. Lett.* **2013**, *8*, 034015. [CrossRef]
21. Marcos-Pérez, M.; Sánchez-Navarro, V.; Zornoza, R. Intercropping Systems between Broccoli and Fava Bean Can Enhance Overall Crop Production and Improve Soil Fertility. *Sci. Hortic.* **2023**, *312*, 111834. [CrossRef]
22. Norhan, M.E.M.; Khashaba, D.A.S.; Abdelkader, M.A.I. Evaluation of Competitive Indices between Caraway and Garlic as Affected by Intercropping System and Potassium Fertilization Level. *IOP Conf. Ser. Earth Environ. Sci.* **2018**, *1214*, 012018. [CrossRef]
23. Maitra, S.; Hossain, A.; Brestic, M.; Skalicky, M.; Ondrisik, P.; Gitari, H.; Brahmachari, K.; Shankar, T.; Bhadra, P.; Palai, J.B.; et al. Intercropping: A Low Input Agricultural Strategy for Food and Environmental Security. *Agronomy* **2021**, *11*, 343. [CrossRef]

24. Boult, C.; Chancellor, W. *Productivity of Australian Broadacre and Dairy Industries, 2018–2019*; Research Report; Australian Bureaus of Agricultural and Resource Economics and Sciences (ABARES): Canberra, Australia, 2020.
25. Bybee-Finley, K.; Ryan, M.R. Advancing Intercropping Research and Practices in Industrialized Agricultural Landscapes. *Agriculture* **2018**, *8*, 80. [CrossRef]
26. Dowling, A.; Sadras, V.O.; Roberts, P.; Doolette, A.; Zhou, Y.; Denton, M.D. Legume-Oilseed Intercropping in Mechanised Broadacre Agriculture—A Review. *Field Crops Res.* **2021**, *260*, e107980. [CrossRef]
27. Rao, M.M.; Edmunds, J.E. A Review of Banana/Plantain Cropping Systems. *Fruits* **1984**, *39*, 79–88.
28. Rodrigues Filho, V.A.; Neves, J.C.; Donato, S.L.R. Model to Estimate Nutritional and Non-Nutritional Limitations of 'Prata-Anã' Banana Crops Grown in Different Environments. *Rev. Caatinga* **2021**, *34*, 58–67. [CrossRef]
29. Sagar, M.; Bharati, P.J.; Pilli, M.; Prasanna, K. Potential of Intercropping System in Sustaining Crop Productivity. *Int. J. Agric. Environ. Biotechnol.* **2019**, *12*, 39–45. [CrossRef]
30. Bakshi, P.; Bhushan, A.; Bali, K.; Kour, K. Intercropping in Fruit Orchards: A Way Forward for Doubling the Farmer's Income. *Int. J. Agric. Sci.* **2019**, *11*, 9274–9276.
31. Olubode, O.O.; Aiyelaagbe, I.O.O.; Bodunde, J.G. Responses of 'Sunrise-Solo' Pawpaw, Okra and Cucumber Components of Pawpaw-Based Cropping System to Time of Intercropping. *Sci. Hortic.* **2012**, *139*, 71–78. [CrossRef]
32. Mattalia, G.; Wezel, A.; Costet, P.; Jagoret, P.; Deheuvels, O.; Migliorini, P.; David, C. Contribution of Cacao Agroforestry Versus Mono-Cropping Systems for Enhanced Sustainability: A Review with a Focus on Yield. *Agrofor. Syst.* **2022**, *96*, 1077–1089. [CrossRef]
33. Gendy, A.S.H.; Walid, N.S.; Dalia, N.A.S. Evaluation of Competitive Indices Between Roselle and Cowpea as Influenced by Intercropping System and Bio-Fertilization Type. *Middle East J. Agric. Res.* **2017**, *6*, 199–207.
34. Nassary, E.K.; Baijukya, F.; Ndakidemi, P.A. Productivity of Intercropping with Maize and Common Bean Over Five Cropping Seasons on Smallholder Farms of Tanzania. *Eur. J. Agron.* **2020**, *113*, 125964. [CrossRef]
35. Diniz, W.J.S.; Silva, T.G.F.; Ferreira, J.M.S.; Santos, D.C.; Moura, M.S.B.; Araújo, G.G.L.; Zolnier, S. Forage Cactus-Sorghum Intercropping at Different Irrigation Water Depths in the Brazilian Semiarid Region. *Pesqui. Agropecu. Bras.* **2017**, *52*, 724–733. [CrossRef]
36. Silva, J.N.; Bezerra Neto, F.; Lima, J.S.S.; Rodrigues, G.S.O.; Barros Júnior, A.P.; Chaves, A.P. Combinations of Coriander and Salad Rocket Cultivars in Bicropping Systems Intercropped with Carrot Cultivars. *Rev. Caatinga* **2017**, *30*, 125–135. [CrossRef]
37. Auzins, A.; Geipele, I.; Stamure, I. Measuring Land-Use Efficiency in Land Management. *Adv. Mater. Res.* **2013**, *804*, 205–210. [CrossRef]
38. Lin, H.C.; Hülsbergen, K.J. A New Method for Analyzing Agricultural Land-Use Efficiency, and Its Application in Organic and Conventional Farming Systems in Southern Germany. *Eur. J. Agron.* **2017**, *83*, 15–27. [CrossRef]
39. Ferreira, M.D.P.; Féres, J.G. Farm Size and Land Use Efficiency in the Brazilian Amazon. *Land Use Policy* **2020**, *99*, 104901. [CrossRef]
40. Silva, J.N.; Bezerra Neto, F.; de Lima, J.S.S.; Chaves, A.P.; Nunes, R.L.C. Sustainability of Carrot-Cowpea Intercropping Systems Through Optimization of Green Manuring and Spatial Arrangements. *Ciênc. Rural* **2021**, *51*, e20190838. [CrossRef]
41. Ditzler, L.; Komarek, A.M.; Chiang, T.W.; Alvarez, S.; Chatterjee, S.A.; Timler, C.; Raneri, J.E.; Carmona, N.E.; Kennedy, G.; Groot, J.C.J. A Model to Examine Farm Household Trade-Offs and Synergies with an Application to Smallholders in Vietnam. *Agric. Syst.* **2019**, *173*, 49–63. [CrossRef]
42. Wortmann, C.S.; Sengooba, T.; Kyamanywa, S. Banana and Bean Intercropping: Factors Affecting Bean Yield and Land Use Efficiency. *Exp. Agric.* **1992**, *28*, 287–294. [CrossRef]
43. Ocimati, W.; Groot, J.C.J.; Blomme, G.; Timler, C.J.; Remans, R.; Taulya, G.; Ntamwira, J.; Tittonell, P. A Multi-Objective Model Exploration of Banana-Canopy Management and Nutrient Input Scenarios for Optimal Banana-Legume Intercrop Performance. *Agronomy* **2021**, *11*, 311. [CrossRef]
44. Van Asten, P.J.A.; Wairegi, L.W.I.; Mukasa, D.; Uringi, N.O. Agronomic and Economic Benefits of Coffee-Banana Intercropping in Uganda's Smallholder Farming Systems. *Agric. Syst.* **2011**, *104*, 326–334. [CrossRef]
45. Sonavane, S.S.; Solia, B.M.; Gonge, A.P.; Patil, S.J.; Gajjar, M.M.; Patil, R.G. Intercropping Options in Autumn Planted Banana. *Trends Biosci.* **2014**, *7*, 1927–1931.
46. Alam, R.; Faruq, O.; Uddin, R.; Zonayet, M.; Syfullah, K. Intercropping of Winter Vegetables with Banana in Khagrachari Hill District of Bangladesh. *J. Glob. Agric. Ecol.* **2021**, *11*, 34–41.
47. de Siqueira, H.M.; de Senna, D.S.; da Silva Araújo, J.B.; da Silva, M.W.; Turbay, E.R.M.G. Análise Econômica de Consórcios do Cafeeiro Conillon com Espécies Perenes e Florestais no Sul do Espírito Santo. *Rev. Bras. Agroecol.* **2020**, *15*, 14–25. [CrossRef]
48. Manju, P.R.; Swain, S.; Pradhan, B.K.; Pushpalatha, P.B.; Patil, P. Effect of Banana (*Musa* spp.) Intercropping with Root and Tuber Crops. *Curr. Hortic.* **2021**, *9*, 26–30. [CrossRef]
49. Teixeira, A.G.; Oliveira, F.L.; Parajara, M.C.; Moises Zucoloto, M.; Dalvi, L.P. Yacon Planting Density in Intercropping with Corn under Banana Orchard. *Hortic. Bras.* **2020**, *38*, 83–88. [CrossRef]
50. Dissanayake, S.M.; Palihakkara, I.R. Effects of Intercropping of Immature Oil Palm (*Elaeis guineensis*) with Banana, Ginger, and Turmeric in the Galle District, Sri Lanka. *Environ. Qual. Manag.* **2024**, *33*, 31–36. [CrossRef]
51. Perdoná, M.J.; Soratto, R.P. Higher Yield and Economic Benefits are Achieved in the Macadamia Crop by Irrigation and Intercropping with Coffee. *Sci. Hortic.* **2015**, *185*, 59–67. [CrossRef]

52. Ouma, G. Intercropping and Its Application to Banana Production in East Africa: A Review. *J. Plant Breed. Crop Sci.* **2009**, *1*, 013–015.
53. Srivastava, A.K.; Huchche, A.D.; Ram, L.; Singh, S. Yield Prediction in Intercropped Versus Monocropped Citrus Orchards. *Sci. Hortic.* **2007**, *114*, 67–70. [CrossRef]
54. Bellamy, A.S. Banana Production Systems: Identification of Alternative Systems for More Sustainable Production. *Ambio* **2013**, *42*, 334–343. [CrossRef] [PubMed]
55. Ashish, K.; Jnanesha, A.C.; Lal, R.K.; Dubey, B.K. Intercropping and Mixed Herb Distillation for High-Quality Oil Yield Using Lemon-Scented Basil (*Ocimum africanum* Lur.) cv. CIM-Jyoti and Lemongrass (*Cymbopogon flexuous* (Nees ex Steud.) cv. Krishna. *Acta Ecol. Sin.* **2021**, *42*, 269–273. [CrossRef]
56. Oliveira, S.G.; Bonfim, F.P.G.; Alves, L.F.; Marques, I.B.; Araújo, E.O. Alelopatia de capim-cidreira na germinação, vigor de sementes e no desenvolvimento inicial do tomate-cereja. *Cad. Ciênc. Agrár.* **2018**, *10*, 7–12.
57. De Oliveira, C.A.A.; Santos, J.S. Active Compounds of Lemon Grass (*Cymbopogon citratus*): A Review. *Res. Soc. Dev.* **2021**, *10*, e26310122081. [CrossRef]
58. Mukarram, M.; Choudhary, S.; Khan, M.A.; Poltronieri, P.; Khan, M.M.A.; Ali, J.; Kurjak, D.; Shahid, M. Lemongrass Essential Oil Components with Antimicrobial and Anticancer Activities. *Antioxidants* **2021**, *11*, 20. [CrossRef] [PubMed]
59. Sun, Y.; Chen, L.; Zhang, S.; Miao, Y.; Zhang, Y.; Li, Z.; Zhao, J.; Zhang, J.; Qin, X.; Yao, Y. Plant Interaction Patterns Shape the Soil Microbial Community and Nutrient Cycling in Different Intercropping Scenarios of Aromatic Plant Species. *Front. Microbiol.* **2022**, *13*, 888789. [CrossRef]
60. Rodrigues de Jesus, P.R.; Leonel, S.; Leonel, M.; Cândido, H.T.; Molha, N.Z.; Domiciano, V.M.; Ouros, L.F.; Tecchio, M.A. Performance and Leaf Nutritional Content of Banana Cultivars Intercropped with Lemongrass. *Rev. Caatinga* **2024**, *37*, e12448. [CrossRef]
61. Spinoni, J.; Marinho Ferreira Barbosa, P.; Bucchignani, E.; Cassano, J.; Cavazos, T.; Christensen, J.; Christensen, O.; Coppola, E.; Evans, J.; Geyer, B.; et al. Future Global Meteorological Drought Hot Spots: A Study Based on CORDEX Data. *J. Clim.* **2020**, *33*, 3635–3661. [CrossRef]
62. Tilman, D. Benefits of Intensive Agricultural Intercropping. *Nat. Plants* **2020**, *6*, 604–605. [CrossRef]
63. Stagnari, F.; Maggio, A.; Galieni, A.; Pisante, M. Multiple Benefits of Legumes for Agriculture Sustainability: An Overview. *Chem. Biol. Technol. Agric.* **2017**, *4*, 2. [CrossRef]
64. Andrade Neto, R.C.; Almeida, U.O.; Lunz, A.M.P.; Oliveira, T.K.; Nogueira, S.R.; Oliveira, J.R. Características Agronômicas de Bananeira Terra, cv. D'Angola, em Consórcio com Açaizeiro (*Euterpe precatoria* Mart.). In *Boletim de Pesquisa*; Embrapa Acre: Rio Branco, Brazil, 2015. Available online: https://ainfo.cnptia.embrapa.br/digital/bitstream/item/136874/1/25853.pdf (accessed on 19 June 2024).
65. Muliele, M.T.; Bielders, C.L.; Van Asten, P.J.A. Short- and Medium-Term Impact of Manual Tillage and No-Tillage with Mulching on Banana Roots and Yields in Banana-Bean Intercropping Systems in the East African Highlands. *Field Crops Res.* **2015**, *171*, 1–10. [CrossRef]
66. Concenço, G.; Motta, I.S.; Correia, I.V.T.; Santos, S.A.; Mariani, A.; Marques, R.F.; Palharini, W.G.; Alves, M.E.S. Infestation of Weed Species in Monocrop Coffee or Intercropped with Banana, Under Agroecological System. *Planta Daninha* **2014**, *32*, 665–674. [CrossRef]
67. Seenivasan, N. Management of Radopholus similis and Helicotylenchus multicinctus in ratoon banana grown under high density planting systems. *Int. J. Fruit Sci.* **2017**, *17*, 41–62. [CrossRef]
68. Ren, X.; Zhou, Z.; Liu, M.; Shen, Z.; Wang, B.; Jousset, A.; Geisen, S.; Ravanbakhsh, M.; Kowalchuk, G.A.; Li, R.; et al. Intercropping with *Trifolium repens* contributes disease suppression of banana Fusarium wilt by reshaping soil protistan communities. *Agric. Ecosyst. Environ.* **2024**, *361*, 108797. [CrossRef]
69. Li, Y.; Lin, J.; Xiao, S.; Feng, D.; Deng, Y.; Xuan, W. Effects of sweet potato intercropping in banana orchard on soil microbial population diversity. *Ann. Microbiol.* **2022**, *72*, 46. [CrossRef]
70. Rahman, M.Z.; Rahman, M.H.; Haqu, M.E.; Ekalar, M.H.; Naber, S.L. Banana-based intercropping system in North-west part of Bangladesh. *J. Agron.* **2006**, *5*, 228–231. [CrossRef]
71. Dassou, A.G.; Carval, D.; Dépgny, S.; Fansi, G.; Tixier, P. Ant abundance and *Cosmopolites sordidus* damage in plantain fields as affected by intercropping. *Biol. Control* **2015**, *81*, 51–57. [CrossRef]
72. Butler, D. Fungus threatens top banana. *Nature* **2013**, *504*, 195–196. [CrossRef]
73. Shen, Z.; Xue, C.; Penton, C.R.; Thomashow, L.S.; Zhang, N.; Wang, B.; Ruan, Y.; Li, R.; Shen, Q. Suppression of banana Panama disease induced by soil microbiome reconstruction through an integrated agricultural strategy. *Soil Biol. Biochem.* **2019**, *128*, 164–174. [CrossRef]
74. Hong, S.; Jv, H.; Lu, M.; Wang, B.; Zhao, Y.; Ruan, Y. Significant decline in banana Fusarium wilt disease is associated with soil microbiome reconstruction under chilli-pepper-banana rotation. *Eur. J. Soil Biol.* **2020**, *97*, 103154. [CrossRef]
75. Zhu, S.; Morel, J.B. Molecular mechanisms underlying microbial disease control in intercropping. *Mol. Plant-Microbe Interact.* **2019**, *32*, 20–24. [CrossRef] [PubMed]
76. Li, Z.; Wang, T.; He, C.; Cheng, K.; Zeng, R.; Song, Y. Control of Panama disease of banana by intercropping with Chinese chive (*Allium tuberosum* Rottler): Cultivar differences. *BMC Plant Biol.* **2020**, *20*, 432. [CrossRef] [PubMed]

77. Yang, J.; Duan, Y.; Liu, X.; Sun, M.; Wang, Y.; Liu, M.; Zhu, Z.; Shen, Z.; Gao, W.; Wang, B.; et al. Reduction of banana Fusarium wilt associated with soil microbiome reconstruction through green manure intercropping. *Agric. Ecosyst. Environ.* **2022**, *337*, 108065. [CrossRef]
78. Li, Z.; Jiao, Y.; Yin, J.; Li, D.; Wang, B.; Zhang, K.; Zheng, X.; Hong, Y.; Zhang, H.; Xie, C.; et al. Productivity and quality of banana in response to chemical fertilizer reduction with bio-organic fertilizer: Insight into soil properties and microbial ecology. *Agric. Ecosyst. Environ.* **2021**, *322*, 107659. [CrossRef]
79. Geisen, S.; Mitchell, E.A.D.; Adl, S.; Bonkowski, M.; Dunthorn, M.; Ekelund, F.; Fernández, L.D.; Jousset, A.; Krashevska, V.; Singer, D.; et al. Soil protists: A fertile frontier in soil biology research. *FEMS Microbiol. Rev.* **2018**, *42*, 293–323. [CrossRef]
80. Amacker, N.; Gao, Z.; Hu, J.; Jousset, A.L.C.; Kowalchuk, G.A.; Geisen, S. Protist feeding patterns and growth rate are related to their predatory impact on soil bacterial communities. *FEMS Microbiol. Ecol.* **2022**, *98*, fiac057. [CrossRef]
81. Gao, Z.; Karlsson, I.; Geisen, S.; Kowalchuk, G.; Jousset, A. Protists: Puppet masters of the rhizosphere microbiome. *Trends Plant Sci.* **2019**, *24*, 165–176. [CrossRef]
82. Guo, S.; Tao, C.; Jousset, A.; Xiong, W.; Wang, Z.; Shen, Z.; Wang, B.; Xu, Z.; Gao, Z.; Liu, S.; et al. Trophic interactions between predatory protists and pathogen-suppressive bacteria impact plant health. *ISME J.* **2022**, *16*, 1932–1943. [CrossRef]
83. Maltais-Landry, G.; Scow, K.; Brennan, E.; Vitousek, P. Long-term effects of compost and cover crops on soil phosphorus in two California agroecosystems. *Soil Sci. Soc. Am. J.* **2015**, *79*, 688–697. [CrossRef]
84. Turrini, A.; Sbrana, C.; Avio, L.; Njeru, E.M.; Bocci, G.; Barberi, P.; Giovannetti, M. Changes in the composition of native root arbuscular mycorrhizal fungal communities during a short-term cover crop-maize succession. *Biol. Fertil. Soils* **2016**, *52*, 643–653. [CrossRef]
85. Ye, X.Q.; Yan, Y.N.; Wu, M.; Yu, F.H. High capacity of nutrient accumulation by invasive *Solidago canadensis* in a coastal grassland. *Front. Plant Sci.* **2019**, *10*, 575. [CrossRef] [PubMed]
86. Akinola, S.A.; Babalola, O.O. The fungal and archaeal community within plant rhizosphere: A review on their contribution to crop safety. *J. Plant Nutr.* **2021**, *44*, 600–618. [CrossRef]
87. Leonel, S.; Bolfarini, A.C.B.; Souza, J.M.A.; Leonel, M.; Ferreira, R.B.; Putti, F.F.; Tecchio, M.A. Agronomic performance of Banana 'FHIA 18' in response to phosphate fertilization. *Agron. J.* **2020**, *112*, 2033–2046. [CrossRef]
88. Maia, A.H.; Rebelatto, B.F.; Reis, D.S.; Trento, M.B.R.; dos Santos Ferreira, L. Crescimento Inicial de Cultivares de Bananeira Consorciadas com Adubos Verdes/Initial Growth of Banana Cultivars Consorted with Green Manures. *Braz. J. Dev.* **2020**, *6*, 20245–20261. [CrossRef]
89. Musongora, M.; Karanja, N.; Kimenju, W.; Kamau, S. Spatio-Temporal Change of Selected Soil Physico-Chemical Properties in Grevillea-Banana Agroforestry Systems. *Heliyon* **2023**, *9*, e16121. [CrossRef]
90. Glaze-Corcoran, S.; Hashemi, M.; Sadeghpour, A.; Jahanzad, E.; Afshar, R.K.; Liu, X.; Hebert, S. Understanding Intercropping to Improve Agricultural Resiliency and Environmental Sustainability. *Adv. Agron.* **2020**, *162*, 199–256. [CrossRef]
91. Huss, C.; Holmes, K.D.; Blubaugh, C. Benefits and Risks of Intercropping for Crop Resilience and Pest Management. *J. Econ. Entomol.* **2022**, *115*, 1350–1362. [CrossRef]
92. Abele, S.; Twine, E.; Legg, C. *Food Security in Eastern Africa and the Great Lakes*; IITA: Ibadan, Nigeria, 2007.
93. Blomme, G.; Ocimati, W.; Groot, J.; Ntamwira, J.; Bahati, L.; Kantungeko, D.; Remans, R.; Tittonell, P. Agroecological Integration of Shade and Drought Tolerant Food/Feed Crops for Year-Round Productivity in Banana-Based Systems Under Rainfed Conditions in Central Africa. *Acta Hortic.* **2018**, *1196*, 41–54. [CrossRef]
94. Ocimati, W.; Ntamwira, J.; Groot, J.C.J.; Taulya, G.; Tittonell, P.; Dhed'a, P.; van Astens, P.; Vanlauwe, B.; Ruhigwa, B.; Blomme, G. Banana Leaf Pruning to Facilitate Annual Legume Intercropping as an Intensification Strategy in the East African Highlands. *Eur. J. Agron.* **2019**, *110*, 125923. [CrossRef]
95. Rodrigo, V.H.L.; Stirling, C.M.; Teklehaimanot, Z.; Nugawela, A. The Effect of Planting Density on Growth and Development of Component Crops in Rubber/Banana Intercropping Systems. *Field Crops Res.* **1997**, *52*, 95–108. [CrossRef]
96. Silva, D.M.N.; Heitor, L.C.; Candido, A.O.; Moraes, B.S.A.; Souza, G.S.; Araújo, J.B.S.; Mendonça, E.S. Carbon Balance in Organic Conilon Coffee Intercropped with Tree Species and Banana. *Rev. Árvore* **2020**, *44*, e4421. [CrossRef]
97. Bebber, D.P. The long road to a sustainable banana trade. *Plants People Planet* **2023**, *5*, 662–671. [CrossRef]

horticulturae

Review

Understanding the Invasion, Ecological Adaptations, and Management Strategies of *Bactrocera dorsalis* in China: A Review

Saleem Jaffar [1], Syed Arif Hussain Rizvi [2] and Yongyue Lu [1,*]

[1] Department of Entomology, College of Plant Protection, South China Agricultural University, Guangzhou 510642, China; saleemjaffar@aup.edu.pk
[2] Xinjiang Institute of Ecology and Geography, Urumqi 830011, China; arifrizvi@aup.edu.pk
* Correspondence: luyongyue@scau.edu.cn

Abstract: *Bactrocera dorsalis* (Hendel, 1912) (Diptera: Tephritidae), commonly known as the oriental fruit fly, is a highly destructive pest that globally infests fruits and vegetables, resulting in significant annual economic losses. Initially detected in Taiwan Island, it has rapidly expanded its distribution range to various regions in mainland China since the 1980s, with a continuous northward spread. To mitigate the damage caused by this pest, extensive efforts have been undertaken to comprehend its ecological and physiological adaptations and develop management strategies. This review article provides an overview of the invasion history of *B. dorsalis* in China, its ecological and physiological mechanisms facilitating its invasion, and the progress made in understanding its major biological characteristics. Moreover, the key approaches for managing *B. dorsalis* that have been or are likely to be implemented in China are presented, including quarantine measures, monitoring procedures, physical controls, biological controls, the sterile insect technique, RNA interference, and CRISPR-Cas-9. Finally, some suggestions for future research directions are provided.

Keywords: *B. dorsalis*; oriental fruit fly; invasion; biology; IPM; pest management; China

Citation: Jaffar, S.; Rizvi, S.A.H.; Lu, Y. Understanding the Invasion, Ecological Adaptations, and Management Strategies of *Bactrocera dorsalis* in China: A Review. *Horticulturae* **2023**, *9*, 1004. https://doi.org/10.3390/horticulturae9091004

Academic Editors: Sergio Ruffo Roberto, Sarita Leonel and Simone Rodrigues da Silva

Received: 18 July 2023
Revised: 25 August 2023
Accepted: 1 September 2023
Published: 5 September 2023

1. Introduction

Tephritid fruit flies are an economically significant pest species globally, including mainland China [1]. They exhibit endophagous feeding behavior, which causes both quantitative and qualitative yield reductions. As a result, they pose significant threats to global fruit and vegetable production [2,3]. The pest affects a broad array of fruit and fleshy vegetable crops in tropical and subtropical regions. The presence of these pests was first observed in Taiwan Island, China, in 1912 [4,5]. The genus *Bactrocera*, which comprises a minimum of 440 species [6], is primarily distributed throughout tropical Asia, Australia, and the South Pacific [7,8]. The wide host range, great climate tolerance, and strong dispersing capacities of these species have led to their spread over the Asia Pacific region in the last century, covering all of South-East Asia from India to Hawaii [7]. The oriental fruit fly, *Bactrocera dorsalis* (Hendel) (Diptera: Tephritidae), is recognized as a destructive and persistent fruit fly pest. *B. dorsalis* has been documented to infest over 250 host plant species [9,10], including mango (*Mangifera indica* L., Anacardiaceae), banana (*Musa* spp., Musaceae), guava (*Psidium guajava* L., Myrtaceae), orange (*Citrus* spp., Rutaceae), papaya (*Carica papaya* L., Caricaceae), peach (*Prunus persica* (L.) Batsch, Rosaceae), grape (*Vitis* spp., Vitaceae), pomegranate (*Punica granatum* L., Lythraceae), lychee (*Litchi chinensis* Sonn., Sapindaceae), and longan (*Dimocarpus longan* Lour., Sapindaceae) [11,12]. Numerous studies have documented the economic damage caused by *B. dorsalis*. For instance, a study carried out in Thailand found that *B. dorsalis* infestation in mango farms caused an average annual yield loss of 15.5% [13]. Similarly, in India, fruit fly infestation led to a reduction in the marketable yield of mango by 25–30% [14]. According to an estimate, guava, sapota, citrus fruits, and mango in India, incurred losses equivalent to USD 356 million [15]. This significant economic loss is attributed to the fruit damage caused by *B. dorsalis*, which can affect 30% to 100% of fruits, depending on the season [16].

In addition to yield reduction, *B. dorsalis* also leads to the quality degradation of fruits, causing phytosanitary issues and triggering trade restrictions, thereby aggravating economic losses. A study conducted in Taiwan revealed that the infestation of fruit flies resulted in trade restrictions on the export of guava to the United States and Japan, leading to an estimated economic loss of USD 2.5 million per year [17]. These studies demonstrate the substantial economic losses caused by *B. dorsalis* and emphasize the necessity for implementing effective management strategies to mitigate the impact of this insect pest on horticultural crops. In China, the economic losses caused by the fruit fly pest species in citrus orchards have been widely reported, especially in Guangdong [18] and Fujian Provinces of China [1]. *B. dorsalis* exhibits three to eleven generations per year in China, with the majority of areas experiencing four to eight generations [8,19]. In the near future, there is the potential for *B. dorsalis* to expand into temperate northern and southern areas of China [20] (Figure 1).

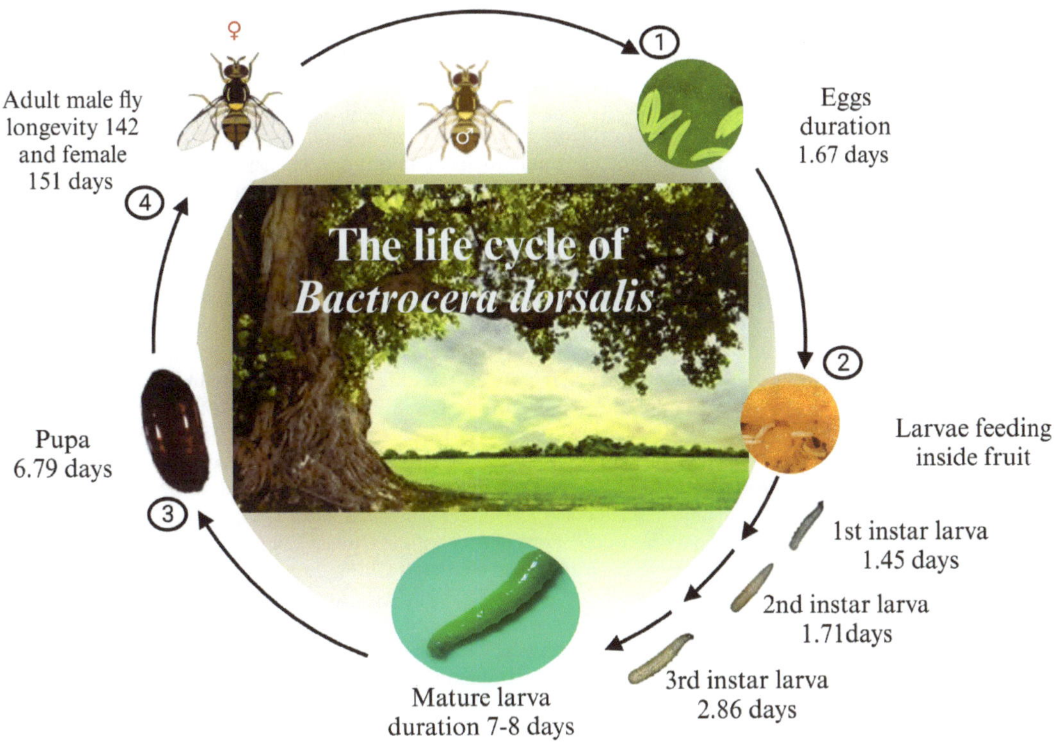

Figure 1. The life cycle of *B. dorsalis*.

Keeping in view the severity of the pest, the key objective of the current study is to review the invasion of *B. dorsalis* from Taiwan to mainland China and its subsequent expansion. The review will also analyze the biological traits of *B. dorsalis* that have facilitated the insect pest's invasion into new areas. Moreover, the review will provide an overview of the integrated pest management (IPM) strategies employed in China to control *B. dorsalis*, encompassing quarantine measures, surveillance techniques, physical and biological control methods, the sterile insect technique, RNA interference, and CRISPR-Cas-9.

2. Dispersion Ecology

2.1. Dispersion of *B. dorsalis* in China

B. dorsalis was first recorded in Taiwan Island and subsequently invaded Hainan Island, China, in 1934 [21]. Prior to the 1970s, the species was only rarely detected in isolated areas in southern China. However, in the 1980s, its population size increased, due to international trade, resistance to pesticides, abundant host plants, and a lack of natural enemies (predators and parasitoids), and its distribution area expanded to encompass most of southern China. In recent years, *B. dorsalis* has continued to spread, crossing the Yangtze River and other neighboring areas [22,23]. To predict the potential geographical distribution of *B. dorsalis* in China, several techniques have been employed. An analysis based on the CLIMEX model has indicated that the species' most suitable habitats are located in Southern China, including Guangdong, Hainan, and Guangxi Provinces, as well as in Eastern China, including Fujian, Zhejiang, and Shanghai [19,24]. The analysis revealed that moderately suitable habitats for the oriental fruit fly were identified in Southwest China, including Yunnan, Sichuan, and Guizhou Provinces. In contrast, Hunan, Hubei, and Jiangxi Provinces were found to exhibit relatively low suitability. Furthermore, areas located to the north of the Yangtze River were deemed unsuitable for the species [7,8]. Subsequent analyses using the genetic algorithm for rule-set prediction (GARP) ecological niche modeling [25] and the emergence rate model combined with ArcGIS yielded similar predictions [7,24]. So far, *B. dorsalis* has effectively colonized the majority of the aforementioned regions and provinces, encompassing territories beyond the Yangtze River and including Henan and Anhui Provinces. Anticipated future developments suggest a substantial northward expansion of its range [26] (Figure 2). The utilization of microsatellite loci analysis unveiled valuable insights into the origin of oriental fruit fly populations within China, Korea, Thailand, and Laos [27,28]. These regions are believed to have been among the initial areas to be invaded by the species. Furthermore, indications suggest that Guangxi Province could potentially represent another early region of *B. dorsalis* invasion. The observed genetic differentiation in the hierarchical model using *nad1*, *nad5*, *Cytb*, and concatenated sequences among the various populations of *B. dorsalis* in China exhibited substantial levels, implying that this invasive pest might have entered China through two distinct invasion routes. The initial route likely originated in Southeast China, leading to subsequent spread into the provinces of Fujian, Taiwan, Guangdong, and Hainan. The second route is thought to have originated in Southwest China, resulting in subsequent dissemination into the provinces of Sichuan, Yunnan, Guangxi, Guizhou, and Hunan, as corroborated by the findings presented in previous studies [29,30].

Figure 2. Distribution of *B. dorsalis* in China. The map information was produced using the geospatial data cloud https://www.gscloud.cn (accessed on 19 August 2023) and datav.aliyun.com (accessed on 19 August 2023). The spatial analysis was accessed through ArcGIS version (10.7) and Mapshaper, which were applied on 19 August.

2.2. Key Factors in the Invasion of B. dorsalis

Over the past few decades, numerous research studies have been published on *B. dorsalis*, with a particular focus on understanding key contributors that help the pest to invade different regions. These studies have investigated various aspects of the insect, including its aggressive reproductive behavior, tolerance to stressful ecological niches, and resilience to chemical pesticides. By examining these factors, researchers aim to understand the mechanisms driving the successful invasion of *B. dorsalis* and develop effective management strategies. Following is a detailed review of all the factors that make *B. dorsalis* a highly invasive insect, particularly in China.

2.2.1. Aggressive Reproductive Behavior

One of the reasons for the successful invasion of *B. dorsalis* is attributed to its reproductive behavior. The pest shows a unique phenomenon of remating in female flies [31], which enables the pest to acquire supplementary material from male accessory gland products E-coniferyl alcohol (ECF) and dimethoxyphenol (DMP), particularly when males have fed on methyl eugenol (ME). These sequestered components can enhance the overall fitness of female flies and contribute to their successful establishment in new environments. The phenomena have also been observed in female *B. dorsalis* [32,33]. Studies have found that female *B. dorsalis*, due to multiple mating, tend to have increased fecundity. One study reported that approximately 50% of the *B. dorsalis* population re-mated and exhibited significantly higher fecundity compared to single-mated females [34]. The remating potential could play a significant role in the rapid population growth of *B. dorsalis* in natural habitats [35,36]. Group housing, which refers to the social living and interaction of insects, is another reproductive behavior that also plays a role in reproductive success. Group housing has been shown to enhance mating frequency and success in arthropods [37]. Recently, studies have shown that group housing has a positive impact on mating and chemical cue sensitization in several insect species, including *B. dorsalis* [38,39]. Studies have reported that group-housed flies exhibit higher mating rates compared to those housed in individual containers. Additionally, group housing creates a more natural social environment for the flies, which stimulates reproductive behavior and increases the chances of successful mating. Furthermore, this social interplay has been found to increase the sensitivity of *B. dorsalis* to chemical cues [40]. This is important for the development of new management options, as chemical cues are often used to attract and trap fruit flies in the field. In group-housed insects, exposure to the pheromones and other volatile chemicals released by conspecifics leads to an increased sensitivity to these cues, which could enhance their efficacy in trapping and monitoring programs. Additional research is needed to fully understand the underlying mechanisms of social interactions, as well as their practical implications for fruit fly management [39]. In another study, the effect of cis-vaccenyl acetate (cVA), a pheromone commonly found in *Drosophila* [41], on the mating behavior of *B. dorsalis* was investigated. These results showed that low levels of cVA had a positive impact on the mating rate of *B. dorsalis* [39], whereas high levels had an inhibitory effect. This finding implies that cVA may play a role in regulating mating behavior not only in *Drosophila* but also in other insect species [42].

Another study has found that group housing conditions had a positive impact on the mating rates and cVA-mediated behaviors of fruit flies [43]. These effects were shown to be dependent on the activity of the CREB (cAMP response element-binding protein) binding protein (CBP), which regulates gene expression and interacts with CREB. In other words, the presence of other insects in the group setting influences the responses of *B. dorsalis* to cVA, leading to altered mating rates and behaviors [44]. This study highlights the potential benefits of group housing for enhancing mating behaviors and cVA-mediated responses in *B. dorsalis* [45]. Further research may be needed to fully understand the mechanisms behind these effects and their practical implications in fruit fly management programs (Figure 3).

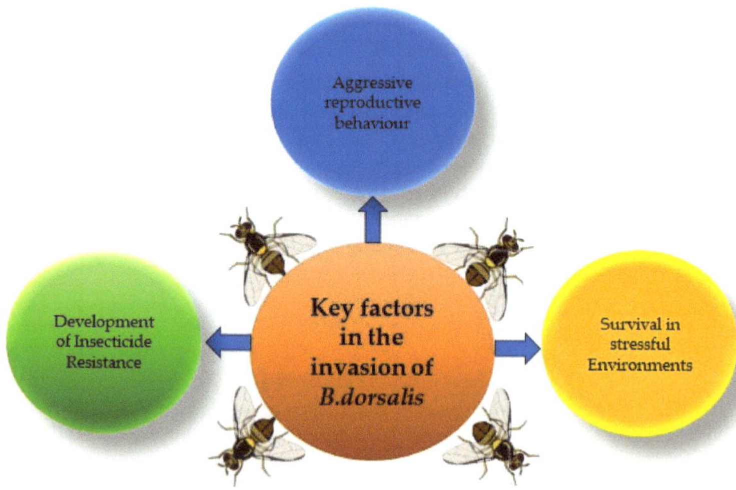

Figure 3. Key factors involved in the invasion biology of *B. dorsalis*.

2.2.2. Survival in Stressful Environments

B. dorsalis is a highly adaptable species, capable of thriving in a broad geographical range and a diverse set of environmental conditions [46]. With reference to the temperature fluctuations, *B. dorsalis* demonstrates a remarkable tolerance, with its eggs displaying a high level of resistance to high temperatures and its pupae exhibiting a remarkable ability to withstand low temperatures, especially at the pre-overwintering stage [8,47]. The cold hardiness of future generations of *B. dorsalis* can be influenced by the host plant they feed on during their larval stage [48]. Physiological mechanisms responsible for the tolerance of *B. dorsalis* to extreme temperatures have been the subject of investigation. Studies suggest that a multitude of oxidoreductases, binding proteins, and transferases are present in large amounts in adults subjected to high and/or low temperatures, providing a form of physiological protection [49,50]. Moreover, antioxidant enzymes, superoxide dismutase (SODs), likely play a key role in mitigating oxidative damage in *B. dorsalis* under thermally stressed conditions [51,52]. Also, *B. dorsalis* can survive in a wide range of humidity levels, with its larvae and pupae thriving and developing in a broad range of moisture content [53]. The larvae of *B. dorsalis* exhibit a remarkable ability to reduce their weight within two hours when exposed to a dry environment, which highlights its resilience under severe conditions. This rapid response to changes in their environment sets them apart from other insect species. On the other hand, the pupae of *B. dorsalis* show an ability to survive in wet environments. In a study, it was found that more than 50% of the *B. dorsalis* pupated in soils with moisture levels between 80 and 100% [54]. Despite the low adult emergence from the highly moist soils, the pupation in highly wet environments shows the resilience of the developmental stages to highly stressful conditions [54]. Thus, the survival of *B. dorsalis* under stressful regimes serves as one of the important reasons for its success.

2.2.3. Development of Insecticide Resistance

The management of *B. dorsalis* is typically achieved through chemical insecticides. However, the extended and frequent use of certain synthetic chemicals has led to the evolution of high levels of insecticide resistance in this species, resulting in more destructive outbreaks [55,56]. For example, studies have shown that populations of the oriental fruit fly in Guangdong, China, as well as in many other parts of the world, developed a high level of resistance to trichlorfon between 2007 and 2020 [8,56,57]. In recent years, resistance to other insecticides such as malathion, β-cypermethrin, and abamectin has also been observed [58,59]. The situation has become particularly critical in Hubei, China, where high resistance to cyantraniliprole, a newly developed anthranilic diamide insecticide, has been reported for many insects [60,61]. Various studies have been conducted to investigate the underlying mechanisms and biochemical processes involved in insecticide resistance development. Experiments have revealed that resistance in *B. dorsalis* is primarily due to increased detoxification mechanisms, including upregulation of cytochrome P450 monooxygenases (P450s), glutathione S-transferases (GSTs), carboxylesterase (CarEs), and insensitivity of acetylcholinesterase (AChE) [62,63]. In addition, physiological resistance, characterized by elevated digestive enzyme activities, has been implicated in resistance development in *B. dorsalis* [55]. Furthermore, the structural features of the cuticle and the interspace between epidermal cells might play a role in the cuticular penetration of insecticides [58]. Advances in molecular techniques, including heterologous expression and RNA interference, have facilitated the functional characterization of various resistance-related genes in *B. dorsalis*. For instance, overexpression of three esterase genes has been linked to malathion resistance in this insect species [64,65]. The rapid development of insecticide resistance in *B. dorsalis* represents one of the important reasons for its success, posing a major challenge to current pest management strategies and necessitating the implementation of innovative approaches for its effective control.

3. Research on *B. dorsalis* Biology in China

Development studies of *B. dorsalis* have focused on various life stages, characterizing the developmental period, growth rate, and morphological changes [66,67]. The reproductive biology of this insect has also been extensively studied, focusing on mating behavior, female remating, egg-laying behavior, and egg survival [68,69]. Regarding chemical resistance, the evolution of high levels of insecticide resistance in *B. dorsalis* has been documented in recent years [8,70]. The biochemical and molecular mechanisms underlying this resistance have been extensively studied. Research has focused on the increased detoxification by cytochrome P450 monooxygenases (P450s), glutathione S-transferases (GSTs), and carboxylesterase (CarEs) [59,71,72]. Physiological resistance, specifically high digestive enzyme activities, has also been implicated in resistance development. Furthermore, recent research has characterized resistance-related genes using heterologous expression and RNA interference techniques [62]. Studies on the associations between *B. dorsalis* and symbionts have also been conducted to determine their effects on the insect's fitness and survival. For instance, research has revealed the benefits of gut symbionts in nutrient acquisition and defense against pathogens [73], expanding our understanding of the complex interactions between *B. dorsalis* and its symbionts. Research on *B. dorsalis* in China has provided valuable insights into the development, reproduction, chemical resistance, communication, and symbiotic associations of this economically important pest.

3.1. *Exploration of Molecular Mechanisms*

To gain insights into the molecular mechanisms underlying this regulation, researchers have cloned several key genes involved in the ecdysone biosynthesis pathway from this insect. Specifically, the cloning of ecdysteroid biosynthetic genes *Cyp302a1*, *Cyp315a1*, and *Cyp314a1*, along with the ecdysone receptor gene *EcR-B1*, has significantly advanced our knowledge in this area [74,75]. The cloning of these genes has facilitated a deeper understanding of how ecdysteroids, specifically 20-hydroxyecdysone (20E), regulate the development of fruit flies. The 20E hormone interacts with the EcR-B1 receptor, which activates a series of downstream target gene expression cascades, leading to cellular processes such as cell proliferation, differentiation, and apoptosis. The genes involved in this process serve as potential molecular targets for the control of this pest [76,77]. The identification of these genes highlights the importance of ecdysteroid signaling in oriental fruit fly development and opens possibilities for utilizing these targets in pest management.

The cloning and characterization of chitin metabolic pathway genes have received significant attention in recent years in *B. dorsalis* research. Key enzymes in the chitin biosynthesis pathway, namely chitin synthases CHS1 and CHS2, have been cloned and their expression profiles analyzed [78]. Among them, a specific splicing variant of CHS1, CHS1a, shows prominent expression during the larval–pupal and pupal–adult transitions. Knockdown of CHS1a expression using RNA interference (RNAi) results in significant phenotypic defects and mortality in treated larvae [78]. Furthermore, the CHS2 transcript is predominantly found in the larval midgut [79], and its expression is positively correlated with the total chitin content during the insect's development. Other crucial enzymes involved in the chitin biosynthesis pathway, such as glucose-6-phosphate isomerase and UDP-*N*-acetylglucosamine pyrophosphorylase, have also been cloned and characterized in *drosophila* [80]. RNAi-mediated knockdown of their expression leads to larval death and abnormal phenotypes. Moreover, chitinase, an enzyme responsible for chitin degradation and starvation, has been identified and characterized in *B. dorsalis* as well [81]. Moreover, the significance of the bHLH transcription factor gene, pipsqueak, in the proper regulation of molting in fruit flies has been demonstrated [82]. In the context of cell growth and metabolism regulation, *B. dorsalis* has an identified ortholog of the target of rapamycin (TOR), a crucial kinase [83]. Furthermore, the evaluation of heat shock proteins 83 and 90 (Hsp 83, 90) has revealed its role in stress response and stress resistance, fecundity, and longevity as its gene expression is upregulated following exposure [84,85]. Furthermore, studies have interpreted the regulation of ecdysteroid hormone biosynthesis in *B. dorsalis*.

The ecdysone biosynthetic gene, cytochrome P450 monooxygenase (Cyp307a1), was identified, and its expression has been shown to be regulated by 20-hydroxyecdysone (20E) and juvenile hormone (JH) [86]. Moreover, the presence of the hormone ecdysone and its receptor, ecdysone receptor, has been confirmed in *B. dorsalis* [87,88].

In addition to these findings, research has demonstrated the crucial role of digestive enzymes in the growth and development of *B. dorsalis*. Fourteen genes encoding digestive enzymes, including alpha-amylases, proteases, and chitinases, have been successfully cloned [89]. It has been observed that these enzymes are regulated by 20E, JH, and insulin signaling pathways, further emphasizing their significance in the development of the mosquito and oriental fruit fly [90,91]. Moreover, comparative transcriptomic analyses have been conducted to reveal differentially expressed genes (DEGs) between *B. dorsalis* and its relatives [47]. These DEGs have yielded essential information concerning the molecular mechanisms of fruit fly development and adaptation, including ecdysone signaling pathways [92], cuticle biosynthesis [78], and insulin signaling [91]. Further studies have uncovered the molecular mechanisms of the *B. dorsalis* immune system, including peptidoglycan recognition proteins (PGRPs) [93,94], the phenoloxidase pathway [95], and toll-like receptor signaling [96,97]. Similarly, recent advancements in CRISPR-Cas9 technology have facilitated precise editing [98] of specific genes in *B. dorsalis* [99,100]. These findings have opened new opportunities for functional genomics research and gene-based control strategies for this insect pest. Furthermore, proteomics and metabolomics analyses have been conducted to comprehend the molecular mechanisms [101] underlying *B. dorsalis* growth and reproduction [102]. These studies have provided vital information on the molecular pathways and metabolites involved in insect development and adaptation. This knowledge could further assist researchers in devising innovative pest management approaches.

3.2. Reproductive Biology

Research has been conducted to understand the molecular mechanisms behind *B. dorsalis* mating behavior. Several neuropeptides and neurohormones involved in regulating insect reproductive behavior were identified in *B. dorsalis*, including corazonin, prothoracicotropic hormone, and allatostatin [103,104]. These neuropeptides play critical roles in regulating various physiological processes such as reproduction, growth, and metabolism. Furthermore, the identification of the gene encoding the alpha subunit of the cGMP-dependent protein kinase in *B. dorsalis* demonstrated its importance in regulating mating behavior [105,106]. Studying the receptor genes responsible for these signaling pathways would offer valuable insights into the molecular basis of *B. dorsalis* mating behavior. Studies focusing on the odorant receptors in *B. dorsalis* have been conducted. The identification of the genes encoding odorant receptors, Orco, was crucial in understanding the molecular mechanisms behind oriental fruit fly olfactory behavior and mate recognition [107,108]. Furthermore, studying the interactions between odorant receptors and their ligands, as well as the mechanisms of signal transduction in the antenna, would pave the way for developing pest management approaches in the future. The role of hormones in *B. dorsalis* female reproduction has been extensively investigated. Juvenile hormone (JH) plays a critical role in regulating reproduction, flight capabilities, and ecdysis in *B. dorsalis*. The biosynthesis, regulation, and function of JH in *B. dorsalis* have been thoroughly studied, and the genes responsible for JH synthesis and degradation have been cloned and characterized [86,109]. Furthermore, the 20-hydroxyecdysone (20E) pathway is another important hormonal pathway in *B. dorsalis*, and it has been found to play a vital role in regulating female reproduction by stimulating the biosynthesis of vitellogenin and other reproductive proteins [90,91].

Furthermore, miRNAs play a critical role in regulating *B. dorsalis* female reproduction through post-transcriptional regulation of target genes. Several miRNAs involved in the regulation of vitellogenin synthesis, oocyte maturation, and egg production have been identified [110,111]. The molecular mechanisms underlying the female maturation process in *B. dorsalis* have been studied. Critical factors in female maturation include genes involved in cuticle formation and sclerotization, ecdysteroid signaling, and ecdysone receptor signaling [87,112]. To date, the molecular mechanisms underlying the female sexual maturation process in *B. dorsalis* have been studied. Critical factors in female sexual maturation include genes involved in cuticle formation and sclerotization, ecdysteroid signaling, and ecdysone receptor signaling. Ongoing research on the reproductive biology and behavior of *B. dorsalis* is expected to yield more important findings in the near future. The discovery of additional genes and molecules involved in the regulation of *B. dorsalis* growth, development, and behavior could offer new strategies for developing effective means of controlling this significant pest [113].

3.3. Functional Analysis of Host Volatile Receptors

The role of volatiles in attracting *B. dorsalis* to crops has been well established, and partial characterization of the mechanisms of volatile detection and processing in the insect has been achieved [107,114]. However, further research is necessary to fully comprehend the complex interplay between the insect and volatile semiochemicals. This will entail a deeper exploration of the expression and function of genes involved in volatile detection, as well as identifying new semiochemicals and developing novel control strategies. The major volatile components of mango fruit that attract *B. dorsalis* have been identified, including hexanal, (E)-2-hexenal, (Z)-3-hexen-1-ol, and (E,E)-2,4-hexadienal [115,116]. The attractant activity of these volatiles was demonstrated through laboratory and field tests [117]. Similarly, other studies have identified attractive volatile components from other fruits, such as banana [118,119]. These findings have important implications for the development of attractive semiochemical-based control actions for *B. dorsalis*.

Volatile organic compounds (VOCs) from ripening fruit play a crucial role in attracting *B. dorsalis*. The interaction between these volatile compounds and the insect olfactory system has been extensively studied. Detection involves the diffusion of volatile compounds into the sensilla located on the antenna, which has numerous pores [120,121]. These pores facilitate the absorption of VOCs by odorant-binding proteins (OBPs) within the sensilla, which are subsequently transferred to the odorant receptors (ORs) located on the sensory neurons [122]. Using RNAi and electrophysiology techniques, researchers have been able to identify a variety of proteins involved in this process, including OBPs, chemosensory proteins (CSPs), ionotropic receptors (IRs), sensory neuron membrane proteins (SNMPs), and the odorant receptor co-receptor (ORCO) [123,124]. Silencing certain genes, such as OBPs, CSPs, and ORCO, has been shown to significantly decrease the electrophysiological response of the antennae [125,126]. In contrast to the genes primarily expressed in the antennae, there are other genes that are expressed both in classical olfactory and non-olfactory organs, or only in non-olfactory organs like the head, legs, and abdomen [127]. Further research is needed to better understand the role of these genes in attracting the oriental fruit fly to ripe fruit, which holds promise as a fruitful area for future study. As the interaction between *B. dorsalis* and the volatile compounds from ripening fruit is a complex and well-studied process, further research into the specific genes and proteins involved could improve our understanding of how insects are attracted to certain fruits and potentially lead to more effective agricultural pest control.

3.4. Gut Microbiota

The gut microbiota of insects is not only important for the health and survival of insects but also has the potential to impact the environment and human health [128]. Recent studies have shed light on the gut bacterial community of *B. dorsalis* and the role they play in insect fitness [129,130]. A study revealed that adult oriental fruit flies harbor a stable gut bacterial community dominated by *Enterobacteriaceae*, *Klebsiella*, *Citrobacter*, *Enterobacter*, and *Pectobacterium* [131,132]. Bacterial diversity was found to be influenced by the type of food. These gut bacteria are believed to indirectly contribute to host fitness by preventing the establishment or proliferation of pathogenic bacteria. Further investigations into the role of the Duox gene in regulating the gut bacterial community homeostasis of

B. dorsalis [133,134] revealed that the suppression of the Duox gene leads to an increased bacterial load and a decreased relative abundance of *Enterobacteriaceae* and *Leuconostocaceae* in the gut. Hence, Duox plays a key role in maintaining the stability of the gut microbiome. Recent studies have demonstrated the significance of gut microbiota in the life history of *B. dorsalis* [135,136]. Research has shown that these bacteria play a crucial role in regulating the gut homeostasis of *B. dorsalis*, preventing the establishment and proliferation of pathogenic bacteria [133,137]. Some bacteria, *Bacillus cereus* and *Enterococcus faecalis*, are able to secrete volatile compounds that are capable of attracting fruit flies, suggesting that these bacteria may be potential biocontrol agents, i.e., microbial attraction baits [138].

Research on the reproductive system of female *B. dorsalis* found *Enterobacter sakazakii* and *Klebsiella oxytoca* as the dominating bacterial species [139]. Culture-dependent (involving isolation and microbial culture) and culture-independent techniques (using molecular techniques—without cultivation) have also been used to survey the gut and reproductive tract of *B. dorsalis*, revealing diverse bacterial communities in fruit flies [140]. These communities are dominated by *Enterobacteriaceae*, *Klebsiella*, *Citrobacter*, *Enterobacter*, and *Pectobacterium* [141]. These findings provide new insights into the role of gut microbiota in insect reproduction. In the reproductive system of female *B. dorsalis*, *Enterobacter sakazakii* and *Klebsiella oxytoca* are the dominant bacterial species [139,142]. These findings provide new insights into the complex interactions between the oriental fruit fly and its associated bacteria, highlighting their importance in the biology of the insect [143]. Further research is needed to better understand the mechanisms underlying these interactions and their role in the survival and fitness of *B. dorsalis*.

3.5. Detection and Infestation of B. dorsalis

B. dorsalis has the potential to expand and infest the northern and southern regions, making it a significant quarantine pest in China [8]. To effectively manage this pest, a comprehensive approach, including quantitative risk assessment, is necessary. Given its ability to spread through the transportation of infested fruit, strict quarantine measures must be implemented to prevent its spread to new areas. Detection of *B. dorsalis* can be accomplished through a combination of morphological identification, acoustic detection, and molecular detection [125,144]. To eliminate infestations, thermal treatment of infected fruit is recommended. Research has shown that treating fruit at 60 °C for 2 h or 45 °C for 5 h effectively kills all fruit flies inside [145,146], and this technique can be implemented at quarantine check points to treat infested fruit. The fly can cause significant damage to a wide range of fruit crops, leading to substantial economic losses for farmers. Its rapid spread has been attributed to its ability to adapt to new environments, high reproductive potential, and resistance to chemical insecticides [147,148]. Strict quarantine measures have been implemented in China to prevent the spread of *B. dorsalis*, including surveillance programs to detect the fly's presence and thermal treatment systems to eliminate fruit infestations. In one study, *B. dorsalis* larvae in navel oranges were exposed to 1.2 °C for 15 days, which resulted in 99.99% larval mortality, hence making it one of the best commercial quarantine treatments [149]. Similarly, the effectiveness of other treatment measures has been studied in recent literature articles [4,8]. Moreover, as a significant quarantine pest in China, *B. dorsalis* requires comprehensive management strategies to control its spread. Quarantine measures, implementation of surveillance programs, and cold and thermal treatment of infested fruit have shown promising results in controlling the spread of *B. dorsalis* [150,151].

4. Pest Management

4.1. Mass Trapping

B. dorsalis could be mass-trapped using pheromone and food-based baits. Various pheromones and scent-based compounds, including synthetic para-pheromones or male lures such as methyl eugenol (ME), have been developed to attract and control *B. dorsalis*. These compounds mimic the natural pheromones produced by melon and oriental fruit flies. These pheromones are used to attract and trap male flies, allowing for population monitoring and infestation assessment [152]. E-coniferyl alcohol (E-CF) has also been found to be effective in attracting female *B. dorsalis* [69]. A comprehensive investigation on current resistance and lure tolerance to fruit flies [153] assessed the response of *B. dorsalis* males to non-ME lures. The experiment evaluated the mating and lure response of non-ME responding (NMR) and non-responding lines (NRLs) of *B. dorsalis* males. Results showed that NMR males had higher mating success rates compared to NRL males and exhibited a greater attraction to non-ME lures, which have been implicated in the development of tolerance mechanisms among *B. dorsalis* populations [153,154].

Another research revealed that *B. dorsalis* causes significant economic losses in the fruit and vegetable industry by laying eggs inside hosts. Chemical controls are not very effective due to the pest's cryptic feeding habits, strong flight ability, and resistance to insecticides. Olfaction-based trapping using ME has been the most cost-effective tool for monitoring and controlling *B. dorsalis* populations for seven decades [69,123]. However, laboratory selection for ME responsiveness has resulted in the non-responsiveness of *B. dorsalis*, which may lead to the recolonization of the pest in some areas [155]. The study aimed to determine the levels of ME responsiveness in *B. dorsalis* field populations in China [153,154]. Results showed that the field populations had lower ME sensitivity compared to the susceptible strain, possibly due to odorant binding protein (BdorOBP2, BdorOBP83b), and P450 gene expressions in olfactory organs [154]. Protein-based baits and food odors, such as yeast, vinegar, and fermentation products, can also attract both male and female oriental fruit flies. These baits can be combined with pheromones to increase trap efficacy [156]. Visual attractants, such as brightly colored sticky traps, can also be used to attract oriental fruit flies, and they can complement other attractants for a more comprehensive monitoring and control approach [114,157]. It is important to note that attractants can be species-specific, and the most effective ones for *B. dorsalis* may vary based on environmental conditions and other factors. Following are the steps involved in bait-based physical control techniques for managing *B. dorsalis* infestations: (a) Monitoring: It includes observations and record-keeping of the presence, distribution, and abundance of *B. dorsalis* in affected areas. (b) Selection of bait material: It includes selection of appropriate bait material, such as food-based baits (fishmeal or yeast hydrolysates, ME, raspberry ketone, cue lure, honey, or molasses) that have been successful in attracting fruit fly species. (c) Formulation of bait: It includes formulating the selected bait material into an attractive and easily dispersible form by adding a food-grade preservative for shelf-life extension and a hydroscopic agent to maintain its moisture content. (d) Deployment of baits: It involves deployment of the baits using various methods, including bait stations, bait trees, or spray applications, depending on the specific circumstances of each situation. (e) Collection and disposal of captured fruit flies: It involves regular monitoring to assess the effectiveness of the bait and removing and disposing of captured fruit flies to prevent escape and further spread. (f) Evaluation: It includes assessing the success of the bait-based physical control technique by monitoring oriental fruit fly population levels over time and comparing pre- and post-treatment populations to determine the reduction in the number of fruit flies.

4.2. Biological Control

Parasitoids, hymenopteran wasps, lay their eggs inside hosts, consuming them from the inside and leading to their death. *Fopius arisanus* (Sonan), a species of egg parasitoid, targets *B. dorsalis* [8]. As a potential biological control agent, *F. arisanus* effectively parasitizes the host eggs and reduces the pest population [158]. It is well-adapted to tropical and subtropical environments, distributed throughout Asia, Africa, and the Pacific region [159]. Utilizing *F. arisanus* offers advantages over chemical pest control, including specificity to the target pest, conservation of beneficial insects, and long-term sustainability [160]. In order to effectively utilize *F. arisanus* for biological control, it is important to understand its biology, behavior, and life cycle, as well as its interactions with the host and other factors that may affect its efficacy. Researchers have also developed mass rearing for *F. arisanus* to produce large numbers of individuals for release into the field. *F. arisanus* is a promising biological control agent for the oriental fruit fly, offering a sustainable and environmentally friendly approach to managing this destructive pest [161]. Another parasitoid, *Spalangia endius* (Walker) (Hymenoptera: Pteromalidae) is a solitary endoparasitoid that attacks fruit fly pupae, including *B. dorsalis*. This wasp lays its eggs inside the pupae, and the emerging larvae consume the host pupae from within, killing the fruit fly [162]. Using *S. endius* for the biological control of *B. dorsalis* has advantages over other methods. It is highly specific in targeting fruit fly pests and does not harm beneficial insects. Field trials have shown that this parasitoid can effectively reduce the number of *B. dorsalis* adults, thereby minimizing crop damage [163,164]. To effectively use *S. endius* for biological control, understanding the biology and behavior of both the wasp and the fruit fly is crucial. The timing of wasp releases is critical in achieving maximum parasitism rates. In general, releases should coincide with the emergence of fruit fly pupae, which is the stage at which *S. endius* lays its eggs. Releasing large numbers of parasitoids can help control fruit fly populations in a targeted area [165,166] (Table 1). Viruses, bacteria, and fungi can also infect and be lethal to the fruit fly adult and larvae [137]. Among the pathogens studied for use against *B. dorsalis*, viruses, especially baculoviruses, have been found to be highly virulent to fruit fly species. They have demonstrated effectiveness in reducing fruit fly populations in both laboratory and field studies.

Baculoviruses are insect-specific viruses that replicate within the insect host and cause death. Among the baculovirus isolates identified and characterized, the nuclear polyhedrosis virus (NPV) has been found to be highly virulent to several insect species [167,168]. NPV studies have demonstrated its ability to reduce the number of fruit fly individuals in laboratory and field settings, thereby decreasing the damage caused by this pest. Moreover, NPV is safe for the environment and non-target organisms, making it a promising option for fruit fly management [169]. However, it is important to note that using pathogenic microorganisms, including viruses, for insect pest management is still in its early stages, and more research is needed to fully understand their potential and limitations. Baculoviruses, particularly NPV, have shown potential for controlling *B. dorsalis*, and further research is needed to integrate them into pest management programs effectively. The entomophagous fungus *Beauveria bassiana* (Sordariomycetes: Clavicipitaceae) is an entomopathogenic fungus that is known to be an effective biological control agent against *B. cucurbitae*. This fungus infects the insects and causes mortality [170]. In China, *B. bassiana* effectively controlled *B. dorsalis*, achieving a mortality rate of over 80% in laboratory experiments. Similarly, another study showed that *B. bassiana* effectively reduced the population density of *B. dorsalis* in the field [171,172]. *B. bassiana* can be used as a biological control agent in several ways: (1) Inoculative releases: This involves releasing large numbers of fungal spores (conidia) into the environment, which then infect the insects. This approach is most effective when used in conjunction with other management strategies, such as the use of pheromones or host-plant resistance. (2) Injection or spraying: This process involves injecting or spraying a suspension of conidia directly onto the insects, causing them to become infected. (3) Formulations: *B. bassiana* can also be formulated into granules or dusts that can be applied to the host plants or environment, where they will encounter the insects. The entomopathogenic bacterium *Bacillus thuringiensis* (Bt) is a naturally occurring soil bacterium that produces a toxic crystal protein effective against many insect pests, including *B. dorsalis* [173]. Entomopathogenic nematodes are parasitic roundworms that can infect and kill fruit fly larvae [174]. Further research and understanding of biology will improve their integration into pest management programs (Table 1).

Table 1. Natural enemies *of B. dorsalis*.

Bio-Control Agents	Name of Species	Host Stages	Reference
Predator	*Oecophylla longinoda*	Pupa/larva	[175]
	Pachycrepoideus vindemmiae	Larva/pupa	[176]
Parasitoids	*Fopius arisanus*	Egg	[177]
	Psyttalia cosyrae	Larva-pupal	[178]
	Diachasmimorpha longicaudata	Larva	[178]
Nematodes	*Heterorhabditis taysearae*	Larva/pupa	[176]
	H. indica	Larva/pupa	[179]
	Steinernema sp	Larva/pupa	[180]

4.3. *Sterile Insect Technique (SIT)*

The sterile insect technique is a promising biological control for *Bactrocera* species, with a proven track record of success in various countries worldwide. It is a sustainable and eco-friendly method of pest management that complements other control strategies, providing long-term control of this economically important insect pest. The technique has been used for decades to manage various insect pests, including *B. dorsalis*, commonly known as the oriental fruit fly. SIT involves mass-rearing and sterilization of male insects, which are then released into the wild to mate with females. Mating with sterilized males leads to the laying of eggs by female insects that do not hatch, ultimately leading to a decline in the pest population [8,181]. This technique was first employed in the 1950s to control the screwworm, *Cochliomyia hominivorax*, in the southern United States and has since been effectively utilized against various other insect pests worldwide. Fruit flies, including *B. dorsalis*, have been successfully managed using SIT in several countries, such as China, Australia, and Hawaii [182,183]. For instance, in Hawaii, SIT was implemented in the early 2000s to manage oriental fruit fly outbreaks in the state's agriculture industry. The program's success resulted in a significant decline in the pest population [184]. In Australia, SIT has been incorporated into integrated pest management to control Mediterranean fruit fly *C. capitata* populations in the country's horticulture industry [185]. The genetic sexing strain is a technique utilized to manipulate the sex ratios of a population, leading to more effective and efficient pest management. It has been successfully applied worldwide, including China, to manage *B. dorsalis*. This technique employs a genetic marker to distinguish between male and female fruit flies. By releasing only sterilized males into the environment, the population growth of the pest can be suppressed without the need for chemical insecticides [186].

In China, researchers have developed a genetic sexing strain for *B. dorsalis* using the temperature-sensitive lethal (tsl) mutation. This mutation causes the death of females at a certain temperature, enabling the separation of male and female fruit flies [187]. The genetic sexing strain has been proven effective in suppressing the population growth of several fruit fly species in field trials [188]. This technique has also been applied to manage fruit flies in other countries, including Australia and Thailand [13,185,189,190]. These studies demonstrate the potential of the genetic sexing strain as an integrated pest management tool for managing tephritid fruit flies.

4.4. *Molecular Control*

The management of *B. dorsalis* is challenging due to its high resistance to insecticides. To overcome this challenge, it is crucial to identify new targets for insect pest control. Transient receptor potential (TRP) channels play a crucial role in various physiological processes in insects, including nociception, thermo-sensation, and olfaction [191,192]. In recent years, there have been extensive studies on the identification and characterization of TRP channels in various insect species, including *B. dorsalis*. In one study [192], 15 TRP channel genes were identified in the genome of *B. dorsalis*. The expression patterns of these genes were analyzed in different tissues, such as the antennae, brain, midgut, Malpighian tubules, and fat body. The results revealed that TRP channels were differentially expressed across various tissues, with some TRP genes being predominantly expressed in specific tissues. Additionally, another study [193] investigated the role of TRP channels in insecticide resistance in insects. They used RNA interference (RNAi) to knock down the expression of TRP channels in *B. dorsalis*. The findings showed that knockdown of TRP channels significantly reduced insecticide resistance in *B. dorsalis*, suggesting the potential utilization of TRP channels as targets for insect pest control [194,195]. The identification, characterization, and expression analysis of TRP channel genes in the oriental fruit fly will provide crucial information for the development of new and effective strategies for the management and control of this pest.

4.5. RNA Interference (RNAi)

RNA interference (RNAi) is a highly effective technique for gene silencing through the use of double-stranded RNA (dsRNA) [196]. It has shown promise in knocking down insect pests as a more environmentally friendly option. Previous studies have demonstrated successful silencing of genes rpl19, v-ATPase-D, noa, and rab11 in adult *B. dorsalis* through the feeding of corresponding dsRNA. Other potential target genes involved in midgut digestion and detoxification have also been identified [197,198]. However, using RNAi for controlling the oriental fruit fly faces challenges, including effectively delivering dsRNA to the insect and potential risks to non-target organisms. The delivery of dsRNA has not been fully implemented yet, and the possible impacts on non-target organisms and host fruits and vegetables must be carefully considered. There is a risk of reducing the expression of genes in natural enemies and other beneficial insects due to the high similarity in rpl19 sequences between these insects and *B. dorsalis*. Therefore, minimizing the impact of dsRNA on non-target insects and host fruits and vegetables is a priority in ongoing efforts to use RNAi for controlling *B. dorsalis*. In a research article addressing the problem of insecticide resistance in *B. dorsalis*, a global pest affecting various crops, researchers focused on the role of UDP-glycosyltransferases (UGTs) in resistance development [199]. These enzymes are involved in metabolically processing both plant secondary metabolites and synthetic insecticides. The study identified 31 UGT genes in the genome of *B. dorsalis*, with 12 of them highly expressed in key tissues such as the antennae, midgut, Malpighian tubules, and fat body. Furthermore, exposure to four different insecticides caused a significant upregulation of 17 UGT genes. To investigate further, RNA interference was used to knock down five selected UGT genes, resulting in reduced oriental fruit fly mortality in response to insecticides from 9.29% to 27.22% [200].

4.6. CRISPR-Cas9

The clustered regularly interspaced palindromic repeat (CRISPR-Cas9) system is a revolutionary tool for precise and efficient genome editing in various organisms [201]. In a study of *B. dorsalis*, researchers targeted a specific gene known as the Sex Peptide Receptor (*Bdspr*) using CRISPR/Cas9 technology [100]. The *Bdspr* gene plays a critical role in the regulation of female reproduction, including ovary development and egg laying. By introducing mutations into this gene, the researchers aimed to examine its effects on female fecundity and reproductive functions in *B. dorsalis*. Several research experiments showed that CRISPR/Cas9-mediated disruption of the *Bdspr* gene, when the insects were fed with the ds-spr gene, led to significant changes in the number and size of ovarioles, a reduction in the number of eggs laid, and a decrease in overall female fecundity. This indicated the importance of the *Bdspr* gene in the normal functioning of the female reproductive system in *B. dorsalis*. The study also demonstrated that the CRISPR/Cas9 system is an effective tool for studying gene function and disrupting specific genes in insects. In the future, this information could potentially be used to develop new strategies for controlling the population of oriental fruit flies, a major agricultural pest causing significant damage to crops worldwide [100,202–204]. The CRISPR/Cas9-induced mutation of the *Bdspr* gene in the oriental fruit fly underscores the significance of this gene in female reproduction and highlights the potential of genome editing technology for advancing the field of insect pest management.

In another study focused on understanding the functional role of the white gene in pigmentation in *B. dorsalis*, the white gene was cloned, and knockout strains were created using the CRISPR/Cas9 genome editing system. The results revealed that the mutants lost pigmentation in the compound eye and their head spots. Further analysis using quantitative reverse-transcription PCR showed lower expression levels of the Bd-yellow1 gene in the head of mutants compared to the wild-type strain, while there were no significant differences in the expression of the other six genes. As the yellow gene is crucial for melanin biosynthesis, the reduced expression of Bd-yellow1 in mutants led to a decrease in dark pigmentation in the head spots. This study provides evidence for the first time that the white gene may play a role in cuticle pigmentation by affecting the expression of the yellow gene [99].

5. Conclusions and Future Perspective

The oriental fruit fly has been the most common and significant orchard pest since its invasion of mainland China. Currently, it is expanding to more suitable regions in North China. Over the past few decades, many tactics have been developed to track its occurrence, spread, and damage, along with numerous studies conducted to understand its invasion process. However, its ability to adapt to different habitats, insecticide resistance, and high reproductive capacity have been helping the pest to spread to wider landscapes. These factors might provide valuable information for the development of new pest management techniques. Currently, the application of synthetic pesticides remains the basis of *B. dorsalis* management worldwide, including China. However, when this pest develops insecticide resistance, the effectiveness of chemical management significantly reduces. To address this issue, it is essential to take preventive measures against the emergence of insecticide resistance and adopt novel pesticide options, such as botanical and microbial pesticides. Gaining a sufficient understanding of the physiological mechanism and molecular roots of mating choice and behavior could lead to new control techniques based on behavioral alteration. SIT (sterile insect technique) and area-wide pest eradication programs have been used as techniques to manage the oriental fruit fly and other insect species. Although it has been widely used in China, it has not yet eradicated the invasive species. However, it holds the potential to successfully eradicate the pest species.

Author Contributions: Conceptualization, S.J. and Y.L.; literature search, S.J.; software, S.J. and S.A.H.R.; validation, Y.L. and S.J.; formal analysis, S.J.; investigation, Y.L.; resources, Y.L.; data curation, S.J.; writing—original draft preparation, S.J.; writing—review and editing, S.J., Y.L.; visualization, Y.L.; funding acquisition, Y.L. All authors have read and agreed to the published version of the manuscript.

Funding: This research was funded by the Common Key Technology R & D Innovation Team Project in Modern Agricultural Industry of Guangdong grant number 2022kj134.

Data Availability Statement: The data used in this study will be made available by the corresponding author on request.

Acknowledgments: The authors thank all anonymous reviewers for their constructive comments, invaluable suggestions, and recommendations.

Conflicts of Interest: The authors declare they have no conflicts of interest.

References

1. He, Y.; Xu, Y.; Chen, X. Biology, Ecology and Management of Tephritid Fruit Flies in China: A Review. *Insects* **2023**, *14*, 196. [CrossRef] [PubMed]
2. Clarke, A.R.; Armstrong, K.F.; Carmichael, A.E.; Milne, J.R.; Raghu, S.; Roderick, G.K.; Yeates, D.K. Invasive phytophagous pests arising through a recent tropical evolutionary radiation: The *Bactrocera dorsalis* complex of fruit flies. *Annu. Rev. Entomol.* **2005**, *50*, 293–319. [CrossRef] [PubMed]
3. Hossain, M.; Sarkar, B.; Hossain, M.; Mian, M.; Rajotte, E.; Muniappan, R.; O'Rourke, M. Comparison of biorational management approaches against mango fruit fly (*Bactrocera dorsalis* Hendel) in Bangladesh. *Crop Prot.* **2020**, *135*, 104807. [CrossRef]
4. Jaffar, S.; Lu, Y. Toxicity of Some Essential Oils Constituents against Oriental Fruit Fly, *Bactrocera dorsalis* (Hendel) (Diptera: Tephritidae). *Insects* **2022**, *13*, 954. [CrossRef] [PubMed]
5. Wan, X.; Liu, Y.; Zhang, B. Invasion history of the oriental fruit fly, *Bactrocera dorsalis*, in the Pacific-Asia region: Two main invasion routes. *PLoS ONE* **2012**, *7*, e36176. [CrossRef] [PubMed]
6. Drew, R.; Hancock, D. Biogeography, Speciation and Taxonomy within the genus *Bactrocera* Macquart with application to the *Bactrocera dorsalis* (Hendel) complex of fruit flies (Diptera: Tephritidae: Dacinae). *Zootaxa* **2022**, *5190*, 333–360. [CrossRef]
7. Zeng, Y.; Reddy, G.V.; Li, Z.; Qin, Y.; Wang, Y.; Pan, X.; Jiang, F.; Gao, F.; Zhao, Z.H. Global distribution and invasion pattern of oriental fruit fly, *Bactrocera dorsalis* (Diptera: Tephritidae). *J. Appl. Entomol.* **2019**, *143*, 165–176. [CrossRef]
8. Liu, H.; Zhang, D.; Xu, Y.; Wang, L.; Cheng, D.; Qi, Y.; Zeng, L.; Lu, Y. Invasion, expansion, and control of *Bactrocera dorsalis* (Hendel) in China. *J. Integr. Agric.* **2019**, *18*, 771–787. [CrossRef]
9. Aketarawong, N.; Guglielmino, C.; Karam, N.; Falchetto, M.; Manni, M.; Scolari, F.; Gomulski, L.; Gasperi, G.; Malacrida, A. The oriental fruitfly *Bactrocera dorsalis* ss in East Asia: Disentangling the different forces promoting the invasion and shaping the genetic make-up of populations. *Genetica* **2014**, *142*, 201–213. [CrossRef]
10. Meng, L.W.; Yuan, G.R.; Lu, X.P.; Jing, T.X.; Zheng, L.S.; Yong, H.X.; Wang, J.J. Two delta class glutathione S-transferases involved in the detoxification of malathion in *Bactrocera dorsalis* (Hendel). *Pest. Manag. Sci.* **2019**, *75*, 1527–1538. [CrossRef]
11. Mutamiswa, R.; Nyamukondiwa, C.; Chikowore, G.; Chidawanyika, F. Overview of oriental fruit fly, *Bactrocera dorsalis* (Hendel) (Diptera: Tephritidae) in Africa: From invasion, bio-ecology to sustainable management. *Crop Prot.* **2021**, *141*, 105492. [CrossRef]
12. Zhu, Y.; Qi, F.; Tan, X.; Zhang, T.; Teng, Z.; Fan, Y.; Wan, F.; Zhou, H. Use of Age-Stage, Two-Sex Life Table to Compare the Fitness of *Bactrocera dorsalis* (Diptera: Tephritidae) on Northern and Southern Host Fruits in China. *Insects* **2022**, *13*, 258. [CrossRef] [PubMed]
13. Orankanok, W.; Chinvinijkul, S.; Thanaphum, S.; Sitilob, P.; Enkerlin, W.R. Area-Wide Integrated Control of Oriental Fruit Fly *Bactrocera dorsalis* and Guava Fruit Fly *Bactrocera correcta* in Thailand. In *Area-Wide Control of Insect Pests*; Springer: Dordrecht, The Netherlands, 2007; pp. 517–526.
14. Bana, J.; Sharma, H.; Kumar, S.; Singh, P. Impact of weather parameters on population dynamics of oriental fruit fly, *Bactrocera dorsalis* (Hendel)(Diptera: Tephritidae) under south Gujarat mango ecosystem. *J. Agrometeorol.* **2017**, *19*, 78–80. [CrossRef]
15. Jitendra, M.; Sandeep, S.; Akilesh, T.; Chaube, M. Population dynamics of oriental fruit fly, *Bactrocera dorsalis* (Hendel) in relation to abiotic factors. *HortFlora Res. Spectr.* **2012**, *1*, 187–189.
16. Dhillon, M.; Singh, R.; Naresh, J.; Sharma, H. The melon fruit fly, *Bactrocera cucurbitae*: A review of its biology and management. *J. Insect Sci.* **2005**, *5*, 40. [CrossRef]
17. Coates, B.S.; Poelchau, M.; Childers, C.; Evans, J.D.; Handler, A.; Guerrero, F.; Skoda, S.; Hopper, K.; Wintermantel, W.M.; Ling, K.-S. Arthropod genomics research in the United States Department of Agriculture-Agricultural Research Service: Current impacts and future prospects. *Trends Entomol.* **2015**, *11*, 12–27.
18. Liu, L.J.; Martinez-Sañudo, I.; Mazzon, L.; Prabhakar, C.S.; Girolami, V.; Deng, Y.L.; Dai, Y.; Li, Z.H. Bacterial communities associated with invasive populations of *Bactrocera dorsalis* (Diptera: Tephritidae) in China. *Bull. Entomol. Res.* **2016**, *106*, 718–728. [CrossRef]
19. Zhu, Y.-F.; Tan, X.-M.; Qi, F.-J.; Teng, Z.-W.; Fan, Y.-J.; Shang, M.-Q.; Lu, Z.-Z.; Wan, F.-H.; Zhou, H.-X. The host shift of *Bactrocera dorsalis*: Early warning of the risk of damage to the fruit industry in northern China. *Entomol. Gen.* **2022**, *42*, 691–699. [CrossRef]
20. Ullah, F.; Gul, H.; Hafeez, M.; Güncan, A.; Tariq, K.; Desneux, N.; Zhao, Z.; Li, Z. Impact of temperature stress on demographic traits and population projection of *Bactrocera dorsalis*. *Entomol. Gen.* **2022**, *42*, 949–957. [CrossRef]
21. Doorenweerd, C.; Leblanc, L.; Hsu, Y.-F.; Huang, C.-L.; Lin, Y.-C.; San Jose, M.; Rubinoff, D. Taiwan's Dacini Fruit Flies: Rare Endemics and Abundant Pests, along Altitudinal Gradients. *Pac. Sci.* **2019**, *73*, 35–59. [CrossRef]
22. Jing, W.; QI, G.-j.; Jun, M.; Ren, Y.; Rui, W.; McKirdy, S. Predicting the potential geographic distribution of *Bactrocera bryoniae* and *Bactrocera neohumeralis* (Diptera: Tephritidae) in China using MaxEnt ecological niche modeling. *J. Integr. Agric.* **2020**, *19*, 2072–2082.
23. Li, X.; Yang, H.; Wang, T.; Wang, J.; Wei, H. Life history and adult dynamics of *Bactrocera dorsalis* in the citrus orchard of Nanchang, a subtropical area from China: Implications for a control timeline. *Scienceasia* **2019**, *45*, 212–220. [CrossRef]
24. Qin, Y.; Wang, C.; Zhao, Z.; Pan, X.; Li, Z. Climate change impacts on the global potential geographical distribution of the agricultural invasive pest, *Bactrocera dorsalis* (Hendel) (Diptera: Tephritidae). *Clim. Chang.* **2019**, *155*, 145–156. [CrossRef]
25. Kim, S.B.; Park, J.-J.; Kim, D.-S. CLIMEX simulated predictions of the potential distribution of *Bactrocera dorsalis* (Hendel) (Diptera: Tephritidae) considering the northern boundary: With special emphasis on Jeju, Korea. *J. Asia-Pac. Entomol.* **2020**, *23*, 797–808. [CrossRef]

26. Zhang, Y.; Liu, S.; De Meyer, M.; Liao, Z.; Zhao, Y.; Virgilio, M.; Feng, S.; Qin, Y.; Singh, S.; Wee, S.L. Genomes of the cosmopolitan fruit pest *Bactrocera dorsalis* (Diptera: Tephritidae) reveal its global invasion history and thermal adaptation. *J. Adv. Res.* **2022**. [CrossRef]
27. Kim, H.; Kim, S.; Kim, S.; Lee, Y.; Lee, H.-S.; Lee, S.-J.; Choi, D.-S.; Jeon, J.; Lee, J.-H. Population genetics for inferring introduction sources of the oriental fruit fly, *Bactrocera dorsalis*: A test for quarantine use in Korea. *Insects* **2021**, *12*, 851. [CrossRef]
28. Ouyang, H.; Wang, X.; Zheng, X.; Lu, W.; Qin, F.; Chen, C. Full-length SMRT transcriptome sequencing and SSR analysis of *Bactrocera dorsalis* (Hendel). *Insects* **2021**, *12*, 938. [CrossRef]
29. Li, Y.; Wu, Y.; Chen, H.; Wu, J.; Li, Z. Population structure and colonization of *Bactrocera dorsalis* (Diptera: Tephritidae) in China, inferred from mtDNA COI sequences. *J. Appl. Entomol.* **2012**, *136*, 241–251. [CrossRef]
30. Liu, J.; Shi, W.; Ye, H. Population genetics analysis of the origin of the Oriental fruit fly, *Bactrocera dorsalis* Hendel (Diptera: Tephritidae), in northern Yunnan Province, China. *Entomol. Sci.* **2007**, *10*, 11–19. [CrossRef]
31. Moschilla, J.A.; Tomkins, J.L.; Simmons, L.W. Males adjust their manipulation of female remating in response to sperm competition risk. *Proc. Biol. Sci.* **2020**, *287*, 20201238. [CrossRef]
32. Wei, D.; Feng, Y.-C.; Wei, D.-D.; Yuan, G.-R.; Dou, W.; Wang, J.-J. Female remating inhibition and fitness of *Bactrocera dorsalis* (Diptera: Tephritidae) associated with male accessory glands. *Fla. Entomol.* **2015**, *98*, 52–58. [CrossRef]
33. Reyes-Hernández, M.; Thimmappa, R.; Abraham, S.; Pagadala Damodaram, K.J.; Pérez-Staples, D. Methyl eugenol effects on *Bactrocera dorsalis* male total body protein, reproductive organs and ejaculate. *J. Appl. Entomol.* **2019**, *143*, 177–186. [CrossRef]
34. Shelly, T.E. Fecundity of Female Oriental Fruit Flies (Diptera: Tephritidae): Effects of Methyl Eugenol-Fed and Multiple Mates. *Ann. Entomol. Soc. Am.* **2000**, *93*, 559–564. [CrossRef]
35. Shelly, T. Ability of sterile males to inhibit female remating in the oriental fruit fly, *Bactrocera dorsalis* (Hendel) (Diptera: Tephritidae). *Proc. Hawaii. Entomol. Soc.* **2020**, *52*, 15–23.
36. Olabiyi, D.; Shrestha, B.; Zaka, S.M.; Neupane, S. Insect Pests of Citrus Production. In *Citrus Production*; CRC Press: Boca Raton, FL, USA, 2022; pp. 287–302.
37. Kota, M.; Olzer, R. Insect Communication. In *Encyclopedia of Animal Cognition and Behavior*; Vonk, J., Shackelford, T., Eds.; Springer International Publishing: Cham, Switzerland, 2019; pp. 1–7. [CrossRef]
38. Meunier, J. Social immunity and the evolution of group living in insects. *Philos. Trans. R. Soc. B Biol. Sci.* **2015**, *370*, 20140102. [CrossRef]
39. Chen, Y.; Zhang, Y.; Yang, L.; Chen, W.; Jiang, Z.; Xiao, Z.; Xie, X.; Zhong, G.; Yi, X. Group housing enhances mating and increases the sensitization of chemical cues in *Bactrocera dorsalis*. *Pest. Manag. Sci.* **2023**, *79*, 391–401. [CrossRef] [PubMed]
40. Kamala Jayanthi, P.D.; Saravan Kumar, P.; Vyas, M. Odour Cues from Fruit Arils of *Artocarpus heterophyllus* Attract both Sexes of Oriental Fruit Flies. *J. Chem. Ecol.* **2021**, *47*, 552–563. [CrossRef]
41. Wang, L.; Anderson, D.J. Identification of an aggression-promoting pheromone and its receptor neurons in *Drosophila*. *Nature* **2010**, *463*, 227–231. [CrossRef]
42. Billeter, J.-C.; Levine, J. The role of cVA and the Odorant binding protein Lush in social and sexual behavior in *Drosophila melanogaster*. *Front. Ecol. Evol.* **2015**, *3*, 75. [CrossRef]
43. Ejima, A.; Smith, B.P.; Lucas, C.; Van Naters, W.v.d.G.; Miller, C.J.; Carlson, J.R.; Levine, J.D.; Griffith, L.C. Generalization of courtship learning in *Drosophila* is mediated by cis-vaccenyl acetate. *Curr. Biol.* **2007**, *17*, 599–605. [CrossRef]
44. Roy, A.; George, S.; Palli, S.R. Multiple functions of CREB-binding protein during postembryonic development: Identification of target genes. *BMC Genom.* **2017**, *18*, 996. [CrossRef]
45. Geib, S.M.; Calla, B.; Hall, B.; Hou, S.; Manoukis, N.C. Characterizing the developmental transcriptome of the oriental fruit fly, *Bactrocera dorsalis* (Diptera: Tephritidae) through comparative genomic analysis with Drosophila melanogaster utilizing modENCODE datasets. *BMC Genom.* **2014**, *15*, 942. [CrossRef]
46. De Villiers, M.; Hattingh, V.; Kriticos, D.J.; Brunel, S.; Vayssières, J.F.; Sinzogan, A.; Billah, M.K.; Mohamed, S.A.; Mwatawala, M.; Abdelgader, H.; et al. The potential distribution of *Bactrocera dorsalis*: Considering phenology and irrigation patterns. *Bull. Entomol. Res.* **2016**, *106*, 19–33. [CrossRef] [PubMed]
47. Gu, X.; Zhao, Y.; Su, Y.; Wu, J.; Wang, Z.; Hu, J.; Liu, L.; Zhao, Z.; Hoffmann, A.A.; Chen, B. A transcriptional and functional analysis of heat hardening in two invasive fruit fly species, *Bactrocera dorsalis* and Bactrocera correcta. *Evol. Appl.* **2019**, *12*, 1147–1163. [CrossRef] [PubMed]
48. Ren, L.; Lu, Y.-Y.; Zeng, L.; Pang, S.-T. Effect of different hostfruits on the cold hardiness of *Bactrocera dorsalis* (Hendel) (Diptera: Tephritidae). *Acta Entomol. Sin.* **2006**, *49*, 447–453.
49. Mutamiswa, R.; Tarusikirwa, V.; Nyamukondiwa, C.; Chidawanyika, F. Fluctuating environments impact thermal tolerance in an invasive insect species *Bactrocera dorsalis* (Diptera: Tephritidae). *J. Appl. Entomol.* **2020**, *144*, 885–896. [CrossRef]
50. Pieterse, W.; Terblanche, J.S.; Addison, P. Do thermal tolerances and rapid thermal responses contribute to the invasion potential of *Bactrocera dorsalis* (Diptera: Tephritidae)? *J. Insect Physiol.* **2017**, *98*, 1–6. [CrossRef] [PubMed]
51. Jia, F.-X.; Dou, W.; Hu, F.; Wang, J.-J. Effects of thermal stress on lipid peroxidation and antioxidant enzyme activities of oriental fruit fly, *Bactrocera dorsalis* (Diptera: Tephritidae). *Fla. Entomol.* **2011**, *94*, 956–963. [CrossRef]
52. Gao, X.M.; Jia, F.X.; Shen, G.M.; Jiang, H.Q.; Dou, W.; Wang, J.J. Involvement of superoxide dismutase in oxidative stress in the oriental fruit fly, *Bactrocera dorsalis*: Molecular cloning and expression profiles. *Pest. Manag. Sci.* **2013**, *69*, 1315–1325. [CrossRef] [PubMed]

53. Wei, D.; Dou, W.; Jiang, M.; Wang, J. Oriental Fruit Fly *Bactrocera dorsalis* (Hendel). In *Biological Invasions and Its Management in China: Volume 1*; Wan, F., Jiang, M., Zhan, A., Eds.; Springer: Dordrecht, The Netherlands, 2017; pp. 267–283. [CrossRef]
54. Hou, B.; Xie, Q.; Zhang, R. Depth of pupation and survival of the Oriental fruit fly, *Bactrocera dorsalis* (Diptera: Tephritidae) pupae at selected soil moistures. *Appl. Entomol. Zool.* **2006**, *41*, 515–520. [CrossRef]
55. Jaffar, S.; Ahmad, S.; Lu, Y. Contribution of insect gut microbiota and their associated enzymes in insect physiology and biodegradation of pesticides. *Front. Microbiol.* **2022**, *13*, 979383. [CrossRef] [PubMed]
56. Jin, T.; Zeng, L.; Lin, Y.; Lu, Y.; Liang, G. Insecticide resistance of the oriental fruit fly, *Bactrocera dorsalis* (Hendel) (Diptera: Tephritidae), in mainland China. *Pest. Manag. Sci.* **2011**, *67*, 370–376. [CrossRef]
57. Abubakar, M.; Ali, H.; Shad, S.A.; Anees, M.; Binyameen, M. Trichlorfon resistance: Its stability and impacts on biological parameters of *Bactrocera zonata* (Diptera: Tephritidae). *Appl. Entomol. Zool.* **2021**, *56*, 473–482. [CrossRef]
58. Lin, Y.; Jin, T.; Zeng, L.; Lu, Y. Cuticular penetration of β-cypermethrin in insecticide-susceptible and resistant strains of *Bactrocera dorsalis*. *Pestic. Biochem. Physiol.* **2012**, *103*, 189–193. [CrossRef]
59. Wang, J.-J.; Wei, D.; Dou, W.; Hu, F.; Liu, W.-F.; Wang, J.-J. Toxicities and Synergistic Effects of Several Insecticides Against the Oriental Fruit Fly (Diptera: Tephritidae). *J. Econ. Entomol.* **2013**, *106*, 970–978. [CrossRef]
60. Zhang, R.; He, S.; Chen, J. Monitoring of *Bactrocera dorsalis* (Diptera: Tephritidae) Resistance to Cyantraniliprole in the South of China. *J. Econ. Entomol.* **2014**, *107*, 1233–1238. [CrossRef] [PubMed]
61. Wang, R.; Wang, J.; Che, W.; Luo, C. First report of field resistance to cyantraniliprole, a new anthranilic diamide insecticide, on *Bemisia tabaci* MED in China. *J. Integr. Agric.* **2018**, *17*, 158–163. [CrossRef]
62. Lu, X.P.; Wang, L.L.; Huang, Y.; Dou, W.; Chen, C.T.; Wei, D.; Wang, J.J. The epsilon glutathione S-transferases contribute to the malathion resistance in the oriental fruit fly, *Bactrocera dorsalis* (Hendel). *Comp. Biochem. Physiol. C Toxicol. Pharmacol.* **2016**, *180*, 40–48. [CrossRef]
63. Jing, T.-X.; Wang, D.-F.; Ma, Y.-P.; Zeng, L.-L.; Meng, L.-W.; Zhang, Q.; Dou, W.; Wang, J.-J. Genome-wide and expression-profiling analyses of the cytochrome P450 genes in *Bactrocera dorsalis* (Hendel) and screening of candidate P450 genes associated with malathion resistance. *Pest. Manag. Sci.* **2020**, *76*, 2932–2943. [CrossRef]
64. He, L.; Huang, Y.; Tang, X. RNAi-based pest control: Production, application and the fate of dsRNA. *Front. Bioeng. Biotechnol.* **2022**, *10*, 1080576. [CrossRef]
65. Yan, S.; Ren, B.-Y.; Shen, J. Nanoparticle-mediated double-stranded RNA delivery system: A promising approach for sustainable pest management. *Insect Sci.* **2021**, *28*, 21–34. [CrossRef] [PubMed]
66. Hou, Q.-L.; Chen, E.-H.; Dou, W.; Wang, J.-J. Assessment of *Bactrocera dorsalis* (Diptera: Tephritidae) Diets on Adult Fecundity and Larval Development: Insights Into Employing the Sterile Insect Technique. *J. Insect Sci.* **2020**, *20*, 1–8. [CrossRef]
67. Jing, T.X.; Zhang, Y.X.; Dou, W.; Jiang, X.Y.; Wang, J.J. First Insights into the Intrapuparial Development of *Bactrocera dorsalis* (Hendel): Application in Predicting Emergence Time for Tephritid Fly Control. *Insects* **2019**, *10*, 283. [CrossRef] [PubMed]
68. Zhao, X.; Zhang, X.; Chen, Z.; Wang, Z.; Lu, Y.; Cheng, D. The Divergence in Bacterial Components Associated with *Bactrocera dorsalis* across Developmental Stages. *Front. Microbiol.* **2018**, *9*, 114. [CrossRef] [PubMed]
69. Liu, H.; Wang, D.D.; Wan, L.; Hu, Z.Y.; He, T.T.; Wang, J.B.; Deng, S.Z.; Wang, X.S. Assessment of attractancy and safeness of (E)-coniferyl alcohol for management of female adults of Oriental fruit fly, *Bactrocera dorsalis* (Hendel). *Pest. Manag. Sci.* **2022**, *78*, 1018–1028. [CrossRef]
70. Cheng, D.; Guo, Z.; Riegler, M.; Xi, Z.; Liang, G.; Xu, Y. Gut symbiont enhances insecticide resistance in a significant pest, the oriental fruit fly *Bactrocera dorsalis* (Hendel). *Microbiome* **2017**, *5*, 13. [CrossRef]
71. Hu, F.; Dou, W.; Wang, J.J.; Jia, F.X.; Wang, J.J. Purification and biochemical characterization of glutathione S-transferases from four field populations of *Bactrocera dorsalis* (Hendel) (Diptera: Tephritidae). *Arch. Insect Biochem. Physiol.* **2011**, *78*, 201–215. [CrossRef]
72. Huang, Y.-B.; Chi, H. Assessing the application of the jackknife and bootstrap techniques to the estimation of the variability of the net reproductive rate and gross reproductive rate: A case study in *Bactrocera cucurbitae* (Coquillett) (Diptera: Tephritidae). *J. Agric. For.* **2012**, *61*, 37–45.
73. Amiri, A.; Bandani, A.R.; Kafil, M. Gut compartments and ovary bacterial symbionts of the Sunn pest. *J. Asia-Pac. Entomol.* **2020**, *23*, 723–730. [CrossRef]
74. Zhou, X.; Ye, Y.-Z.; Ogihara, M.H.; Takeshima, M.; Fujinaga, D.; Liu, C.-W.; Zhu, Z.; Kataoka, H.; Bao, Y.-Y. Functional analysis of ecdysteroid biosynthetic enzymes of the rice planthopper, *Nilaparvata lugens*. *Insect Biochem. Mol. Biol.* **2020**, *123*, 103428. [CrossRef]
75. Dhadialla, T.S.; Retnakaran, A.; Smagghe, G. 6.3—Insect Growth- and Development-Disrupting Insecticides. In *Comprehensive Molecular Insect Science*; Gilbert, L.I., Ed.; Elsevier: Amsterdam, The Netherlands, 2005; pp. 55–115. [CrossRef]
76. Baehrecke, E.H. Ecdysone signaling cascade and regulation of *Drosophila metamorphosis*. *Arch. Insect Biochem. Physiol.* **1996**, *33*, 231–244. [CrossRef]
77. Uyehara, C.M.; McKay, D.J. Direct and widespread role for the nuclear receptor EcR in mediating the response to ecdysone in *Drosophila*. *Proc. Natl. Acad. Sci. USA* **2019**, *116*, 9893–9902. [CrossRef]
78. Yang, W.-J.; Xu, K.-K.; Cong, L.; Wang, J.-J. Identification, mRNA Expression, and Functional Analysis of Chitin Synthase 1 Gene and Its Two Alternative Splicing Variants in Oriental Fruit Fly, *Bactrocera dorsalis*. *Int. J. Biol. Sci.* **2013**, *9*, 331–342. [CrossRef] [PubMed]

79. Fujii, S.; Kobayashi, S.; Tettey, P.A.; Lisdiana, L.; Saito, H.; Kawai, K.; Sambongi, Y. Transcriptome analysis of *Chironomus sulfurosus* larvae living in acidic environments: Insights into molecular mechanisms for acid tolerance. *J. Insect Physiol.* **2021**, *133*, 104288. [CrossRef] [PubMed]
80. Laurie-Ahlberg, C.C.; Williamson, J.H.; Cochrane, B.J.; Wilton, A.N.; Chasalow, F.I. Autosomal factors with correlated effects on the activities of the glucose 6-phosphate and 6-phosphogluconate dehydrogenases in *Drosophila melanogaster*. *Genetics* **1981**, *99*, 127–150. [CrossRef]
81. Yang, W.J.; Xu, K.K.; Zhang, R.Y.; Dou, W.; Wang, J.J. Transcriptional regulation of a chitinase gene by 20-hydroxyecdysone and starvation in the oriental fruit fly, *Bactrocera dorsalis*. *Int. J. Mol. Sci.* **2013**, *14*, 20048–20063. [CrossRef]
82. Yu, L.; Zhou, Q.; Zhang, C.; Pignoni, F. Identification of *Bombyx* atonal and functional comparison with the *Drosophila* atonal proneural factor in the developing fly eye. *Genesis* **2012**, *50*, 393–403. [CrossRef]
83. Bakshi, A.; Moin, M.; Madhav, M.S.; Kirti, P.B. Target of rapamycin, a master regulator of multiple signalling pathways and a potential candidate gene for crop improvement. *Plant Biol.* **2019**, *21*, 190–205. [CrossRef]
84. Yoveva, A.; Sawarkar, R. Chromatin Immunoprecipitation (ChIP) of Heat Shock Protein 90 (Hsp90). *Methods Mol. Biol.* **2018**, *1709*, 221–231. [CrossRef]
85. Chen, B.; Wagner, A. Hsp90 is important for fecundity, longevity, and buffering of cryptic deleterious variation in wild fly populations. *BMC Evol. Biol.* **2012**, *12*, 25. [CrossRef]
86. Chen, P.; Chen, M.; Ye, H.; Yuan, R.; Du, C.; Ong, S.P. Juvenile hormone regulation on the flight capability of *Bactrocera dorsalis* (Diptera: Tephritidae). *J. Asia-Pac. Entomol.* **2020**, *23*, 1120–1129. [CrossRef]
87. Shi, Y.; Liu, T.Y.; Jiang, H.B.; Liu, X.Q.; Dou, W.; Park, Y.; Smagghe, G.; Wang, J.J. The Ecdysis Triggering Hormone System, via ETH/ETHR-B, Is Essential for Successful Reproduction of a Major Pest Insect, *Bactrocera dorsalis* (Hendel). *Front. Physiol.* **2019**, *10*, 151. [CrossRef] [PubMed]
88. Liu, S.-H.; Xia, Y.-D.; Zhang, Q.; Li, W.; Li, R.-Y.; Liu, Y.; Chen, E.-H.; Dou, W.; Stelinski, L.L.; Wang, J.-J. Potential targets for controlling *Bactrocera dorsalis* using cuticle- and hormone-related genes revealed by a developmental transcriptome analysis. *Pest. Manag. Sci.* **2020**, *76*, 2127–2143. [CrossRef] [PubMed]
89. Vaidya, S.; Srivastava, P.; Rathore, D.P.; Pandey, A.K. Amylases: A Prospective Enzyme in the Field of Biotechnology. *J. Appl. Biosci.* **2015**, *41*, 1–18.
90. Ekoka, E.; Maharaj, S.; Nardini, L.; Dahan-Moss, Y.; Koekemoer, L.L. 20-Hydroxyecdysone (20E) signaling as a promising target for the chemical control of malaria vectors. *Parasites Vectors* **2021**, *14*, 86. [CrossRef]
91. Xu, K.-K.; Yang, W.-J.; Tian, Y.; Wu, Y.-B.; Wang, J.-J. Insulin signaling pathway in the oriental fruit fly: The role of insulin receptor substrate in ovarian development. *General. Comp. Endocrinol.* **2015**, *216*, 125–133. [CrossRef]
92. Kannangara, J.R.; Mirth, C.K.; Warr, C.G. Regulation of ecdysone production in *Drosophila* by neuropeptides and peptide hormones. *Open Biol.* **2021**, *11*, 200373. [CrossRef]
93. Dong, W.; Zhe, W.; XU, H.-q.; NIU, J.-z.; WANG, J.-j. Cloning and functional characterization of two peptidoglycan recognition protein isoforms (PGRP-LC) in *Bactrocera dorsalis* (Diptera: Tephritidae). *J. Integr. Agric.* **2020**, *19*, 3025–3034.
94. Kampouraki, A.; Tsakireli, D.; Koidou, V.; Stavrakaki, M.; Kaili, S.; Livadaras, I.; Grigoraki, L.; Ioannidis, P.; Roditakis, E.; Vontas, J. Functional characterization of cytochrome P450s associated with pyrethroid resistance in the olive fruit fly *Bactrocera oleae*. *Pestic. Biochem. Physiol.* **2023**, *191*, 105374. [CrossRef]
95. Bai, P.-P.; Chen, E.-H.; Shen, G.-M.; Wei, D.; Wei, D.-D.; Wang, J.-J. Inhibition of phenoloxidase activity delays development in *Bactrocera dorsalis* (Diptera: Tephritidae). *Fla. Entomol.* **2014**, *97*, 477–485. [CrossRef]
96. Huang, Y.; Yu, Y.; Zhan, S.; Tomberlin, J.K.; Huang, D.; Cai, M.; Zheng, L.; Yu, Z.; Zhang, J. Dual oxidase Duox and Toll-like receptor 3 TLR3 in the Toll pathway suppress zoonotic pathogens through regulating the intestinal bacterial community homeostasis in *Hermetia illucens* L. *PLoS ONE* **2020**, *15*, e0225873. [CrossRef]
97. Lima, L.F.; Torres, A.Q.; Jardim, R.; Mesquita, R.D.; Schama, R. Evolution of Toll, Spatzle and MyD88 in insects: The problem of the Diptera bias. *BMC Genom.* **2021**, *22*, 562. [CrossRef]
98. Hsu, P.D.; Lander, E.S.; Zhang, F. Development and applications of CRISPR-Cas9 for genome engineering. *Cell* **2014**, *157*, 1262–1278. [CrossRef]
99. Bai, X.; Zeng, T.; Ni, X.Y.; Su, H.A.; Huang, J.; Ye, G.Y.; Lu, Y.Y.; Qi, Y.X. CRISPR/Cas9-mediated knockout of the eye pigmentation gene white leads to alterations in colour of head spots in the oriental fruit fly, *Bactrocera dorsalis*. *Insect Mol. Biol.* **2019**, *28*, 837–849. [CrossRef]
100. Chen, H.; Sun, H.; Xie, J.; Yao, Z.; Zheng, W.; Li, Z.; Deng, Z.; Li, X.; Zhang, H. CRISPR/Cas9-induced Mutation of Sex Peptide Receptor Gene Bdspr Affects Ovary, Egg Laying, and Female Fecundity in *Bactrocera dorsalis* (Hendel) (Diptera: Tephritidae). *J. Insect Sci.* **2023**, *23*, 2. [CrossRef]
101. Pitt, A.R.; Kolch, W. 24 Chapter Genomics, Proteomics, and Metabolomics. In *Medical Biochemistry-E-Book*; Elsevier Health Sciences: Barcelona, Spain, 2022; Volume 333.
102. Yang, Y.; Jiang, H.-B.; Liang, C.-H.; Ma, Y.-P.; Dou, W.; Wang, J.-J. Chromosome-level genome assembly reveals potential epigenetic mechanisms of the thermal tolerance in the oriental fruit fly, *Bactrocera dorsalis*. *Int. J. Biol. Macromol.* **2023**, *225*, 430–441. [CrossRef]
103. Altstein, M.; Nässel, D.R. Neuropeptide Signaling in Insects. In *Neuropeptide Systems as Targets for Parasite and Pest Control*; Geary, T.G., Maule, A.G., Eds.; Springer: Boston, MA, USA, 2010; pp. 155–165. [CrossRef]

104. Hou, Q.L.; Jiang, H.B.; Gui, S.H.; Chen, E.H.; Wei, D.D.; Li, H.M.; Wang, J.J.; Smagghe, G. A Role of Corazonin Receptor in Larval-Pupal Transition and Pupariation in the Oriental Fruit Fly *Bactrocera dorsalis* (Hendel) (Diptera: Tephritidae). *Front. Physiol.* **2017**, *8*, 77. [CrossRef]
105. Gui, S.-H.; Jiang, H.-B.; Smagghe, G.; Wang, J.-J. The neuropeptides and protein hormones of the agricultural pest fruit fly *Bactrocera dorsalis*: What do we learn from the genome sequencing and tissue-specific transcriptomes? *Peptides* **2017**, *98*, 29–34. [CrossRef]
106. Caplan, S.L.; Milton, S.L.; Dawson-Scully, K. A cGMP-dependent protein kinase (PKG) controls synaptic transmission tolerance to acute oxidative stress at the *Drosophila* larval neuromuscular junction. *J. Neurophysiol.* **2013**, *109*, 649–658. [CrossRef]
107. Miyazaki, H.; Otake, J.; Mitsuno, H.; Ozaki, K.; Kanzaki, R.; Chieng, A.C.-T.; Hee, A.K.-W.; Nishida, R.; Ono, H. Functional characterization of olfactory receptors in the oriental fruit fly *Bactrocera dorsalis* that respond to plant volatiles. *Insect Biochem. Mol. Biol.* **2018**, *101*, 32–46. [CrossRef]
108. Xu, L.; Jiang, H.-B.; Tang, K.-Y.; Yan, Y.; Schetelig, M.F.; Wang, J.-J. CRISPR-mediated mutagenesis of the odorant receptor co-receptor (Orco) gene disrupts olfaction-mediated behaviors in *Bactrocera dorsalis*. *Insect Sci.* **2022**, *29*, 1275–1286. [CrossRef] [PubMed]
109. Zhou, Q.H.; Zhang, Q.; Yang, R.L.; Yuan, G.R.; Wang, J.J.; Dou, W. RNAi-mediated knockdown of juvenile hormone acid O-methyltransferase disrupts larval development in the oriental fruit fly, *Bactrocera dorsalis* (Hendel). *Pestic. Biochem. Physiol.* **2022**, *188*, 105285. [CrossRef] [PubMed]
110. Zhang, R.; Zhang, S.; Li, T.; Li, H.; Zhang, H.; Zheng, W. RNA sequencing identifies an ovary-enriched microRNA, miR-311-3p, involved in ovarian development and fecundity by targeting Endophilin B1 in *Bactrocera dorsalis*. *Pest. Manag. Sci.* **2023**, *79*, 688–700. [CrossRef]
111. Wu, Z.; Yang, L.; He, Q.; Zhou, S. Regulatory Mechanisms of Vitellogenesis in Insects. *Front. Cell Dev. Biol.* **2020**, *8*, 593613. [CrossRef] [PubMed]
112. Yamanaka, N. Chapter One—Ecdysteroid signalling in insects—From biosynthesis to gene expression regulation. In *Advances in Insect Physiology*; Adams, M.E., Ed.; Academic Press: Cambridge, MA, USA, 2021; Volume 60, pp. 1–36.
113. Roh, G.-H.; Kendra, P.E.; Cha, D.H. Preferential attraction of oviposition-ready oriental fruit flies to host fruit odor over protein food odor. *Insects* **2021**, *12*, 909. [CrossRef]
114. Kimbokota, F.; Hassanali, A.; Njagi, P.G.N. Potential attractants from three host plants of *Bactrocera dorsalis* (Hendel) (Diptera: Tephritidae). *Int. J. Trop. Insect Sci.* **2023**, *43*, 107–114. [CrossRef]
115. Ren, L.; Zeng, L.; Lin, J.; Lu, Y.; Liu, C. Biological Control of *Bactrocera dorsalis* (Hendel) (Diptera: Tephritidae). *Biol. Invasions Biol. Control Theory Pract.* **2008**, 413–440.
116. Ren, L.; Ma, J.; Lv, Y.; Tong, Q.; Guo, H. Characterization of key off-odor compounds in thermal duck egg gels by GC-olfactometry-MS, odor activity values, and aroma recombination. *LWT* **2021**, *143*, 111182. [CrossRef]
117. Xu, L.; Yu, X.; Li, M.; Chen, J.; Wang, X. Monitoring oxidative stability and changes in key volatile compounds in edible oils during ambient storage through HS-SPME/GC–MS. *Int. J. Food Prop.* **2017**, *20*, S2926–S2938. [CrossRef]
118. Cornelius, M.L.; Duan, J.J.; Messing, R.H. Volatile host fruit odors as attractants for the oriental fruit fly (Diptera: Tephritidae). *J. Econ. Entomol.* **2000**, *93*, 93–100. [CrossRef]
119. Zhu, X.; Li, Q.; Li, J.; Luo, J.; Chen, W.; Li, X. Comparative Study of Volatile Compounds in the Fruit of Two Banana Cultivars at Different Ripening Stages. *Molecules* **2018**, *23*, 2456. [CrossRef]
120. Li, Y.-Y.; Liu, D.; Wen, P.; Chen, L. Detection of Volatile Organic Compounds by Antennal Lamellae of a Scarab Beetle. *Front. Ecol. Evol.* **2021**, *9*, 392–401. [CrossRef]
121. Steinbrecht, R.A. Pore structures in insect olfactory sensilla: A review of data and concepts. *Int. J. Insect Morphol. Embryol.* **1997**, *26*, 229–245. [CrossRef]
122. Francia, S.; Lodovichi, C. The role of the odorant receptors in the formation of the sensory map. *BMC Biol.* **2021**, *19*, 174. [CrossRef] [PubMed]
123. Liu, H.; Chen, Z.-S.; Zhang, D.-J.; Lu, Y.-Y. BdorOR88a Modulates the Responsiveness to Methyl Eugenol in Mature Males of *Bactrocera dorsalis* (Hendel). *Front. Physiol.* **2018**, *9*, 987. [CrossRef]
124. Agnihotri, A.; Liu, N.; Xu, W. Chemosensory proteins (CSPs) in the cotton bollworm *Helicoverpa armigera*. *Insects* **2021**, *13*, 29. [CrossRef]
125. Liu, Z.; Liang, X.-F.; Xu, L.; Keesey, I.W.; Lei, Z.-R.; Smagghe, G.; Wang, J.-J. An Antennae-Specific Odorant-Binding Protein Is Involved in *Bactrocera dorsalis* Olfaction. *Front. Ecol. Evol.* **2020**, *8*, 63. [CrossRef]
126. Chen, J.; Li, S.-S.; Fang, S.-M.; Zhang, Z.; Yu, Q.-Y. Olfactory dysfunction and potential mechanisms caused by volatile organophosphate dichlorvos in the silkworm as a model animal. *J. Hazard. Mater.* **2022**, *425*, 127940. [CrossRef]
127. Liu, Z.; Xie, Q.-P.; Guo, H.-W.; Xu, W.; Wang, J.-J. An odorant binding protein mediates *Bactrocera dorsalis* olfactory sensitivity to host plant volatiles and male attractant compounds. *Int. J. Biol. Macromol.* **2022**, *219*, 538–544. [CrossRef]
128. Jang, S.; Kikuchi, Y. Impact of the insect gut microbiota on ecology, evolution, and industry. *Curr. Opin. Insect Sci.* **2020**, *41*, 33–39. [CrossRef]
129. Ren, X.; Cao, S.; Akami, M.; Mansour, A.; Yang, Y.; Jiang, N.; Wang, H.; Zhang, G.; Qi, X.; Xu, P.; et al. Gut symbiotic bacteria are involved in nitrogen recycling in the tephritid fruit fly *Bactrocera dorsalis*. *BMC Biol.* **2022**, *20*, 201. [CrossRef] [PubMed]

130. Jing, T.-Z.; Qi, F.-H.; Wang, Z.-Y. Most dominant roles of insect gut bacteria: Digestion, detoxification, or essential nutrient provision? *Microbiome* **2020**, *8*, 38. [CrossRef] [PubMed]
131. Guentzel, M.N. *Escherichia, Klebsiella, Enterobacter, Serratia, Citrobacter*, and Proteus. In *Medical Microbiology*; Baron, S., Ed.; The University of Texas Medical Branch at Galveston: Galveston, TX, USA, 1996.
132. Wang, H.; Jin, L.; Zhang, H. Comparison of the diversity of the bacterial communities in the intestinal tract of adult *Bactrocera dorsalis* from three different populations. *J. Appl. Microbiol.* **2011**, *110*, 1390–1401. [CrossRef] [PubMed]
133. Zeng, T.; Jaffar, S.; Xu, Y.; Qi, Y. The Intestinal Immune Defense System in Insects. *Int. J. Mol. Sci.* **2022**, *23*, 15132. [CrossRef] [PubMed]
134. Yao, Z.; Wang, A.; Li, Y.; Cai, Z.; Lemaitre, B.; Zhang, H. The dual oxidase gene BdDuox regulates the intestinal bacterial community homeostasis of *Bactrocera dorsalis*. *ISME J.* **2016**, *10*, 1037–1050. [CrossRef] [PubMed]
135. Raza, M.F.; Wang, Y.; Cai, Z.; Bai, S.; Yao, Z.; Awan, U.A.; Zhang, Z.; Zheng, W.; Zhang, H. Gut microbiota promotes host resistance to low-temperature stress by stimulating its arginine and proline metabolism pathway in adult *Bactrocera dorsalis*. *PLoS Pathog.* **2020**, *16*, e1008441. [CrossRef]
136. Akami, M.; Ren, X.; Wang, Y.; Mansour, A.; Cao, S.; Qi, X.; Ngakou, A.; Ngane, R.A.N.; Niu, C.-Y. Host fruits shape the changes in the gut microbiota and development of *Bactrocera dorsalis* (Diptera: Tephritidae) larvae. *Int. J. Trop. Insect Sci.* **2022**, *42*, 2127–2141. [CrossRef]
137. Zeng, T.; Bai, X.; Liu, Y.L.; Li, J.F.; Lu, Y.Y.; Qi, Y.X. Intestinal responses of the oriental fruit fly *Bactrocera dorsalis* (Hendel) after ingestion of an entomopathogenic bacterium strain. *Pest. Manag. Sci.* **2020**, *76*, 653–664. [CrossRef]
138. Koul, B.; Chopra, M.; Lamba, S. Chapter 3—Microorganisms as biocontrol agents for sustainable agriculture. In *Relationship Between Microbes and the Environment for Sustainable Ecosystem Services, Volume 1*; Samuel, J., Kumar, A., Singh, J., Eds.; Elsevier: Amsterdam, The Netherlands, 2022; pp. 45–68. [CrossRef]
139. Shi, Z.; Wang, L.; Zhang, H. Low Diversity Bacterial Community and the Trapping Activity of Metabolites from Cultivable Bacteria Species in the Female Reproductive System of the Oriental Fruit Fly, *Bactrocera dorsalis* Hendel (Diptera: Tephritidae). *Int. J. Mol. Sci.* **2012**, *13*, 6266–6278. [CrossRef]
140. Liu, S.H.; Chen, Y.; Li, W.; Tang, G.H.; Yang, Y.; Jiang, H.B.; Dou, W.; Wang, J.J. Diversity of Bacterial Communities in the Intestinal Tracts of Two Geographically Distant Populations of *Bactrocera dorsalis* (Diptera: Tephritidae). *J. Econ. Entomol.* **2018**, *111*, 2861–2868. [CrossRef]
141. Noman, M.S.; Liu, L.; Bai, Z.; Li, Z. Tephritidae bacterial symbionts: Potentials for pest management. *Bull. Entomol. Res.* **2020**, *110*, 1–14. [CrossRef]
142. Wang, M.; Cao, B.; Gao, Q.; Sun, Y.; Liu, P.; Feng, L.; Wang, L. Detection of *Enterobacter sakazakii* and other pathogens associated with infant formula powder by use of a DNA microarray. *J. Clin. Microbiol.* **2009**, *47*, 3178–3184. [CrossRef]
143. Li, H.; Ren, L.; Xie, M.; Gao, Y.; He, M.; Hassan, B.; Lu, Y.; Cheng, D. Egg-surface bacteria are indirectly associated with oviposition aversion in *Bactrocera dorsalis*. *Curr. Biol.* **2020**, *30*, 4432–4440.e4434. [CrossRef]
144. Taddei, A.; Reisenzein, H.; Mouttet, R.; Lethmayer, C.; Egartner, A.; Gottsberger, R.A.; Blümel, S.; Heiss, C.; Pohn, C.; Reynaud, P. Morphological and Molecular Identification Protocols for *Bactrocera dorsalis*: A Joint Validation Study. *PhytoFrontiers™* **2023**, *3*, 186–198. [CrossRef]
145. Choi, K.S.; Samayoa, A.C.; Hwang, S.-Y.; Huang, Y.-B.; Ahn, J.J. Thermal effect on the fecundity and longevity of *Bactrocera dorsalis* adults and their improved oviposition model. *PLoS ONE* **2020**, *15*, e0235910. [CrossRef] [PubMed]
146. Fiaboe, K.K.; Kekeunou, S.; Nanga, S.N.; Kuate, A.F.; Tonnang, H.E.; Gnanvossou, D.; Hanna, R. Temperature-based phenology model to predict the development, survival, and reproduction of the oriental fruit fly *Bactrocera dorsalis*. *J. Therm. Biol.* **2021**, *97*, 102877.
147. Jiang, F.; Liang, L.; Wang, J.; Zhu, S. Chromosome-level genome assembly of *Bactrocera dorsalis* reveals its adaptation and invasion mechanisms. *Commun. Biol.* **2022**, *5*, 25. [CrossRef] [PubMed]
148. Wu, S.-X.; Chen, Y.; Lei, Q.; Peng, Y.-Y.; Jiang, H.-B. Sublethal Dose of β-Cypermethrin Impairs the Olfaction of *Bactrocera dorsalis* by Suppressing the Expression of Chemosensory Genes. *Insects* **2022**, *13*, 721. [CrossRef] [PubMed]
149. Fang, Y.; Kang, F.; Zhan, G.; Ma, C.; Li, Y.; Wang, L.; Wei, Y.; Gao, X.; Li, Z.; Wang, Y. The Effects of a Cold Disinfestation on *Bactrocera dorsalis* Survival and Navel Orange Quality. *Insects* **2019**, *10*, 452. [CrossRef]
150. Lin, K.W.; Lin, H.L.; Shiesh, C.C.; Hsu, Y.L.; Lin, C.H.; Chen, S.C.; Yeh, W.B. Cold treatment for guava fruits infested with oriental fruit fly, *Bactrocera dorsalis* (Diptera: Tephritidae). *Appl. Entomol. Zool.* **2020**, *55*, 37–44. [CrossRef]
151. Hsu, Y.-L.; Chen, S.-C.; Lin, K.-W.; Shiesh, C.-C.; Lin, C.-H.; Yeh, W.-B. Quarantine vapor heat treatment of papaya fruit for *Bactrocera dorsalis* and *Bactrocera cucurbitae* (Diptera: Tephritidae). *J. Econ. Entomol.* **2018**, *111*, 2101–2109. [CrossRef] [PubMed]
152. Shelly, T. Effects of methyl eugenol and raspberry ketone/cue lure on the sexual behavior of *Bactrocera* species (Diptera: Tephritidae). *Appl. Entomol. Zool.* **2010**, *45*, 349–361. [CrossRef]
153. Mandanayake, M.A.R.A.; Hee, A.K.W. Assessment of non-methyl eugenol-responding lines of *Bactrocera dorsalis* (Diptera: Tephritidae) males on lure response and mating. *Bull. Entomol. Res.* **2023**, *113*, 396–401. [CrossRef]
154. Fan, Y.; Zhang, C.; Qin, Y.; Yin, X.; Dong, X.; Desneux, N.; Zhou, H. Monitoring the Methyl Eugenol Response and Non-Responsiveness Mechanisms in Oriental Fruit Fly *Bactrocera dorsalis* in China. *Insects* **2022**, *13*, 1004. [CrossRef]
155. Mandanayake, M.A.R.A.; Shohaimi, S.; Ghani, N.I.A.; Hee, A.K.W. Establishment of non-methyl eugenol-responding lines from feral Oriental fruit fly, *Bactrocera dorsalis* (Hendel) (Diptera Tephritidae). *Phytoparasitica* **2023**, *51*, 425–436. [CrossRef]

156. Raza, A.B.M.; Majeed, M.Z.; Ismail, M.; Qasim, Z. Field evaluation of IPM strategies against *Bactrocera dorsalis* PUJZ. *Punjab Univ. J. Zool.* **2023**, *38*, 59–64. [CrossRef]
157. Tan, K.-H.; Wee, S.-L.; Nishida, R.; Shelly, T.E. Attraction of feral *Bactrocera dorsalis* males (Diptera: Tephritidae) to natural versus commercial sources of methyl eugenol. *J. Asia-Pac. Entomol.* **2021**, *24*, 1095–1100. [CrossRef]
158. Bautista, R.C.; Mochizuki, N.; Spencer, J.P.; Harris, E.J.; Ichimura, D.M. Mass-Rearing of the Tephritid Fruit Fly Parasitoid *Fopius arisanus* (Hymenoptera: Braconidae)1. *Biol. Control* **1999**, *15*, 137–144. [CrossRef]
159. Lahiri, S.; Orr, D. Chapter 11—Biological Control in Tomato Production Systems: Theory and Practice. In *Sustainable Management of Arthropod Pests of Tomato*; Wakil, W., Brust, G.E., Perring, T.M., Eds.; Academic Press: San Diego, CA, USA, 2018; pp. 253–267. [CrossRef]
160. Nanga Nanga, S.; Hanna, R.; Gnanvossou, D.; Fotso Kuate, A.; Fiaboe, K.K.M.; Djieto-Lordon, C. Fruit Preference, Parasitism, and Offspring Fitness of *Fopius arisanus* (Hymenoptera: Braconidae) Exposed to *Bactrocera dorsalis'* (Diptera: Tephritidae) Infested Fruit Species. *Environ. Entomol.* **2019**, *48*, 1286–1296. [CrossRef]
161. Sime, K.R.; Daane, K.M.; Wang, X.G.; Johnson, M.W.; Messing, R.H. Evaluation of *Fopius arisanus* as a biological control agent for the olive fruit fly in California. *Agric. For. Entomol.* **2008**, *10*, 423–431. [CrossRef]
162. Zheng, Y.; Song, Z.-W.; Zhang, Y.-P.; Li, D.-S. Ability of *Spalangia endius* (Hymenoptera: Pteromalidae) to Parasitize *Bactrocera dorsalis* (Diptera: Tephritidae) after Switching Hosts. *Insects* **2021**, *12*, 613. [CrossRef] [PubMed]
163. Tang, L.-D.; Ji, X.-C.; Han, Y.; Fu, B.-L.; Liu, K. Parasitism, Emergence, and Development of *Spalangia endius* (Hymenoptera: Pteromalidae) in Pupae of Different Ages of *Bactrocera cucurbitae* (Diptera: Tephritidae). *J. Insect Sci.* **2015**, *15*, 15. [CrossRef] [PubMed]
164. Li, L.; Chen, J.; Niu, L.; Han, D.; Zhang, F.; Fu, Y. Field evaluation of *Spalangia endius* Walker in controlling *Zeugodacus cucurbitae* (Coquillett). *Int. J. Pest. Manag.* **2021**, *67*, 216–221. [CrossRef]
165. van Lenteren, J.C. The state of commercial augmentative biological control: Plenty of natural enemies, but a frustrating lack of uptake. *BioControl* **2012**, *57*, 1–20. [CrossRef]
166. Liang, J.; Tang, S.; Cheke, R.A.; Wu, J. Adaptive release of natural enemies in a pest-natural enemy system with pesticide resistance. *Bull. Math. Biol.* **2013**, *75*, 2167–2195. [CrossRef]
167. Blissard, G.W.; Theilmann, D.A. Baculovirus Entry and Egress from Insect Cells. *Annu. Rev. Virol.* **2018**, *5*, 113–139. [CrossRef]
168. Prasad, V.; Srivastava, S. Chapter 13—Insect Viruses. In *Ecofriendly Pest Management for Food Security*; Omkar, Ed.; Academic Press: San Diego, CA, USA, 2016; pp. 411–442. [CrossRef]
169. Zhang, W.; Gu, Q.; Niu, J.; Wang, J.J. The RNA Virome and Its Dynamics in an Invasive Fruit Fly, *Bactrocera dorsalis*, Imply Interactions Between Host and Viruses. *Microb. Ecol.* **2020**, *80*, 423–434. [CrossRef]
170. Iqbal, M.; Gogi, M.D.; Atta, B.; Nisar, M.J.; Arif, M.J.; Javed, N. Assessment of pathogenicity of *Beauveria bassiana*, *Metarhizium anisopliae*, *Verticillium lecanii* and *Bacillus thuringiensis* var. kurstaki against *Bactrocera cucurbitae* Coquillett (Diptera: Tephritidae) via diet-bioassay technique under controlled conditions. *Int. J. Trop. Insect Sci.* **2021**, *41*, 1129–1145. [CrossRef]
171. Wang, D.; Liang, Q.; Chen, M.; Ye, H.; Liao, Y.; Yin, J.; Lü, L.; Lei, Y.; Cai, D.; Jaleel, W.; et al. Susceptibility of oriental fruit fly, *Bactrocera dorsalis* (Diptera: Tephritidae) pupae to entomopathogenic fungi. *Appl. Entomol. Zool.* **2021**, *56*, 269–275. [CrossRef]
172. Martínez-Barrera, O.Y.; Toledo, J.; Cancino, J.; Liedo, P.; Gómez, J.; Valle-Mora, J.; Montoya, P. Interaction Between *Beauveria bassiana* (Hypocreales: Cordycipitaceae) and *Coptera haywardi* (Hymenoptera: Diapriidae) for the Management of *Anastrepha obliqua* (Diptera: Tephritidae). *J. Insect Sci.* **2020**, *20*, 1–10. [CrossRef]
173. Shishir, M.A.; Akter, A.; Bodiuzzaman, M.; Hossain, M.A.; Alam, M.M.; Khan, S.A.; Khan, S.N.; Hoq, M.M. Novel toxicity of *Bacillus thuringiensis* strains against the melon fruit fly, *Bactrocera cucurbitae* (Diptera: Tephritidae). *Biocontrol Sci.* **2015**, *20*, 115–123. [CrossRef] [PubMed]
174. Wakil, W.; Usman, M.; Piñero, J.C.; Wu, S.; Toews, M.D.; Shapiro-Ilan, D.I. Combined application of entomopathogenic nematodes and fungi against fruit flies, *Bactrocera zonata* and B. dorsalis (Diptera: Tephritidae): Laboratory cups to field study. *Pest. Manag. Sci.* **2022**, *78*, 2779–2791. [CrossRef]
175. Heve, W.K.; Adjadeh, T.A.; Billah, M.K. Overview and future research needs for development of effective biocontrol strategies for management of *Bactrocera dorsalis* Hendel (Diptera: Tephritidae) in sub-Saharan Africa. *Pest. Manag. Sci.* **2021**, *77*, 4224–4237. [CrossRef]
176. Godjo, A.; Chabi, N.; Zadji, L.; Dossou, P.; Batcho, O.; Baimey, H.; Bonou, W.; Sinzogan, A.A.C.; Bokonon-Ganta, A.; Decraemer, W.; et al. Evaluation of the ability of indigenous nematode isolates of *Heterorhabditis taysearae* and *Steinernema kandii* to control mango fruit fly *Bactrocera dorsalis* under laboratory, semi-field and field conditions in Northern Benin. *Crop Prot.* **2021**, *149*, 105754. [CrossRef]
177. Agboka, K.M.; Tonnang, H.E.Z.; Abdel-Rahman, E.M.; Kimathi, E.; Mutanga, O.; Odindi, J.; Niassy, S.; Mohamed, S.A.; Ekesi, S. A systematic methodological approach to estimate the impacts of a classical biological control agent's dispersal at landscape: Application to fruit fly *Bactrocera dorsalis* and its endoparasitoid *Fopius arisanus*. *Biol. Control* **2022**, *175*, 105053. [CrossRef]
178. Ndlela, S.; Mohamed, S.A.; Azrag, A.G.A.; Ndegwa, P.N.; Ong'amo, G.O.; Ekesi, S. Interactions between Two Parasitoids of Tephritidae: *Diachasmimorpha longicaudata* (Ashmead) and *Psyttalia cosyrae* (Wilkinson) (Hymenoptera: Braconidae), under Laboratory Conditions. *Insects* **2020**, *11*, 671. [CrossRef] [PubMed]
179. Labaude, S.; Griffin, C.T. Transmission Success of Entomopathogenic Nematodes Used in Pest Control. *Insects* **2018**, *9*, 72. [CrossRef] [PubMed]

180. Toledo, J.; Morán-Aceves, B.M.; Ibarra, J.E.; Liedo, P. Can Entomopathogenic Nematodes and Their Symbiotic Bacteria Suppress Fruit Fly Pests? A Review. *Microorganisms* **2023**, *11*, 1682. [CrossRef]
181. Diouf, E.G.; Brévault, T.; Ndiaye, S.; Faye, E.; Chailleux, A.; Diatta, P.; Piou, C. An agent-based model to simulate the boosted Sterile Insect Technique for fruit fly management. *Ecol. Model.* **2022**, *468*, 109951. [CrossRef]
182. Pérez-Staples, D.; Díaz-Fleischer, F.; Montoya, P. The Sterile Insect Technique: Success and Perspectives in the Neotropics. *Neotrop. Entomol.* **2021**, *50*, 172–185. [CrossRef]
183. Cladera, J.L.; Vilardi, J.C.; Juri, M.; Paulin, L.E.; Giardini, M.C.; Gómez Cendra, P.V.; Segura, D.F.; Lanzavecchia, S.B. Genetics and biology of *Anastrepha fraterculus*: Research supporting the use of the sterile insect technique (SIT) to control this pest in Argentina. *BMC Genet.* **2014**, *15* (Suppl. S2), S12. [CrossRef] [PubMed]
184. Mau, R.; Jang, E.; Vargas, R. The Hawaii area-wide fruit fly pest management programme. In *Area-Wide Control of Insect Pests*; Springer: Berlin/Heidelberg, Germany, 2007; pp. 671–683.
185. Haq, I.U.; Abd-Alla, A.; Tomas, U.S.; Meza, J.S.; Bourtzis, K.; Cáceres, C. Cryopreservation of the Mediterranean fruit fly (Diptera: Tephritidae) VIENNA 8 genetic sexing strain: No effect on large scale production of high quality sterile males for SIT applications. *PLoS ONE* **2019**, *14*, e0211259. [CrossRef] [PubMed]
186. Ramírez-Santos, E.; Rendon, P.; Gouvi, G.; Zacharopoulou, A.; Bourtzis, K.; Cáceres, C.; Bloem, K. A Novel Genetic Sexing Strain of *Anastrepha ludens* for Cost-Effective Sterile Insect Technique Applications: Improved Genetic Stability and Rearing Efficiency. *Insects* **2021**, *12*, 499. [CrossRef]
187. Sollazzo, G.; Gouvi, G.; Nikolouli, K.; Cancio Martinez, E.I.; Schetelig, M.F.; Bourtzis, K. Temperature Sensitivity of Wild-Type, Mutant and Genetic Sexing Strains of *Ceratitis capitata*. *Insects* **2022**, *13*, 943. [CrossRef]
188. Robinson, A.S. Genetic sexing strains in medfly, *Ceratitis capitata*, sterile insect technique programmes. *Genetica* **2002**, *116*, 5–13. [CrossRef] [PubMed]
189. Aluja, M.; Norrbom, A. *Fruit Flies (Tephritidae): Phylogeny and Evolution of Behavior*; CRC Press: Boca Raton, FL, USA, 1999.
190. Hendrichs, J.; Robinson, A.; Cayol, J.; Enkerlin, W. Medfly areawide sterile insect technique programmes for prevention, suppression or eradication: The importance of mating behavior studies. *Fla. Entomol.* **2002**, *85*, 1–13. [CrossRef]
191. Kaneko, Y.; Szallasi, A. Transient receptor potential (TRP) channels: A clinical perspective. *Br. J. Pharmacol.* **2014**, *171*, 2474–2507. [CrossRef]
192. Su, H.; Bai, X.; Zeng, T.; Lu, Y.; Qi, Y. Identification, characterization and expression analysis of transient receptor potential channel genes in the oriental fruit fly, *Bactrocera dorsalis*. *BMC Genom.* **2018**, *19*, 674. [CrossRef]
193. Nesterov, A.; Spalthoff, C.; Kandasamy, R.; Katana, R.; Rankl, N.B.; Andrés, M.; Jähde, P.; Dorsch, J.A.; Stam, L.F.; Braun, F.-J.; et al. TRP Channels in Insect Stretch Receptors as Insecticide Targets. *Neuron* **2015**, *86*, 665–671. [CrossRef]
194. Zhang, Y.; Zhang, Y.-J.; Guo, D.; Wang, L.-X.; Niu, C.-D.; Wu, S.-F.; Zhang, Y.V.; Gao, C.-F. Function of Transient Receptor Potential-Like Channel in Insect Egg Laying. *Front. Mol. Neurosci.* **2022**, *15*, 823563. [CrossRef]
195. Wang, R.; Gao, B.; Zhang, Q.; Qu, C.; Luo, C. Knockdown of TRPV gene Nanchung decreases resistance to the novel pyropene insecticide, afidopyropen, in *Bemisia tabaci*. *Int. J. Biol. Macromol.* **2023**, *224*, 1566–1575. [CrossRef] [PubMed]
196. Xie, Y.F.; Niu, J.Z.; Jiang, X.Z.; Yang, W.J.; Shen, G.M.; Wei, D.; Smagghe, G.; Wang, J.J. Influence of various stressors on the expression of core genes of the small interfering RNA pathway in the oriental fruit fly, *Bactrocera dorsalis*. *Insect Sci.* **2017**, *24*, 418–430. [CrossRef] [PubMed]
197. Li, X.; Zhang, M.; Zhang, H. RNA interference of four genes in adult *Bactrocera dorsalis* by feeding their dsRNAs. *PLoS ONE* **2011**, *6*, e17788. [CrossRef]
198. Dong, Y.-C.; Wang, Z.-J.; Chen, Z.-Z.; Clarke, A.R.; Niu, C.-Y. *Bactrocera dorsalis* male sterilization by targeted RNA interference of spermatogenesis: Empowering sterile insect technique programs. *Sci. Rep.* **2016**, *6*, 35750. [CrossRef] [PubMed]
199. Zhou, Y.; Fu, W.-B.; Si, F.-L.; Yan, Z.-T.; Zhang, Y.-J.; He, Q.-Y.; Chen, B. UDP-glycosyltransferase genes and their association and mutations associated with pyrethroid resistance in *Anopheles sinensis* (Diptera: Culicidae). *Malar. J.* **2019**, *18*, 62. [CrossRef]
200. Chen, M.-L.; Zhang, S.-X.; Guo, P.-Y.; Qin, Q.-S.; Meng, L.-W.; Yuan, G.-R.; Wang, J.-J. Identification and characterization of UDP-glycosyltransferase genes and the potential role in response to insecticides exposure in *Bactrocera dorsalis*. *Pest. Manag. Sci.* **2023**, *79*, 666–677. [CrossRef]
201. Moon, T.T.; Maliha, I.J.; Khan, A.A.M.; Chakraborty, M.; Uddin, M.S.; Amin, M.R.; Islam, T. CRISPR-Cas Genome Editing for Insect Pest Stress Management in Crop Plants. *Stresses* **2022**, *2*, 493–514. [CrossRef]
202. Wang, X.-F.; Chen, Z.; Wang, X.-B.; Xu, J.; Chen, P.; Ye, H. Bacterial-mediated RNAi and functional analysis of Natalisin in a moth. *Sci. Rep.* **2021**, *11*, 4662. [CrossRef]
203. Kim, Y.-J.; Bartalska, K.; Audsley, N.; Yamanaka, N.; Yapici, N.; Lee, J.-Y.; Kim, Y.-C.; Markovic, M.; Isaac, E.; Tanaka, Y.; et al. MIPs are ancestral ligands for the sex peptide receptor. *Proc. Natl. Acad. Sci. USA* **2010**, *107*, 6520–6525. [CrossRef]
204. Zheng, W.; Liu, Y.; Zheng, W.; Xiao, Y.; Zhang, H. Influence of the silencing sex-peptide receptor on *Bactrocera dorsalis* adults and offspring by feeding with ds-spr. *J. Asia-Pac. Entomol.* **2015**, *18*, 477–481. [CrossRef]

MDPI AG
Grosspeteranlage 5
4052 Basel
Switzerland
Tel.: +41 61 683 77 34

Horticulturae Editorial Office
E-mail: horticulturae@mdpi.com
www.mdpi.com/journal/horticulturae

www.ingramcontent.com/pod-product-compliance
Lightning Source LLC
Chambersburg PA
CBHW041350210926
51850CB00013B/3038
* 9 7 8 3 7 2 5 8 3 3 3 7 5 *